100 Years of General Relativity – Vol. 4

LOOP QUANTUM GRAVITY

100 Years of General Relativity

ISSN: 2424-8223

Series Editor: Abhay Ashtekar *(Pennsylvania State University, USA)*

This series is to publish about two dozen excellent monographs written by top-notch authors from the international gravitational community covering various aspects of the field, ranging from mathematical general relativity through observational ramifications in cosmology, relativistic astrophysics and gravitational waves, to quantum aspects.

Published

Vol. 1   Numerical Relativity
         by Masaru Shibata (Kyoto University, Japan)

Vol. 2   Chern–Simons (Super)Gravity
         by Mokhtar Hassaine (Universidad de Talca, Chile) &
         Jorge Zanelli (Centro de Estudios Científicos, Chile)

Vol. 3   An Introduction to Covariant Quantum Gravity and Asymptotic Safety
         by Roberto Percacci
         (SISSA, Italy & Perimeter Institute for Theoretical Physics, Canada)

Vol. 4   Loop Quantum Gravity: The First 30 Years
         edited by Abhay Ashtekar (Pennsylvania State University, USA) &
         Jorge Pullin (Louisiana State University, USA)

100 Years of General Relativity – Vol. 4

LOOP QUANTUM GRAVITY

Editors

Abhay Ashtekar
Pennsylvania State University, USA

Jorge Pullin
Louisiana State University, USA

World Scientific

NEW JERSEY · LONDON · SINGAPORE · BEIJING · SHANGHAI · HONG KONG · TAIPEI · CHENNAI · TOKYO

Published by

World Scientific Publishing Co. Pte. Ltd.
5 Toh Tuck Link, Singapore 596224
USA office: 27 Warren Street, Suite 401-402, Hackensack, NJ 07601
UK office: 57 Shelton Street, Covent Garden, London WC2H 9HE

Library of Congress Cataloging-in-Publication Data
Names: Ashtekar, Abhay, editor. | Pullin, Jorge, editor.
Title: Loop quantum gravity : the first 30 years / editors: Abhay Ashtekar
  (The Pennsylvania State University, USA), Jorge Pullin (Louisiana State University, USA).
Other titles: 100 years of general relativity ; v. 4.
Description: Singapore ; Hackensack, NJ : World Scientific, [2017] |
  Series: 100 years of general relativity, ISSN 2424-8223 ; vol. 4 |
  Includes bibliographical references and index.
Identifiers: LCCN 2017005542| ISBN 9789813209923 (hardcover ; alk. paper) |
  ISBN 9813209925 (hardcover ; alk. paper) | ISBN 9789813209930 (pbk. ; alk. paper) |
  ISBN 9813209933 (pbk. ; alk. paper)
Subjects: LCSH: Quantum gravity. | Quantum cosmology. | Quantum theory.
Classification: LCC QC178 .L87 2017 | DDC 539.7/54--dc23
LC record available at https://lccn.loc.gov/2017005542

British Library Cataloguing-in-Publication Data
A catalogue record for this book is available from the British Library.

Cover credit: Dr. Andreas Müller (Technical University of Munich)

Copyright © 2017 by World Scientific Publishing Co. Pte. Ltd.

All rights reserved. This book, or parts thereof, may not be reproduced in any form or by any means, electronic or mechanical, including photocopying, recording or any information storage and retrieval system now known or to be invented, without written permission from the publisher.

For photocopying of material in this volume, please pay a copying fee through the Copyright Clearance Center, Inc., 222 Rosewood Drive, Danvers, MA 01923, USA. In this case permission to photocopy is not required from the publisher.

Printed in Singapore

Preface

A hundred years ago, Einstein unveiled his final formulation of general relativity. Perhaps the deepest feature of this theory is the encoding of the gravitational interaction in the very geometry of spacetime. Thus, general relativity required that we describe the physical universe using a new syntax — that of the pseudo-Riemannian geometry. As we venture beyond Einstein to unify general relativity with quantum physics, then, we are led to seek a fresh syntax to compose the new conceptual paradigm. Loop quantum gravity (LQG) takes the central lesson of general relativity seriously, adopting the view that the natural home of Planck scale physics would be in *quantum* Riemannian geometry.

Creation of this syntax is challenging because all of twentieth century physics presupposes a classical spacetime, with its rigid metric, sharp light cones, and ensuing micro-causality. It is quite unsettling to lift the anchor that tied us to these safe havens and learn to sail the open seas, foregoing the comfort of an underlying spacetime continuum. Yet, hundreds of researchers in LQG chose to abandon this comfort zone, venture forth, and toil together to make this courageous leap. Given that it took astronomers and physicists many decades to come to grips with the dynamical nature of spacetime enshrined in general relativity, it is not surprising that it has taken a while to unfold the new syntax and use it to address the age-old, central problems of quantum gravity. In the 1990s, they laid down the foundations of quantum geometry. The central idea was to use a reformulation of general relativity in the language of gauge theories without, however, any reference to a background spacetime geometry. The ensuing framework turned out not only to be natural in intuitive, geometric terms, but it is also supported by precise uniqueness theorems that reveal the astonishing power of background independence. This kinematic framework provides the foundation for the current efforts to describe non-perturbative, background independent quantum dynamics. Over the last fifteen years, these efforts have provided a rich set of concrete results, creating new paradigms to describe the very early universe and quantum properties of black holes. In particular, the quantum nature of spacetime geometry has been shown to be directly responsible for resolution of the most important singularities that plague general relativity, for an unforeseen interplay between the ultraviolet and the infrared in the very early universe, and for creating avenues to recover the information that is apparently lost behind black hole horizons. The maturity of the field is reflected in the fact that now there are several research programs aimed at bridging observations and the Planck scale physics emerging from quantum geometry, especially in the context of the early universe.

The eight Chapters of this volume, authored by leading younger experts in various areas, provide a comprehensive overview of the current state of the field. Thus, the volume complements the existing excellent texts on LQG, in that it is up to date and covers all the main areas, rather than focusing on a few. We envisage the volume to serve three purposes. *First*, it provides a detailed but concise introduction to the field for beginning researchers. Part **2** introduces the conceptual, mathematical and physical foundations, while Part **3** summarizes the applications to black holes and the early universe, bringing out a detailed interplay between the theory and observations. The material contained in these Chapters is sufficiently detailed to provide a thorough understanding of the basic structure that is needed to pursue research projects specialized in one area or another. *Second*, the volume should also be helpful to researchers who are already working in LQG. The subject has grown immensely, with tens of thousands of journal articles. Therefore, it has become increasingly difficult to follow, even in the International LQG seminars, the current lines of research in areas of LQG that one is not actively working on. Parts **2** and **3** provide the material that is necessary first to understand the current research, and then to contribute to these other areas. Part **1** should help in understanding the interrelations between diverse ideas, distinguishing what is well-established from what remains open, and grasping the healthy tensions that are fertile grounds for further research. *Third*, the monograph should also be useful to physicists, mathematicians and cosmologists outside LQG, who are interested in issues at the interface of general relativity and quantum physics. They will see the opportunities and challenges that accompany the notion of 'background independence', the deep interplay between geometry and physics, and fertile areas where Planck scale physics can have observational implications in the foreseeable future. Finally, senior experts can also use this monograph as a text for an advanced course.

Since the volume is part of the '100 Years of General Relativity' series of monographs, we will conclude by comparing and contrasting LQG with other approaches to physics beyond Einstein. A key feature of LQG is that it is rooted in well-established physics: principles of general relativity and quantum mechanics. The underlying viewpoint is that ideas that have no observational support should not constitute an integral part of the foundation of quantum gravity, even when they can lead to rich mathematical structures. In particular, a negative cosmological constant, extended objects, supersymmetry and specific matter content involving towers of fields and particles do not feature in the fundamentals of LQG. The starting point is general relativity coupled with matter — just as the starting point in QED is the classical Maxwell theory with charged sources. However, LQG is also radical in important ways. The fundamental quanta of geometry are one dimensional, polymer-like excitations over nothing, rather than gravitons, the wavy undulations over a continuum background. In particular, classical general relativity is recovered only in an appropriate coarse-grained limit. This balance between well-established principles and radical ideas is a hallmark of LQG. Another key feature is the prominent role of the quantum nature of spacetime geometry. In particular, in LQG the area operator has a lowest non-zero eigenvalue, called the area gap. It is a basic microscopic parameter of the theory that dictates the macroscopic parameters, such as the maximum value that matter density and curvature can attain in the very early universe. Thus, quantum geometry provides a natural, built-in ultraviolet cutoff. In this respect it differs from other approaches such as Asymptotic Safety or Dynamical Triangulations that are more closely aligned with standard quantum field theories. Because of its emphasis on quantum geometry and non-perturbative techniques, LQG is well placed to address the long standing problems of quantum gravity, such as the resolution of physically important singularities, the so-called 'trans-Planckian issues', the 'problem of time' and diffeomorphism covariance. Indeed, over the past decade and a half, these issues have been at the center of investigations in LQG. On the other hand, the very emphasis on quantum geometry and physics at the Planck scale has made it difficult for LQG to make rapid progress on establishing a detailed contact with low energy effective theories, and on finding implications of quantum gravity on matter couplings. By contrast, the Asymptotic Safety program, for example, has made significant progress in both directions. In String Theory, advances have occurred in yet other directions. Since the advent of AdS/CFT some two decades ago, focus of research has slowly moved away from quintessentially quantum gravity issues to applications of techniques from general relativity and supergravity to problems in an array of non-gravitational areas of physics. Thus, because leading approaches use diverse points of departure, reflecting the striking differences on what should be regarded as fundamental, they have led to new insights in different directions, reflecting their complementary strengths. Given the difficulty of the task, this diversity is both healthy and essential.

State College and Baton Rouge
December 31st 2016

Acknowledgments

We are grateful to all authors for graciously agreeing to contribute, to make revisions, and especially for their patience with the slow editing process. They have provided a fresh perspective on the current status of the field that will be useful not only for beginning researchers but also experts, since the field has now grown so much. We would also like to thank numerous referees who reviewed the eight Chapters as well as the Introduction and made valuable comments. Throughout this project, we received efficient help from the World Scientific staff. We would especially like to thank our Editor, Ms Lakshmi Narayan, for her diligence, care, and flexibility. Finally, the idea of commemorating *100 Years of General Relativity* with a series of monographs, including this volume, came from Dr. K. K. Phua, who also provided all the support we needed for this project.

This work was supported in part by NSF grants PHY-1505411, PHY-1305000, PHY-16030630, the Eberly research funds of Penn State and the Hearne Institute and the Center for Computation and Technology of Louisiana State University.

Contents

Preface   v

List of Symbols   1

Part 1: Introduction   3

    An Overview   5
    Abhay Ashtekar and Jorge Pullin

Part 2: Foundations of loop quantum gravity   29

1. Quantum Geometry   31
   Kristina Giesel

2. Quantum Dynamics   69
   Alok Laddha and Madhavan Varadarajan

3. Spinfoam Gravity   97
   Eugenio Bianchi

4. Group Field Theory and Loop Quantum Gravity   125
   Daniele Oriti

5. The Continuum Limit of Loop Quantum Gravity: A Framework for Solving the Theory   153
   Bianca Dittrich

Part 3: Applications of loop quantum gravity   **181**

6.   Loop Quantum Cosmology   183
   Ivan Agullo and Parampreet Singh

7.   Quantum Geometry and Black Holes   241
   J. Fernando Barbero G. and Alejandro Perez

8.   Loop Quantum Gravity and Observations   281
   Aurélien Barrau and Julien Grain

Subject Index   307

List of Symbols

$a, b, \ldots$ — spacial indices in the 3-manifold M
$i, j, \ldots$ — internal indices in so(3) = su(2)
A^i_a — a connection 1-form on M
$A(e)$ — the holonomy transport along a curve e defined by a connection A
$\mathcal{A}$ — the space of the connections on M with a given gauge group G
$\bar{A}(e)$ — the holonomy along a curve e defined by a generalized connection $\bar{A}$
$\hat{A}(e)$ — the corresponding quantum operator
$\bar{\mathcal{A}}$ — the quantum configuration space of the generalized connections
A_S — the classical area of a 2-surface S
$\hat{A}_S$ — the corresponding quantum operator
$\mathcal{C}_{\text{Diff}}(\vec{N})$ — the diffeomorphism constraint corresponding to a shift vector field $\vec{N}$
$\mathcal{C}(N)$ — the scalar constraint corresponding to a laps function N
$\mathcal{C}_{\text{G}}(\Lambda)$ — the Gauss constraint corresponding to a Lagrange multiplier Λ^i
$C^{(n)}$ — a class of differentiability
Cyl — the algebra of the cylindrical functions on $\mathcal{A}$
Cyl$_\alpha$ — the algebra of the cylindrical functions compatible with a given graph α
Cyl* — the space of linear functionals defined on Cyl
Cyl$^\star_{\text{diff}}$ — the image of Cyl upon the diffeomorphism averaging map
Diff(M) — the space of diffeomorphisms of M
E^a_i — 3-vector density, representing geometry on M
ϵ_{ijk} — the structure constants of $\mathfrak{g}$ and su(2) in particular
η_{ij} — the Killing form in $\mathfrak{g}$ and in su(2) in particular
η_{abc}, η^{abc} — the anti-symmetric pseudo tensor densities, defined by $\eta_{123} = 1 = \eta^{123}$ in arbitrary coordinate system
η — the diffeomorphism averaging map
F^i_{ab} — the curvature of A^i_a
G — a fixed compact Lie group
G_{N} — Newton's constant
$\mathfrak{g}$ — its Lie algebra

List of Symbols

γ — Immirzi parameter
$\mathcal{H}$ — the Hilbert space of the cylindrical functions
$\mathcal{H}_\alpha$ — the Hilbert subspace of cylindrical functions compatible with α
$\mathcal{I}_E, \mathcal{I}_V$ — maps from the space of connections and, respectively, gauge transformations on a graph, into the Cartesian products G^n
$\hat{J}_i^{(v,e)}$ — the spin operator associated to an edge e of α and its vertex v
k — $8\pi G_\mathrm{N}$
κ — the black hole surface gravity
$\kappa(S,e)$ — a $-1, 1, 0$ number assigned suitably to a surface S and an oriented edge e
ℓ_Pl — the Planck length
L^2 — the space of square integrable functions
$\hat{L}_i$ — a frame of left invariant vector fields on the group G multiplied by $\sqrt{-1}$
M — a 3-dimensional manifold, a spacial slice of space-time
P_a^i — the momentum field canonically conjugate to A_a^i
$P(S,f)$ — the flux of $P_a^i f_i$ across a two surface S
$\hat{P}(S,f)$ — the quantum flux operator
$\hat{e}M(g)$ — the quantum Maxwell electric field smeared against g
q_{ab} — a metric tensor defined on M
$\hat{q}_{v,\alpha}$ — the quantum operator representing determinant of $q_{ab}(v)$, restricted to Cyl_α
$\hat{R}_i$ — a frame of right invariant vector fields on the group G multiplied by $\sqrt{-1}$
Tr — trace
$V_\mathcal{R}$ — the volume of a region $\mathcal{R}$ defined by q_{ab}
$\hat{V}_\mathcal{R}$ — the corresponding quantum volume

Part 1
Introduction

... a really new field of experience will always lead to crystallization of a new system of scientific concepts and laws ... when faced with essentially new intellectual challenges, we continually follow the example of Columbus who possessed the courage to leave the known world in almost insane hope of finding land beyond the sea.

– Werner Heisenberg (Changes in the Foundation of Exact Science)

An Overview

Abhay Ashtekar[*] and Jorge Pullin[†]

[*]*Institute for Gravitation and the Cosmos and Department of Physics, Pennsylvania State University, University Park, PA 16802, USA*
[†]*Department of Physics and Astronomy, Louisiana State University, Baton Rouge, LA 70803-4001, USA*

The quote on the last page from Heisenberg's essay *Changes in the Foundation of Exact Science* succinctly captures the spirit that drives Loop Quantum Gravity (LQG). One leaves behind the *terra firma* of a rigid spacetime continuum in the hope of finding a more supple and richer habitat for physics beyond Einstein. In Columbus' case, while the vision and hope that led to the expedition were indeed almost 'insane', he was well aware of the risks. Therefore he embarked on the voyage well prepared, equipped with the most reliable navigational charts and tools then available. Similarly, in LQG one starts with principles of general relativity and quantum mechanics that are firmly rooted in observations, knowing fully well that one will encounter surprises along the way and habitats at the destination will not look anything like those on these charts. Yet the known charts are essential at the point of departure to ensure that the sails are properly aligned and one does not drift into a fantasy landscape with little relation to the physical world we inhabit.

The five chapters in Part **2** of this volume describe current status of this voyage and the new habitats it has already led to. One starts with well established general relativity coupled to matter and uses proven tools from quantum mechanics insisting, however, that there be no fields in the background, not even a spacetime metric. This insistence leads one to a rigorous mathematical framework whose conceptual implications are deep [1–3]. Quantum spacetime does not look like a 4-dimensional continuum at all; fundamental excitations of geometry — and hence of gravity — are polymer like; geometric observables have purely discrete eigenvalues; local curvature in the classical theory is replaced by non-local holonomies of a spin connection; and, quantum dynamics inherits a natural, built-in ultraviolet cut-off. There is no 'objective time' to describe quantum dynamics; there are no rigid light cones to formulate causality. Nonetheless, time evolution can be described in detail through *relational* dynamics in the cosmological setting, where familiar causality emerges through *qualitatively new* effective descriptions that are valid all the way to the full Planck regime. Thus, the final landscape is very different from that of general relativity and quantum mechanics, although both provided guiding principles at the point of departure.

The three Chapters of Part **3** illustrate the Planck scale 'flora and fauna' that inhabits this new landscape. The ultraviolet properties of geometry naturally tames the most important singularities of general relativity — in particular, the big bang is replaced by a quantum bounce. In the very early universe, cosmological perturbations propagate on these regular, bouncing *quantum* geometries, giving rise to effects that are within observational reach. Quantum geometry has also opened a new window on the microscopic degrees of freedom of horizons. Singularity resolution gives rise to a quantum extension of classical spacetimes, creating new paradigms for black hole evaporation in which the evolution is unitary. Finally, several possibilities have been proposed to test LQG ideas in astrophysics and cosmology. They all involve additional assumptions/hypotheses beyond mainstream LQG. Nonetheless, the very fact that relation to observations can be contemplated, sometimes through detailed calculations, provides a measure of the maturity of the subject.

The purpose of this Introduction is twofold: (i) To provide a global overview to aid the beginning researcher navigate through Parts II and III, especially by comparing and contrasting ideas in individual Chapters; and, (ii) Supplement the detailed discussions in these Parts with a brief discussion of a few general, conceptually important points. To keep the bibliography to a manageable size, we will refer only to reviews and monographs (rather than research articles) where further details can be found. *We urge the beginning readers to read this Introduction first, as it spells out the overall viewpoint and motivation that is often taken for granted in individual Chapters.*

The Setting

In LQG one adopts the viewpoint that among fundamental forces of Nature, gravity is special: it is encoded in the very geometry of spacetime. This is a central feature of GR, a crystallization of the equivalence principle that lies at the heart of the theory. Therefore, one argues, it should be incorporated at a fundamental level in a viable quantum theory.

The perturbative treatments which dominated the field since the 1960s ignored this aspect of gravity. They assumed that the underlying spacetime can be taken to be a continuum, endowed with a smooth background geometry, and the quantum gravitational field can be treated as any other quantum field on this background. But the resulting perturbation theory around Minkowski spacetime turned out to be non-renormalizable; the strategy failed to achieve the initial goals. The new strategy is to free oneself of the background spacetime continuum that seemed indispensable for formulating and addressing physical questions. In particular, in contrast to approaches developed by particle physicists, one does not begin with quantum matter on a background geometry and then use perturbation theory to incorporate quantum effects of gravity. Matter *and geometry* are both quantum-mechanical at birth. There is often an underlying manifold but no metric, or indeed any other physical fields, in the background.[a]

In classical gravity, Riemannian geometry provides the appropriate mathematical language to formulate the physical, kinematical notions as well as the final dynamical equations. This role is now taken by *quantum* Riemannian geometry. In the classical domain, general relativity stands out as the best available theory of gravity, some of whose predictions have been tested to an amazing degree of accuracy, surpassing even the legendary tests of quantum electrodynamics. Therefore, it is natural to ask: *Does quantum general relativity, coupled to suitable matter exist as a consistent theory non-perturbatively?* There is no implication that such a theory would be the final, complete description of Nature. Nonetheless, this is a fascinating and important open question in its own right.

In particle physics circles the answer to this question is often assumed to be in the negative, not because there is concrete evidence against non-perturbative quantum gravity, but because of the analogy to the theory of weak interactions. There, one first had a 4-point interaction model due to Fermi which works quite well at low energies but which fails to be renormalizable. Progress occurred not by looking for non-perturbative formulations of the Fermi model but by replacing the model by the Glashow-Salam-Weinberg renormalizable theory of electro-weak interactions, in which the 4-point interaction is replaced by $W^\pm$ and Z propagators. Therefore, it is often assumed that perturbative non-renormalizability of quantum general relativity points in a similar direction. However this argument overlooks the crucial fact that, in the case of general relativity, there is a qualitatively new element. Perturbative treatments pre-suppose that the spacetime can be assumed to be a continuum *at all scales of interest to physics under consideration*. This assumption is safe for weak interactions. In the gravitational case, on the other hand, the scale of interest is *the Planck length* $\ell_{\rm Pl}$ and there is no physical basis to pre-suppose that the continuum picture should be valid down to that scale. The failure of the standard perturbative treatments may largely be due to this grossly incorrect assumption and a non-perturbative treatment which correctly incorporates the physical micro-structure of geometry may well be free of these inconsistencies.

Are there any situations, outside loop quantum gravity, where such physical expectations are borne out in detail mathematically? The answer is in the affirmative. There exist quantum field theories (such as the Gross-Neveu model in three dimensions) in which the standard perturbation expansion is not renormalizable although the theory is *exactly soluble* [5]! Failure of the standard perturbation expansion can occur because one insists on perturbing around the trivial, Gaussian

---

[a]In 2+1 dimensions, although one begins in a completely analogous fashion, in the final picture one can get rid of the background manifold as well. Thus, the fundamental theory can be formulated combinatorially [4]. In 3+1 dimensions, combinatorial descriptions emerge in several approaches to dynamics but one does not yet have a complete theory.

point rather than the more physical, non-trivial fixed point of the renormalization group (RG) flow. Interestingly, thanks to developments in the Asymptotic Safety program there is now growing evidence that situation may be similar in quantum general relativity [6]. Although there are some basic differences [7] between LQG and the Asymptotic Safety program, these results provide concrete support to the idea that non-perturbative treatments of quantum general relativity can lead to an ultraviolet regular theory.[b]

However, even if the LQG program could be carried out to completion, there is no a priori reason to assume that the result would be the 'final' theory of all known physics. In particular, as is the case with classical general relativity, while requirements of background independence and general covariance do restrict the form of interactions between gravity and matter fields and among matter fields themselves, the theory would not have a built-in principle which *determines* these interactions. Put differently, such a theory may not be a satisfactory candidate for unification of all known forces. However, just as general relativity has had powerful implications in spite of this limitation in the classical domain, LQG should have qualitatively new predictions, pushing further the existing frontiers of physics. Indeed, unification does not appear to be an essential criterion for usefulness of a theory even in other interactions. QCD, for example, is a powerful theory even though it does not unify strong interactions with electro-weak ones. Furthermore, the fact that we do not yet have a viable candidate for grand unified theory does not make QCD any less useful. Finally, as the three Chapters in Part **3** illustrate, LQG has already made interesting predictions for quantum physics of black holes and the very early universe, some of which are detailed and make direct contact with observations.

Quantum Riemannian Geometry

Since the basic dynamical variable in general relativity is the spacetime metric, Wheeler advocated the view that we should regard it as *geometrodynamics*, a dynamical theory of 3-metrics q_{ab} that constitute the configuration variable. For the three other basic forces of Nature, on the other hand, the dynamical variable is a connection 1-form that takes values in the Lie algebra of the appropriate internal group. In QED the connection enables one to parallel transport electrons and positrons while in QCD it serves as the vehicle to parallel transport quarks. Now the configuration variable is a spatial connections A_a^i; we have theories of *connection-dynamics*. Weinberg, in particular, has emphasized that this difference has driven a 'wedge between general relativity and the theory of elementary particles' [9].

As described in *Chapter 1*, the starting point in LQG is a reformulation of general relativity as a dynamical theory of spin connections [4]. We now know that

---

[b]In the Asymptotic Safety program, spacetime geometry in the Planck regime is effectively 2-dimensional as in LQG and the 4-dimensional continuum arises only in the low energy limit [6].

the idea can be traced back to Einstein and Schrödinger who, among others, had recast general relativity as a theory of connections already in the fifties. (For a brief account of this fascinating history, see [10].) However, they used the 'Levi-Civita connection' that features in the parallel transport of vectors and found that the theory becomes rather complicated. The situation is very different with self-dual (or anti-self-dual) spin connections. For example, the dynamical evolution dictated by Einstein's equations can now be visualized simply as a *geodesic motion* on the 'superspace' of spin-connections (with respect to a natural metric extracted from the constraint equations). Furthermore, the (anti-)self-dual connections have a direct physical interpretation: they are the vehicles used to parallel transport spinors with definite helicities of the standard model. With this formulation of general relativity, Weinberg's 'wedge' disappears. In particular, phase-space of general relativity is now the same as that of gauge theories of the other three forces of Nature [4, 11]. However, in the Lorentzian signature, the (anti-)self-dual connections are complex-valued and, so far, this fact has been a road-block in the construction of a rigorous mathematical framework in the passage to the quantum theory.[c] Therefore, the strategy is to pass to real connection variables by performing a canonical transformation [1–3]. The canonical transformation introduces a real, dimensionless constant γ, referred to as the Barbero-Immirzi parameter; the (anti-) self-dual Hamiltonian framework is recovered by formally setting $\gamma = \pm i$.

For real connection variables the 'internal gauge group' reduces to SU(2), which is compact. Therefore, as explained in *Chapter 1*, it is possible to introduce integral and differential calculus on the infinite-dimensional space of (generalized) connections rigorously without having to introduce background geometrical fields, such as a metric. Since this space serves as the quantum configuration space, one can introduce a Hilbert space of square integrable functions and physically interesting self-adjoint operators thereon. This setup then serves as the kinematical framework in LQG.[d]

The most important features of this background independent framework are the following. *First,* just as one has the von-Neumann uniqueness theorem in quantum

---

[c]At a formal level in which one often works in quantum field theory (QFT), one can carry out calculations with complex connections. The road block refers to rigorous mathematical constructions. Specifically, with complex connections, it has not been possible to introduce measures and develop differential geometry on the infinite dimensional space of generalized connections that serve as the quantum configuration space. However, as is the standard practice in constructive QFT, it may be possible to work in the Riemannian signature where (anti-)self-dual connections are real and then pass to the Lorentzian theory through a *generalized* Wick transform that does exist [3]. This is an opportunity for future research.

[d]In rigorous QFT in Minkowski space, while the set of smooth fields that constitute the classical configuration space is topologically dense in the quantum configuration space on which wave functions live, the quantum measure is concentrated on genuinely distributional fields. In LQG the situation is similar. Smooth connections are topologically dense in the space of generalized ones but contained in a set of measure zero. This level of rigor is essential if the kinematical framework is to serve as the point of departure for quantum dynamics. It was never attained in geometrodynamics — the Wheeler-DeWitt framework has remained entirely formal.

mechanics of a finite number of degrees of freedom that singles out the standard Schrödinger representation, there are uniqueness theorems that single out this kinematical framework. This is a highly non-trivial result because the system has an infinite number of degrees of freedom, made possible only because the requirement of background independence is very stringent. *Second,* in this framework, geometric operators describing the quantum Riemannian geometry have purely discrete eigenvalues. This is a striking and unforeseen outcome, given that the point of departure is standard general relativity. For example, this discreteness is *not* shared by quantum geometrodynamics (i.e., by the Wheeler-DeWitt theory).

There is a basis in the LQG kinematical Hilbert space which is well-suited to analyze properties of these geometric operators: the *spin network basis* [1–3]. The notion of spin networks was introduced by Penrose already in 1971 in a combinatorial approach to the Euclidean 3-geometry. This notion was generalized in LQG. Now spin networks are labelled by graphs in which any number of links can meet at nodes and both inks and nodes carry certain information — called 'decorations' or 'colors'. This kinematics brings out the precise sense in which the fundamental excitations of (spatial) geometry in LQG are 1-dimensional, polymer like. Consider for simplicity graphs which have only 4-valent nodes — i.e. in which precisely 4 links meet at each node. Then, one can introduce a simplicial decomposition of the 3-manifold which is dual to the graph: Each node of the graph is contained in a single tetrahedron and each link of the graph associated with that node intersects precisely one face of that tetrahedron. The 'decoration' on the links assigns specific quantized areas to faces of tetrahedra and the 'decoration' at the node determines the volume of that tetrahedron. Thus, each topological tetrahedron can be regarded as an 'atom of space', characterized by the decorations. The volumes of tetrahedra and areas of its faces endow them with geometrical properties. But these geometries are not induced by smooth metrics on the 3-manifold. One thinks of the familiar continuum Riemannian geometries as 'emergent', arising from a coarse graining of the fundamental quantum geometry.

Of particular interest is the smallest non-zero eigenvalue of the area operator, called the *area gap* whose value in Planck units is denoted by Δ. It turns out that Δ depends linearly on the Barbero-Immirzi parameter: $\Delta = 4\sqrt{3}\pi\gamma$. Therefore one can trade one for the other. Conceptually, it is appropriate to regard γ as a 'mathematical parameter' that features in the transition from classical to quantum theory, and Δ as the 'physical parameter' that sets the scale at which quantum geometry effects become important. Thus, from the perspective of the final quantum theory Δ is the fundamental physical parameter. For example, in Loop Quantum Cosmology (LQC) of homogeneous isotropic models, energy density has a maximum value given by $\rho_{\max} = (18\pi/\Delta^3)\rho_{\rm Pl}$ and as $\Delta \to 0$, i.e., as we ignore quantum geometry effects, we recover the classical result $\rho_{\max} \to \infty$. This is completely analogous to the situation for the energy spectrum of the Hydrogen atom in quantum mechanics: the ground state energy is given by $E_0 = -(me^4/2\hbar^2)$ and $E_0 \to -\infty$, the classical value, as $\hbar \to 0$. More generally, one should formulate physical questions in terms of Δ, replacing γ by $\Delta/(4\sqrt{3}\pi)$ in expressions of interest. In the older LQC literature, some confusion arose because one took the limit $\Delta \to 0$, keeping γ fixed in expressions that also involved γ. Similarly, the discussion of black hole entropy becomes significantly clearer if everything is formulated in terms of Δ. In particular, since it is physically clear that the microcanonical calculations of entropy using Planck scale quantum geometry should have a dependence on the value of the area gap, one is not tempted to find arguments to make it independent of γ.[e]

The quantum geometry framework is described in detail in *Chapter 1* and lies at the foundation of issues discussed in later Chapters.

Non-perturbative, Background-independent Dynamics

Chapters 2-5 summarize the current status of dynamics in full LQG from both the Hamiltonian and path integral perspectives.

In the early years of LQG the primary focus was on considering general relativity as a constrained Hamiltonian theory — but not in terms of 3-metrics and extrinsic curvature, but rather in terms of spin connections and their canonically conjugate momenta, the spatial triads (with density weight 1). Because the phase space is the same as in an SU(2) Yang-Mills theory, one could import into gravity the well-developed techniques from gauge theories. However, unlike the familiar Yang-Mills theory in Minkowski spacetime, now there is no metric or any other field in the background. In particular, expressions of general relativity constraints involve only the *dynamical* phase space variables. Since there are no fiducial geometric structures, and coordinates themselves can be rescaled arbitrarily without affecting underlying physics, techniques used in Minkowskian quantum field theories to regulate products of field operators are no longer useful. Therefore a number of novel and astute techniques had to be developed to construct the physical sector of the theory by imposing quantum constraints a la Dirac in the background independent kinematical framework of *Chapter 1*. (In the Hamiltonian framework this is equivalent to solving quantum Einstein's equations.) These developments are discussed in *Chapter 2*. However, as discussed there, to make the constraint operators well-defined in the rigorous setting provided by the kinematical framework, one had to introduce a number of auxiliary structures. This is not surprising by itself; such constructions are also needed to regulate products of operators in Minkowskian quantum field theories. However, in LQG the *final* physical sector of the theory depended on the scheme chosen and the physical meaning of these differences remained opaque.

---

[e]In semi-classical considerations, arguments are restricted to states representing 'near horizon, classical geometries' and one calculates entropy using canonical and micro-canonical ensembles. Then the final result can be insensitive to the area gap Δ.

What was needed was a principle to streamline the calculations and reduce the available freedom. Perhaps the most natural and most attractive of these principles is to demand that the quantum constraint algebra should be closed not only on quantum states that satisfy all or some of the constraints, but on the full kinematical state space. The regularization strategies used in the 1990s could not shed light on this issue of 'off-shell closure' of quantum constraints. Over the past five years or so, this idea is being systematically implemented in models with increasing complexity. The task is technically difficult. A key idea is to use the fact that, in connection dynamics, the difficult Hamiltonian constraint can be regarded as a diffeomorphism constraint in which the shift fields themselves have certain specific dependence on the phase space variables i.e., so called 'q-number' quantities. (This simplification does not occur in geometrodynamics, where the two constraints have entirely different forms.)

Therefore effort is directed to constructing quantum operators that generate infinitesimal (generalized) diffeomorphisms. (In the early LQG works one only had operators implementing *finite* and 'c-number' diffeomorphisms.) The strategy works in simpler models such as parametrized field theories, which were invented to mimic features associated with background independence of general relativity. In these theories, the new LQG techniques have enabled one to overcome some long standing obstacles and construct a satisfactory quantum theory with fundamental discreteness as well as covariance. These ideas have also been successfully applied to certain models that arise from simplifications of general relativity. In these systems, the freedom in quantization is neatly streamlined by the requirement of 'off-shell' closure of quantum constraints. These systematic advances, summarized in *Chapter 2*, have opened concrete directions to complete the Dirac program in the framework of connection dynamics.

Chapter 3 introduces the basics of spin foams, the sum over histories approach which has been a primary focus of recent work on quantum dynamics [12, 13]. Recall that in his original derivation of path integrals, Feynman began with the expressions of transition amplitudes in Schrödinger quantum mechanics and *reformulated* them as an integral over all kinematically allowed paths [14]. In background independent theories, on the other hand, we have a constrained Hamiltonian framework. As we saw above, in the Dirac program, physical states simply solve the quantum constraints and one has to tease out dynamics, e.g., by introducing a relational time variable. Therefore, on formally mimicking the Feynman procedure starting from the Hamiltonian framework, one finds that the analog of the transition amplitude is an *extraction amplitude*. This is a Green's function that extracts from 'incoming' (or 'outgoing') kinematical states, solutions to quantum constraints and also provides the physical inner product between them. Thus, path integrals provide an alternate, covariant avenue to construct the physical Hilbert space of the theory. If the theory can be deparametrized, it inherits a relational time variable and then the extraction amplitude can be re-interpreted as a transition amplitude with respect to that time.[f] Irrespective of whether this is possible, the basic object that encodes quantum dynamics is the extraction amplitude.

In heuristic treatments of sum over histories, kinematical paths are generally represented by smooth classical fields. However, in rigorous QFT it is well known that these paths are contained in a set of measure zero; the measure is concentrated on genuinely distributional fields. Situation is very similar in LQG: the measure is concentrated on generalized connections rather than smooth ones (see footnote (d)). Consequently in spin foams the sum is over *quantum* spacetime geometries rather than smooth metrics. These are represented by 'decorated' 2-complexes that can be heuristically thought of as 'time evolution' of spin networks. Quantum geometries associated with a given 2-complex can be regarded as paths that interpolate between given 'incoming' and 'outgoing' spin networks, the 'decorations' providing specific spacetime quantum geometries which are described in some detail in *Chapter 3*.

In any given 2-complex, (zero-dimensional) nodes of the 'incoming' spin network 'evolve' into (one-dimensional) edges, but every now and then a vertex is created characterizing a 'non-trivial happening'. There is no time, yet 'happenings' are objectively coded in each quantum history. Each 'decorated' 2-complex carries a fixed number of such happenings but can represent many different quantum geometries depending on the choice of decorations. In *any* choice of decorations, areas of 2-dimensional faces are quantized and there is a minimum non-zero area — the area gap. Thus, properties of the underlying quantum geometry provide a natural ultraviolet cut off. To carry out the path integral, one has to assign amplitudes to the faces and vertices of the 2-complex. The non-trivial part turns out to be the specification of the vertex amplitude. The first concrete prescription was given in the Barrett-Crane model which opened up the field of spin foams. However, later attempts to calculate the graviton propagator in Minkowski spacetime starting from non-perturbative spin foams revealed some important limitations of this model. It was replaced by the Engle–Pereira-Rovelli-Livine (EPRL) and (the closely related) Freidel-Krasnov (FK) models. In these models, the built-in, natural cut off makes the integral over 'decorations' ultraviolet finite in any given 2-complex. It is unlikely that there is an analogous infrared finiteness in general relativity with zero cosmological constant. However, in presence of a positive cosmological constant, one is naturally led to replace the SU(2) group of internal rotations (of spinors) with its quantum analog SU(2)$_q$ and the amplitudes have been shown to be infrared finite as well.

---

[f]In LQC these steps have been carried out rigorously in the Friedmann-Lemaître-Robertson-Walker (FLRW) as well as Bianchi I models. That is, one can begin with the well-defined Hamiltonian quantum theory and arrive at the covariant cosmological spin foam, obtain the exact Green's function for the extraction amplitude, show that it admits a 'vertex expansion' that is convergent, and interpret the extraction amplitude as a transition amplitude by an appropriate deparameterization [15].

In full LQG, of course, one must go beyond a single 2-complex and consider 'all possible' 2-complexes interpolating between the two spin networks, allowing for an arbitrary number of vertices. *Chapters 4 and 5* summarize two complementary but different approaches to fulfill this task.

Chapter 4 presents a more general view of sum over histories which, at first, seems very different from spin foams but in fact first arose as a 'generalized Fourier transform' from spin foam models. The point of departure is neither a Hamiltonian formulation of general relativity as in the canonical approach, nor a 'constrained topological theory' with which one starts in spin foams. Not only are there no background fields such as a metric but there is no spacetime manifold at a fundamental level. The underlying idea is that gravity is *to emerge* from a more fundamental theory based on abstract structures that, to begin with, have nothing to do with spacetime geometry. Drawing inspiration from the matrix models (in 2 spacetime dimensions) and especially the 'Boulatov-model' (in 3 spacetime dimensions) the fundamental object is a QFT but formulated on a *group manifold, rather than spacetime*. Therefore the framework is aptly called 'Group Field Theory' (GFT). As in familiar field theories, the Lagrangian has a free and an interaction term, with a coupling constant λ. Even though the point of departure appears to be so different, one can again use the LQG kinematics and represent the 'in' and the 'out' states by spin networks as in spin foams.[g] Remarkably, for a certain choice of the Lagrangian, the n-th term in the perturbation expansion — i.e., the coefficient of λ^n — is the same as the contribution to the extraction amplitude obtained by fixing a 2-complex with precisely n vertices, and summing over the decorations in the EPRL model. But because it arises in a standard perturbation expansion — albeit on a group manifold, not spacetime — one can now borrow techniques from standard QFT. This is especially important in order to (i) go beyond a single 2-complex (i.e., a fixed number of vertices); and (ii) analyze potentially distinct phases, as in standard QFT.

In this summary we have presented GFT from the perspective of Hamiltonian and spin foam LQG (although, as explained in *Chapter 4*, there are also some technical differences). However, GFT offers greater generality. For example, GFT naturally suggests quantum LQG dynamics of a more general, 'grandcanonical' type in which the number of vertices is allowed to vary. It also opens avenues to discuss the 'continuum limit' using ideas in the spirit of the thermodynamic limit in quantum field theories. In addition, it suggests that even if one begins with GFT actions corresponding to, say, the EPRL model, under the RG flow quantum dynamics

---

[g]Each spin network captures only a finite number of degrees of freedom of the quantum gravitational field which can be interpreted as a 'twisted geometry'. In GFT, one often says that they represent a 'first quantized' theory. The full GFT has operators that create spin network states and is therefore regarded as a 'second quantized' theory. This terminology is *not* used outside GFT and 'second quantization' should not be interpreted as going beyond LQG: LQG has the same underlying mathematical structures.

will generate many more terms at different scales, providing new scale dependent physics. Finally, the framework is so general that it may even allow a departure from a fundamental tenet of the rest of LQG that there are no degrees of freedom beyond the Planck scale. Thus, like QFT in Minkowski spacetime, GFT offers a general paradigm, rather than a physical theory. To specify a physical theory, one has to choose a set of fields — now on a group manifold — and fix interactions between them. Hence its scope differs from the more focused approaches to quantum dynamics discussed in *Chapters 2 and 3*. In recent years the emphasis has been on exploiting the generality it offers by applying the ideas to specific models. As in *Chapter 2*, it has been successfully applied to simpler systems. The state of the art can be summarized by saying that GFT has opened a number of new avenues that have the potential to resolve the key open issues in 4-dimensional spin foams.

Finally, in *Chapter 5* we return to the important open issue of the continuum limit in spin foams, which is now taken using a generalization of the standard RG flow induced by refinements in the spacetime manifold in the spirit of lattice QCD, rather than through a field theory on a group manifold. Here, one uses methods from lattice QFT, functional analysis and tensor networks, rather than perturbative expansions in coupling constants together with techniques from non-perturbative QFTs, used in GFT. More precisely, the physical states — i.e., solutions of the quantum constraint — are to be constructed by taking the refinement limit in spacetime.

Let us begin with the kinematical setup of *Chapter 1*. The strategy employed there is to first define structures, such as the scalar product between states and action of geometric and holonomy operators on them, using the finite degrees of freedom that are captured in a single graph. These structures then naturally extend to the full state space in the continuum that captures all the infinitely many degrees of freedom *provided* they satisfy stringent consistency conditions as one coarse grains or refines graphs by adding new nodes and links. These are the so-called *cylindrical consistency conditions* [1]. Quantum geometry has been successfully constructed in the kinematical setting precisely because these cylindrical consistency requirements were met. The idea now is to promote these consistency requirements *to dynamics*, using spin foams of *Chapter 3*.

One can start with a fixed simplicial decomposition of the spacetime manifold together with its dual 2-complex. The 2-complex induces spin network states on the initial and final 3-manifolds which can be regarded as the 'in' and 'out' quantum states of geometry (or, more precisely, simply kinematic states out of which one wishes to extract physical ones). As noted above, a spin foam model provides a 'transition amplitude' between them (or, more precisely, a Green's function that extracts a physical state from the two given kinematical ones). The idea is that one should consider only those refinements (and coarse grainings) of the simplicial decomposition — and hence of the dual 2-complex — that make the procedure cylindrically consistent, so that physics at different 'scales' is appropriately related to constitute a coherent scheme. This is the sense in which the notion of cylindrical consistency is to be promoted from the kinematical to the dynamical setting. A refinement procedure that meets the consistency conditions would then provide a generalization of the RG flow ideas to a setting in which there is no background metric to define the scale. The question of whether such a refinement limit exists is similar to the asymptotic safety conjecture that an ultraviolet fixed point exists. In the same spirit as asymptotic safety, once the program is developed beyond pure gravity, the hope is that cylindrical consistency will severely constrain matter couplings since it is a stringent requirement.

Chapter 5 explains these consistency conditions and also the inductive limit that is to provide the final quantum dynamics in the continuum limit through an admissible refinement. Again, as in *Chapters 2 and 4*, the procedure has been successfully applied to simpler models, now involving decorated tensor networks and an iterative procedure. The general viewpoint in LQG is that the so called Ashtekar-Lewandowski (AL) representation (of the fundamental LQG quantum algebra) underlying the kinematic setup correctly captures the essential features of quantum geometry at Planck scale. But by loosening the requirements that led to the uniqueness of this representation, one can construct a 'dual' description, called the BF representation. General arguments have been put forward to suggest that it may be more directly useful for describing the phase of the theory containing macroscopic, continuum geometries. Each of these representations has a vacuum state and the two vacua are very different from one another. An important feature of the overall strategy is that the truncation scheme is to be determined by dynamics. 'Coarse states' will have few excitations while 'fine states' will have many excitations with respect to a vacuum that is also determined by dynamics.

This approach to continuum limit has several attractive features. First, the procedure brings out a close relation between the the continuum quantum dynamics and spacetime diffeomorphism symmetry in systems with (auxiliary) discrete structures, reflecting the intuitive idea that the diffeomorphism symmetry allows one to refine or coarse grain any region. Second, it provides a background independent analog of a 'complete renormalization trajectory' through the notion of cylindrically consistent amplitudes. Finally, the procedure already has the necessary ingredients in place to lead to the running of coupling constants, once the system is extended to allow for matter sources. As in *Chapters 2 and 4*, this is an ongoing program; now the open question is whether cylindrically consistent amplitudes exist in full 4-dimensional LQG.

Applications

While important issues remain in full LQG, the basic underlying ideas have been successfully applied to two physical sectors of full gravity: black holes and the very early universe. Part **3** of this volume summarize these advances.

Cosmic microwave background (CMB) observations have established that the early universe is spatially homogeneous and isotropic to one part in 10^5. Therefore the current paradigms of the early universe assume that spacetime geometry is well described by a Friedmann-Lemaître-Robertson-Walker (FLRW) geometry, together with first order perturbations prior to the CMB epoch. Although we do not yet have a definitive paradigm, the inflationary scenario has emerged as the leading candidate. In particular, it has been very successful in accounting for the one part in 10^5 inhomogeneities observed in the CMB. Known physics and astrophysics show that these inhomogeneities serve as seeds for formation of the large scale structure we see in the universe. Thus, inflation pushes the issue of the origin of the observed large scale structure further back from the CMB epoch — in fact to *very* early times, when the spacetime curvature was some 10^{62} times that at the surface of a solar mass black hole! While this is truly impressive, from the viewpoint of quantum gravity this epoch lies in the classical general relativity regime since the curvature is still some 10^{-14} times the Planck curvature. That is why it is consistent — as is done in all current paradigms of the early universe scenario — to describe the universe using a classical FLRW background and represent the cosmological perturbations by quantum fields propagating on it. However, this strategy is inadequate if one wishes to go to still earlier times and describe what happened when the matter density and curvature were of Planck scale. For this, one needs a quantum theory of gravity. The inadequacy of standard inflation is brought out by two facts: (i) the big-bang singularity persists in this theory, and all physics comes to a halt there; and (ii) quantum field theory on FLRW backgrounds, used to describe the dynamics of cosmological perturbations, becomes inadequate because even modes that can be observed in the CMB acquire trans-Planckian frequencies in the early epoch. Thus, a challenge to any candidate quantum gravity theory is to provide a completion of the inflationary scenario over the 11 orders of magnitude in matter density and curvature that separate it from the Planck scale and successfully address these issues. As *Chapter 6* describes, in LQG there have been remarkable advances in this direction.

Let us begin with the first issue — that of the resolution of the big-bang singularity in the background spacetime. In LQG cosmological singularities are resolved in all models that have been studied so far. These include the flat and closed FLRW models with and without a cosmological constant (of either sign); the anisotropic Bianchi models that contain non-linear gravitational waves; and the inhomogeneous Gowdy models which also contain non-linear gravitational waves [15]. The resolution does not come about by introducing matter that violates energy conditions or by some fine tuning. The origin of the mechanism can be traced back directly to the underlying quantum geometry — particularly the emergence of the area gap — described in *Chapter 1*. Quantum geometry creates a brand new repulsive force. It is completely negligible already at the onset of inflation and thereafter. However if we evolve back in time, it grows rapidly, overwhelms the classical attraction in the Planck regime, and causes the universe to bounce. Thus the big bang is replaced by a quantum bounce. More generally, the following picture succinctly summarizes the salient features of quantum dynamics in cosmological models: anytime a curvature invariant starts to grow in general relativity signaling approach to a singularity, the repulsive force grows to dilute it, preventing its formation.

What is behind this singularity resolution? In Loop Quantum Cosmology (LQC) one applies the LQG techniques to the cosmological setting described above. It turns out that even after standard gauge fixing there is still some residual diffeomorphism freedom in cosmological models. The requirement of covariance under this freedom — i.e., background dependence in the cosmological context — again leads to a unique representation of the fundamental quantum algebra. As in the AL (or BF) representation for full LQG, the connection operator is not well defined in LQC; only its exponential, the holonomy, is a well defined (unitary) operator. In this LQC representation of quantum states, then, (as in full LQG) one has to express the curvature term in the Hamiltonian constraint in terms of holonomies, now around a loop which encloses the minimum possible physical area defined by the quantum state. As a result the quantum Hamiltonian constraint depends on the area gap and the operator reduces to the Wheeler DeWitt operator of geometrodynamics only in the limit in which the area gap goes to zero. At a fundamental level, there is no time. But in presence of suitable matter, one can deparameterize the theory and interpret the Hamiltonian constraint as an evolution equation in a relational time variable provided, e.g., by matter. One can show that physical (Dirac) observables — such as the energy density or curvature at a given instant of relational time — that diverge at the big bang in general relativity have a finite upper bound on the entire space of physical states. Thus the singularity is resolved in a very direct, physical sense: There are simply *no* states in the physical Hilbert space in which matter density diverges. This resolution has been analyzed using Hamiltonian and path integral methods and has also been discussed in the 'decoherent histories' framework. *Chapter 6* summarizes most of these results and provides references for topics that could not be covered.

Thus, in LQC, the classical FLRW backgrounds $(a(t), \phi(t))$ used in the inflationary scenario are replaced by a wave function $\Psi(a, \phi)$ in the physical Hilbert space. Quantum fields representing the scalar (curvature) and tensor perturbations now propagate on the quantum FLRW geometry $\Psi(a, \phi)$. Over the last several years, quantum field theory on classical FLRW spacetimes was systematically generalized to quantum field theory on these *quantum* FLRW spacetimes. This generalization enables one to face the 'trans-Planckian issues' squarely and has therefore been used to analyze dynamics of the scalar and tensor modes through the Planck regime, all the way to the quantum bounce. This analysis, as well as other methods described in *Chapter 8*, have led to interesting phenomenological predictions that can be tested against observations.

Of particular interest is an unforeseen interplay between the ultraviolet and the infrared. Quantum geometry effects that tame the singularity provide a new ultraviolet LQC length scale $\ell_{\rm LQC}$, the minimum curvature radius — corresponding to the maximum scalar curvature at the bounce. Modes of perturbations whose physical wavelength $\lambda_{\rm phys}$ is *larger* than $\ell_{\rm LQC}$ experience curvature during their evolution in the Planck regime near the bounce and are excited. These turn out to have the longest wavelength among the modes seen in the CMB observations. Thus, dynamics in the Planck regime can leave imprints at the largest angular scales in the sky. Detailed calculations have been performed to use this effect to account for the anomalies that the PLANCK and WMAP teams have found at the largest angular scales in CMB. Furthermore, predictions have been made for other correlation functions that should soon be reported by the PLANCK team for the largest angular scales. Thus, LQC has been a fertile ground within LQG. It has led to a concrete and detailed realization of several key underlying ideas — use of relational time, construction of a complete set of Dirac observables, relation between the canonical framework and spin foams, and the hope that the vertex expansion in spin foams is convergent. At the same time, it has brought quantum gravity from lofty heights of mathematical physics to concrete, observational issues in phenomenology.

Chapter 7 summarizes results on the black hole sector of LQG. Just as the early universe provides us with possibly the best opportunity of directly observing quantum gravity effects, black holes provide us with possibly the best arena to test quantum gravity theories at a conceptual level. Specifically any viable quantum gravity has to address two issues in the black hole sector. The *first* arises from black hole thermodynamics. Einstein equations within classical general relativity inform us that black holes obey certain laws: the zeroth refers to equilibrium configurations, the first to transition from an equilibrium state to a nearby one, and the second to full dynamical situations. Remarkably, these laws become the zeroth, the first and the second laws of thermodynamics if one identifies a multiple of the surface gravity κ of the horizon with temperature T, and $(1/(8\pi G)$ times) the reciprocal multiple of the horizon area $A_{\rm H}$ with (Bekenstein-Hawking) entropy $S_{\rm BH}$. However, purely from dimensional grounds one finds that the multiple must have the same dimensions as $\hbar$, bringing in quantum mechanics in a totally unforeseen fashion. Subsequent calculation, using quantum field theory in the Schwarzschild black hole spacetime, established the Hawking effect: Black holes evaporate quantum mechanically and at late times the outgoing state is extremely well approximated by the thermal radiation from a black body at the (Hawking) temperature $T_H = \kappa\hbar/2\pi$. These discoveries, made over 4 decades ago, make it clear that black holes hold a key to bring together the three pillars of physics — general relativity, thermodynamics and quantum mechanics. They imply that a solar mass black hole has $\exp 10^{77}$ microstates, an enormous number even on statistical mechanics standards. One is thus immediately led to the question: Can we account for the enormous horizon entropy $S_{\rm BH}$ using more fundamental statistical mechanical considerations? The *second* issue concerns the dynamics of the evaporation process. What happens in a self-consistent treatment when one includes back reaction so that, by energy conservation, the black hole mass decreases? Does the horizon evaporate completely? The black hole may have been formed by sending in a variety of, say, pure states. Are we left with particles in a thermal state at the end of evaporation? If so, the evaporation process would not be unitary and information would be lost. Even after four decades, there are no clear answers to this second set of questions in any approach to quantum gravity. Therefore *Chapter 7* focuses only to the first set of issues in this section.

Event horizons that one normally associates with black holes are *extremely* global as well as teleological notions. For example, they can form and grow in a flat, Minkowskian region of a spacetime. Similarly, they can be gotten rid of simply by changing spacetime geometry in a tiny neighborhood of the singularity. Furthermore, it is not unlikely that the spacetime describing an evaporating black hole has an event horizon. Calculations with 2-dimensional black holes that include the back reaction and solve for geometry using a mean field approximation strongly indicate that the full spacetime will not have an event horizon. Therefore, as explained in detail in *Chapter 7*, in LQG one uses the quasi-local notion of a (weakly) *isolated* horizon for which the zeroth and the first laws of black hole mechanics do hold [16]. An additional advantage is that one can incorporate *both* the black hole and the cosmological horizons in one swoop, and one does *not* have to restrict oneself to near extremal black holes.[h]

In LQG, one focuses on the quantum geometry of isolated horizons. The 'quanta of geometry' provide the microstates of the quantum horizon that are used in a statistical mechanical entropy calculation. The subject is mathematically rich. It involves quantum groups, Chern-Simons theory on punctured spheres, mapping class groups and sophisticated techniques from number theory. Detailed calculations show that the number of microstates has interesting properties, especially for small black holes. For large black holes, the entropy $S_{\rm BH}$ is indeed proportional to the horizon area $A_{\rm H}$ but this coefficient is inversely proportional to the area gap Δ of LQG. This is just as one would expect on general grounds. So, as the area gap goes to zero, i.e., we ignore quantum geometry effects, the entropy becomes infinite. However, as a result, the Bekenstein-Hawking formula $S_{\rm BH} = A_{\rm H}/4G\hbar$ is recovered only for a certain value of the area gap Δ. Put differently, by demanding that LQG have the correct semi-classical limit to leading order, one can fix the

---

[h]To obtain the second law, one needs *dynamical* horizons which are also quasi-local notions. While event horizons are null hypersurfaces, dynamical horizons are space-like when the black hole is growing during collapse and time-like when it is evaporating. Unlike event horizons, they are *not* one way membranes. This fact removes considerable confusion in the literature on the evaporation process. There is strong indication that dynamical horizons do exist for evaporating black holes. (When matter is neither falling into the black hole nor leaving it, the dynamical horizon becomes null and an isolated horizon.) [16]

value of the area gap, and hence remove the a priori 1-parameter ambiguity in the kinematical setup of LQG. This general picture is well-established. However, as the discussion in *Chapter 7* shows, there are ambiguities in the precise definition of what constitutes a quantum state of the horizon geometry, primarily because the surface states of the horizon quantum geometry are entangled with states of the bulk quantum geometry. Therefore, at a conceptual level, the subject is not fully settled. However, the ambiguities are all very small to make a difference in practice: The area gap varies only between 5.16 and 5.96 (in Planck units).

These considerations refer to quantum geometry and Planck scale eigenvalues of the area operator. In recent years, there has been a significant advance in relating LQG to the vast literature on black hole thermodynamics based on semi-classical considerations. The main idea is to consider the near horizon geometry corresponding to that of a stationary black hole solution and shift the perspective to that of a suitable family of near horizon stationary observers. In such geometries, one can relate the energy E, as seen by stationary observers at a distance ℓ from the horizon, to the area A_H of the horizon, the low energy Newton constant G_N and ℓ. The 'semi-classical input' is the assumption that the physical sector of LQG contains states that are peaked around solutions admitting isolated horizons with such near horizon geometries. Under this assumption, one can use the expression of E and the LQG expression of the area operator $\hat{A}_H$ to construct the Hamiltonian operator. In this step, the Planck length ℓ_{Pl} in the expression of the area operator is assumed to be given by $\ell_{Pl}^2 = G\hbar$ where G is the gravitational constant in the Planck regime (which one would expect to run to G_N in the low energy regime). Using the Hamiltonian operator (which features G, G_N and ℓ) one can construct a canonical ensemble at the Unruh temperature $T = 2\hbar/\ell$ associated with the stationary observers and calculate the entropy. The leading term in the result turns out to be precisely the Bekenstein-Hawking entropy, with no extra factor. The dependence on the auxiliary parameter ℓ as well as on the area gap Δ drops out of this semi-classical result based on the canonical ensemble tailored to stationary, near horizon observers. The calculation also sheds new light on how the possible running of gravitational constant can be naturally accommodated, even when the Hawking-Bekenstein formula uses the low energy value G_N of the gravitational constant and the area operator from quantum geometry uses the Planck scale value G.

Finally, there have been a number of other developments. First, using spin foams, entanglement entropy associated with horizons has been re-examined. It has been shown that if one changes focus from entropy to variation of entropy, a lot of unnecessary conceptual and computational baggage is removed. In particular the species problem is alleviated. It has also been suggested that it is the entanglement entropy that features in a precise statement of the first law in semi-classical gravity and the idea is under more detailed investigation. Another — and major — direction is a systematic investigation of the quantum geometry underlying a black hole spacetime. This has been successfully carried out under the assumption of spherical symmetry. As in cosmological models, the singularity is naturally resolved by quantum geometry effects. Then it is natural to study quantum fields propagating on this *quantum* black hole geometry. These investigations bring out how the underlying quantum geometry provides a natural ultraviolet regularization in the QFT calculations. Finally, there are calculations of Hawking effect on the *quantum* geometry representing a spherically symmetric black hole, again in the test field approximation. The quantum nature of the underlying geometry leads to ultraviolet modifications of the quantum vacua, eliminating the trans-Planckian modes close to the horizon. As a result, expectation value of the stress energy tensor are ultraviolet finite. Hawking radiation at infinity has been computed and, apart from an ultraviolet cutoff, the outgoing state is the same as the conventional one. In particular, this analysis provides a concrete quantum gravity calculation showing that Hawking radiation is not seriously contaminated by the apparent trans-Planckian problems of QFT in classical, curved spacetimes. Currently, these calculations are being revisited and extended to study the evaporation process.

Chapter 8 covers a broad range of issues related to phenomenology and observations. They include direct and indirect probes into the very early universe, potential for Lorentz invariance violations, modifications of the spectrum of an evaporating black hole due to the discreteness of area eigenvalues, Planck stars and the possibility that the emission from them is related to Gamma ray bursts (GRBs). In the discussion of cosmological probes there is an inevitable, small overlap with *Chapter 6* but the rest of the material is presented for the first time in this volume. The material covered here illustrates the wide scope of current research in LQG, in particular the fact that now there is a significant community that is looking beyond foundational, conceptual and mathematical issues into the interface with observations.

Conceptually, perhaps the most ambitious of these ideas is that of 'Planck stars' which is now superseded by the more recent work on 'fireworks'. One knows from quantum cosmology of closed FLRW models that quantum geometry effects dominate and cause a quantum bounce once the curvature reaches the Planck scale. The curvature at the surface of a solar mass collapsing star reaches the Planck scale when its radius is ~ 13 orders of magnitude larger than the Planck length. Thus the view that quantum gravity can drastically modify classical dynamics only when the star has shrunk to the Planck size is unsubstantiated. It is reasonable to suppose that something like a quantum bounce would occur when the curvature at the surface of the star reaches the Planck scale. The question is: What is the time scale — as measured at infinity — that is involved in this bounce? The time scale for the Hawking evaporation goes as M^3. If the time scale for the bounce is smaller, say $\sim M^2$, then the bounce would dominate, making Hawking evaporation irrelevant for large black holes. The hope in the program is that something like this does happen. This is a bold conjecture. The LQG group pursuing these ideas is well aware of the fact that extraordinary conjectures require extraordinary proofs and is therefore engaged in building up the necessary evidence through detailed calculations.

Concepts and results that are needed as prerequisites to understand the contents of *Chapter 8* have already been discussed in this Introduction. Therefore, we will conclude with just one remark. As in other approaches to quantum gravity, to make contact with observations, one has to make additional assumptions that are not part of LQG proper, since situations of interest to observations are too complex to be systematically arrived at starting from a fundamental quantum gravity theory. Therefore, observations probe the package consisting of the fundamental theory being used in calculations, *together with* the additional assumptions that are made to arrive at phenomenological predictions. This is also the case in other areas of physics, such as QCD. Much of the work on quark-gluon plasmas, for example, makes hydrodynamical assumptions that are physically motivated but are not known to be direct consequences of fundamental QCD. Therefore, if a prediction is falsified, one cannot conclude that there is a problem with QCD; it is *much* more likely that the problem lies with the additional assumptions. Of course, QCD has absolutely huge observational support compared to any quantum gravity theory. But the general spirit of this research in LQG is the same as that in these other areas: use observations to refine paradigms, assumptions and strategies. If a particularly clean prediction is verified, it would give confidence in the underlying assumptions and encourage more detailed analysis within that paradigm, leading to further predictions. If observations contradict a prediction, they provide guidance as to which of the assumptions are suspect and need to be weeded out. However, because observational evidence for any quantum gravity theory is scant, one constantly keeps an eye on the possibility that the observations are suggesting or even requiring a change in the fundamentals of LQG and LQC.

Closing Remarks

Perhaps the most distinguishing feature of LQG is the underlying quantum geometry. Intuitively one expects some sort of discreteness at the Planck scale. Much of the heuristic literature as well as some of the systematic quantum gravity approaches assume that quanta of geometry have a simple structure with lengths that are integral multiples of ℓ_{Pl}, and/or areas that are integral multiples of ℓ_{Pl}^2 and/or volumes that are integral multiples of ℓ_{Pl}^3. The quantum Riemannian geometry of LQG is much more subtle. In particular, the spectra of various geometric operators are quite different from one another; for example, one cannot deduce the eigenvalues of the volume operator from those of the area operator. Secondly, while there is an area gap, higher eigenvalues of the area operator crowd and the level spacing decreases exponentially! Consequently the continuum geometry is approached very quickly. These specific features were first discovered in the 1990s but remained mathematical curiosities until recently. Now such details are turning out to be important in the interplay between theory and observations.

For example, in the Bekenstein-Mukhanov approach to black hole entropy, area eigenvalues are of the type $g\, N\ell_{\rm Pl}^2$ where $g \sim \mathcal{O}(1)$ can be thought of the degeneracy factor — or, the number of 'bits' — in the basic quantum of area. This means that when a black hole is perturbed it makes a transition from an initial state of the horizon characterized by an integer N_1, to a final state characterized by another integer N_2. Therefore, the change ΔA in the horizon area occurs in discrete steps: $\Delta A = g\,(N_2 - N_1)\ell_{\rm Pl}^2$. Recently, these ideas were re-examined using the relation between area, mass and spin of Kerr black holes, and frequency of quasi-normal modes that describe their 'ringing' under perturbations. Specifically, it was argued that the Bekenstein-Mukhanov proposal leads to a different description of ringing and therefore it may soon be under stress once LIGO observes quasi-normal ringing from a sufficient number of spinning black holes. By contrast, for black holes of interest to LIGO, the LQG area spectrum becomes almost continuous due to the exponential crowding of eigenvalues. Therefore, there is no reason for predictions to be observationally distinct from those of classical general relativity. Another example comes from cosmology, where one can constrain the value of the area gap Δ from observations. More precisely, one can leave Δ as a free parameter in LQC calculations and obtain the value that makes the predicted power spectrum fit best with the PLANCK mission observations. Recent investigations in LQC have carried out this task, providing a completely independent way of arriving at the value of area gap Δ (and hence of the Barbero-Immirzi parameter γ). The best-fit value of Δ agrees with that obtained from black hole entropy calculations within the 68% confidence level used by the PLANCK mission to report its data. LQG researchers who first investigated geometric operators in the 1990s did not anticipate that the details they carefully worked out would start making a difference on observational fronts within just two decades.

For quantum dynamics in full LQC, a number of important issues remain. The conceptual framework and mathematical techniques introduced in the early discussions were crucial to get the program rolling. But, with hindsight, we now see that these discussions failed to take into account key issues such as physics at different scales, renormalization group flows, and consistency requirements in the continuum limit procedure. That is why the program has not made as much contact with low energy physics as one would have hoped. Now the community is well aware of these limitations and is actively engaged in overcoming them. *Chapters 2-5* summarize the three directions that are being pursued: Hamiltonian theory a la Dirac, spin foams and their continuum limit, and GFT. While the final goal is essentially the same — uncovering dynamics through solutions to the quantum constraints — there are also some differences and healthy tensions. The Hamiltonian methods of *Chapter 2* focus on canonical gravity while methods used in *Chapters 3-5* come from path integrals. They have complementary strengths and limitations. For example, in terms of concrete results in 4 dimensions, the Hamiltonian methods have been much more useful in the study of the early universe and black hole entropy, while path integral methods have shown how one can obtain the graviton 2-point function in Minkowski space in a background with independent setting. Even within the path integral approaches there are differences. Roughly, GFT of *Chapter 4* is modeled more on methods from QFT, while the continuum limit of spin foams discussed in *Chapter 5* generalizes ideas and techniques used in condensed matter physics.

Given the variety of difficult issues one encounters in non-perturbative quantum dynamics, we believe this diversity is essential. It represents a key strength of the program. Existing tensions between these sets of ideas will bring to forefront deeper issues and launch new investigations leading to a more coherent overall framework. In particular, they should lead to advances on two key open issues:

(i) Contact with the standard *low energy* effective theory. The derivation of the graviton 2-point function, for example, has been a major achievement but one needs to make sure that the current calculations are not changed in the leading order by including spin foams with a large number of vertices, or in an appropriate continuum limit. It is also important to obtain higher order corrections in a reliable fashion; and,

(ii) Matter couplings. We know that there is no conceptual difficulty in incorporating matter either in the Hamiltonian framework [3, 4] or in spin foams [2]. But details have not been worked out. In particular, a satisfactory derivation of the classical limit for gravity interacting with matter is still lacking. It is important to understand whether consistency requirements — such as those used in *Chapters 2 and 5* — strongly constrain matter couplings. In these investigations, recent results in the asymptotic safety program may provide guidance.

Finally, results on symmetry reduced models in the study of the early universe and black holes provide important checks on viability of the main ideas underlying LQG. The fact that the program can be completed in these models and that the results successfully address long standing questions of quantum gravity is a nontrivial indication that the program is well-founded.

However, there is always the issue of whether an important aspect of physics is overlooked by first focusing on a symmetry reduced sector of the classical theory and then passing to the quantum theory. The Dirac model of the hydrogen atom sheds interesting light on this issue. Here one considers the proton-electron system in quantum electrodynamics (QED), truncates the theory to its *spherical symmetric sector* and then carries out quantization. From the perspective of full QED, the strategy seems to introduce a *drastic* oversimplification since it banishes all the photons right from the start! Indeed, conceptually, the truncated theory does ignore most of the rich possibilities one can envisage in full QED. And yet the Dirac theory provides an excellent description of the hydrogen atom. Indeed, to see its limitations, one has to carry out *very* accurate measurements and carefully examine the hyperfine structure due to QED effects such as the Lamb shift! Returning to quantum gravity, it is not unreasonable to expect that the symmetry reduction strategy would be appropriate *provided* it keeps an eye on the structure of the full theory, i.e., provided the quantum theory of the reduced model is not constructed in a manner that is specifically engineered just to fit that model. In LQG this view is taken seriously: quantum theories of reduced models pay due attention to the quantum geometry that emerged from *full* LQG. Therefore, for the limited class of observables that are needed to describe the very early universe and spherically symmetric black holes, the varied and rich possibilities of full quantum gravity could just refine, rather than alter, the predictions of the reduced models. Of course there is no a priori guarantee that this will be the case. Therefore, over the past 2-3 years there has been an increased effort on finding the precise relation between the symmetry reduced quantum theories and full LQG. In particular, there are strong results showing that the LQC faithfully captures the LQG kinematics in the homogeneous, isotropic sector. At the dynamical level, there are partial results that go in the same direction.

We will conclude with a few general remarks about this volume. The list of mathematical symbols that appears in the beginning is intended to help readers as they navigate through detailed arguments. It was sent to all authors and by and large they have followed the conventions laid down in that list. There are several senior figures in LQG who have made invaluable contributions and driven major advances over the past two decades or more. But to provide a fresh perspective that emphasizes future directions, we thought it would be best to invite some of the younger researchers who are currently leading research programs in new directions. We urged them to express their outlook on where we stand and what the strategies are best suited for future advances. Consequently various Chapters express personal visions that sometimes boldly venture beyond the general consensus in the field. This is especially the case on some issues of quantum dynamics and at the interface of theory and observations. In this respect, our intention is captured in the best spirit of the motto: 'let a thousand flowers bloom'. The Editors do not subscribe to all the ideas and views expressed in the 8 Chapters. Rather they feel it is important that beginning researchers be guided by younger leaders who will shape the future of LQG.

References

[1] A. Ashtekar A and J. Lewandowski, Background independent quantum gravity: A status report, *Class. Quant. Grav.* **21** R53-R152 (2004).
[2] C. Rovelli, *Quantum Gravity* (Cambridge University Press, Cambridge (2004)).
[3] T. Thiemann, *Introduction to Modern Canonical Quantum General Relativity* (Cambridge University Press, Cambridge (2005)).

[4] A. Ashtekar, *Lectures on Non-perturbative Canonical Gravity,* Notes prepared in collaboration with R. S. Tate (World Scientific, Singapore (1991)).
[5] J. Fröhlich, *Non-perturbative Quantum Field Theory: Mathematical Aspects and Applications*, pages 630-31 (World Scientific, Singapore (1992)).
[6] M. Niedermaier and M. Reuter, The asymptotic safety scenario in quantum gravity, *Living Rev. Rel.* **9** (2006);
M. Reuter and F. Saueressig, *New. J. Phys.* **14** 055022 (2012).
[7] A. Ashtekar, M. Reuter and C. Rovelli, From general relativity to Quantum Gravity, in *General Relativity and Gravitation: A Centennial Perspective* (Cambridge University Press, Cambridge (2015)).
[8] J. A. Wheeler, *Geometrodynamics*, (Academic Press, New York, (1962)).
[9] S. Weinberg, *Gravitation and Cosmology* (John Wiley, New York, (1972)).
[10] A. Ashtekar, Quantum mechanics of geometry, in *The Universe: Visions and Perspectives*, eds Dadhich N and Kembhavi A (Kluwer Academic, Dordretch, (2000)); gr-qc/9901023.
[11] J. Baez and J. P. Muniain *Gauge Fields, Knots and Gravity* (World Scientific, Singapore (1994)).
[12] A. Perez, The spin foam approach to quantum gravity, *Living Rev. Rel.* **16** (2013).
[13] C. Rovelli and F. Vidotto, *Covariant Loop Quantum Gravity* (Cambridge University Press, Cambridge (2014)).
[14] R. P. Feynman and M. R. Hibbs, *Quantum Mechanics and Path Integrals* Emended Edition (McGraw Hill, New York (2005)).
[15] A. Ashtekar and P. Singh, Loop quantum cosmology: A status report, *Class. Quant. Grav.* **28** 213001 (2011).
[16] A. Ashtekar and B. Krishnan, Isolated and dynamical horizons and their applications, *Living Rev. Rel.* **7** 191 (2004).

Part 2
Foundations

... I think it must be the case that the all-pervading use of the continuum in physics stems from its mathematical utility rather than from any essential physical reality that it may possess. However, it is not even quite clear that such use of the continuum is not, to some extent, a historical accident. ... My own view is that ultimately physical laws should find their most natural expressions in terms of essentially combinatorial principles ... Thus in accordance with such a view, [there] should emerge some form of discrete or combinatorial space-time.

– Roger Penrose (On the Nature of Quantum Geometry)

Chapter 1

Quantum Geometry

Kristina Giesel

*Institute for Quantum Gravity (IQG), FAU-Erlangen-Nürnberg,
Staudtstr. 7, 91058 Erlangen, Germany*

1. Canonical Quantization of General Relativity

With their seminal work in 1960 on the canonical formulation of general relativity, which has come to be called the ADM-formalism, Arnowitt, Deser and Misner [1] provided the background for research which focused on the question how general relativity can be quantized using the technique of canonical quantization. In the covariant formulation of general relativity the elementary variable is a Lorentzian metric $g_{\mu\nu}$, where we use the signature $(3,1)$, on a four-dimensional differentiable manifold $\mathcal{M}$. The equation of motion for $g_{\mu\nu}$ are given by Einstein's equations and encode the dynamics of general relativity. In the context of the ADM-formalism the four-dimensional space time $(\mathcal{M}, g_{\mu\nu})$ is replaced by a 3+1-dimensional picture, using the fact that the topology of globally hyperbolic space times is of the form $\mathcal{M} \simeq \mathbb{R} \times M$ [2]. One associates $\mathbb{R}$ with time and M with space. Hence, the four-dimensional manifold $\mathcal{M}$ is considered as a foliation of three-dimensional space-like hypersurfaces $X_t(M)$ labeled by a parameter $t \in \mathbb{R}$, where $X_t : M \to \mathcal{M}$ is an embedding of the spatial manifold M into $\mathcal{M}$. A particular choice of time and space would break diffeomorphism invariance and therefore in the framework of the ADM-formalism one does not choose a particular foliation but considers *all possible ones*. In the canonical framework the elementary configuration variables are the pull back of the metric $g_{\mu\nu}$ onto M denoted by q_{ab} from now on, also called the ADM 3-metric. The conjugate momenta, denoted by p^{ab}, are related to the extrinsic curvature of the space-like hypersurfaces $X_t(M)$. The diffeomorphism invariance of the theory has the consequence that general relativity is a constrained Hamiltonian theory meaning that in addition to the Hamiltonian equation of motion for q_{ab} and p^{ab} the theory possesses constraints, which are additional equations on phase space, that q_{ab} and p^{ab} have to satisfy. Therefore the constraints select out of the kinematical degrees of freedom (q_{ab}, p^{ab}), which still include gauge degrees of freedom, the physical degrees of freedom. In the case of the ADM-formalism these constraints are called Hamiltonian and (spatial) diffeomorphism constraint.

The latter generates diffeomorphisms within the spatial hypersurface M and the Hamiltonian constraint is generating diffeomorphisms orthogonal to the hypersurface. Note that in the case of the Hamiltonian constraint this is only true on shell, that is when the constraints are satisfied, and in addition when the equation of motion are fulfilled. Furthermore, general relativity is in this sense special as its Hamiltonian consists entirely of a linear combination of the constraints and therefore general relativity is called a fully constrained theory. This property has to be taken into account when one discusses the notion of observables, that is gauge invariant quantities, in the context of general relativity, see for instance Refs. [3] and [4].

For theories with constraints there exist two different approaches to formulate the quantum theory. First, known as Dirac [5] quantization, is based on the idea of quantizing the entire kinematical phase space, including the gauge degrees of freedom, yielding the so called kinematical Hilbert space $\mathcal{H}_{\text{kin}}$. Let us denote the set of classical constraints by $\{\mathcal{C}_I\}$ with $I \in \mathcal{I}$ where $\mathcal{I}$ denotes some arbitrary index set. Then the physical sector of the theory is constructed in the quantum theory by implementing all classical constraints $\{\mathcal{C}_I\}$ as operators $\{\hat{C}_I\}$ on $\mathcal{H}_{\text{kin}}$ and requiring that physical states ψ be annihilated by all constraints operators, that is $\hat{\mathcal{C}}_I \psi = 0$ for all $I \in \mathcal{I}$. These physical states are elements of the so called physical Hilbert space $\mathcal{H}_{\text{phys}}$. The second approach called reduced quantization follows the strategy to solve the constraints already at the classical level. In doing so, one obtains the reduced — also called physical — phase space whose elementary variables are called observables because there are gauge invariant quantities and do not include gauge degrees of freedom any longer. Then one quantizes the physical phase space, which corresponds to the task of finding suitable representations of the algebra of observables leading directly to the physical Hilbert space $\mathcal{H}_{\text{phys}}$. In addition one is only interested in those representations which also allow to implement the dynamics of those observables in the quantum theory.

Now in practice one often does not exclusively follow Dirac or reduced quantization but often combines both approaches. If for example the classical constraints are complicated to solve, Dirac quantization might be of advantage as long one is able to solve the corresponding quantum constraint equations. On the other hand, if one is able to reduce the constraints at the classical level, one quantizes only the physical phase space and has thus a direct access to the physical Hilbert space. Therefore one is interested in both approaches. The technical difficulty in the reduced quantization occurs when the resulting algebra of observables has a much more complicated structure than the corresponding kinematical one because it might be impossible to find representations of the algebra and hence to formulate the quantum theory at all.

As far as the ADM-variables are concerned one has mainly followed the Dirac quantization procedure and used standard Schrödinger quantization techniques, known from other quantum field theories, to construct the corresponding kinematical Hilbert space $\mathcal{H}_{\text{kin}}$ for general relativity [6]. If we denote the diffeomorphism constraint by $\mathcal{C}_{\text{Diff}}$ and the Hamiltonian constraint by $\mathcal{C}$ then one needs to find quantum states $\psi(q_{ab})$ that satisfy $\hat{\mathcal{C}}_{\text{Diff}}\psi(q_{ab}) = 0$ and $\hat{\mathcal{C}}\psi(q_{ab}) = 0$. The latter equation involving the Hamiltonian constraint is also known as the Wheeler–DeWitt equation. However, this quantization of general relativity has to be understood rather at a formal level because not all details about the measure underlying $\mathcal{H}_{\text{kin}}$ have been worked out. Also, using ADM-variables and the standard Schrödinger representation it has up to now not been shown that the Hamiltonian constraint operator can be implemented on $\mathcal{H}_{\text{kin}}$ as a well defined operator. Exactly these difficulties have been the starting point for reconsidering the canonical quantization of general relativity from a different angle. We will see in the next section that a different choice of elementary variables, called connection or Ashtekar variables, to describe the canonical formulation of general relativity will allow to formulate the kinematical Hilbert space $\mathcal{H}_{\text{kin}}$ for general relativity not only at the formal level but will allow to implement all constraints of general relativity as operators on $\mathcal{H}_{\text{kin}}$.

2. General Relativity in Connection Variables

The motivation for deriving a formulation of general relativity in terms of connection variables is that it allows to describe general relativity in a language very close to the language that is used in other quantum field theories for which already powerful quantization techniques exist.

The starting point for the connection formulation is to describe general relativity in terms of frames. A frame field denoted by $e_I := e_I^\mu \partial_\mu$ with $I = 0, 1, 2, 3$ defines a basis of the tangent space $T_p\mathcal{M}$ at each point p of $\mathcal{M}$. Here we will discuss the connection formulation by starting already with the ADM 3+1-split of the spacetime and therefore work with frame fields, which are at a point dependent basis for the tangent space T_pM associated with the 3-dimensional manifold M. Usually one works with orthonormal frames, meaning that $e_j := e_j^a \partial_a$ with $j = 1, 2, 3$ satisfy

$$\langle e_j, e_k \rangle = e_j^a e_k^b q_{ab} = \eta_{ij}, \qquad (1)$$

where η_{jk} denotes the components of the Euclidean metric on $\mathbb{R}^3$ and e_j^a is called triad or 3-bein respectively. Given a frame field, we can define the (inverse) 3-metric in terms of the triads

$$q^{ab} = e_j^a e_k^b \eta^{jk}. \qquad (2)$$

Conversely, q^{ab} defines a triad, however only up to $SO(3)$-rotations. Likewise to a frame we can also introduce a co-frame field $e^j := e_a^j dx^a$ being a point dependent basis for the co-tangent space T_p^*M. At each point in M we can view e_j^a and e_a^j as non-singular matrices. Using the isomorphism between the Lie-algebras of $su(2)$ and $so(3)$ we can regard e_a^j as an $su(2)$-valued one-form. When we take this point of view we have to replace η_{ij} by the Killing metric of $su(2)$, which we will also

denote by η_{ij}. Since the co-frame is at each point the dual basis of the frame we have
$$\eta_k^j = e^j(e_k) = e_a^j e_k^b dx^a(\partial_b) = e_a^j e_k^b \delta_b^a = e_a^j e_k^a. \tag{3}$$
Furthermore, we have
$$\delta_b^a = q^{ac}q_{cb} = e_j^a e_k^c \eta^{jk} e_c^m e_n^b \delta_{mn} = e_j^a e_n^b \eta^{jk} \delta_k^m \delta_{mn} = e_j^a e_b^j. \tag{4}$$
Due to the additional $SO(3)$ freedom encoded in the triads the passage from the ADM-phase space to the frame formulation is not a canonical transformation but an extension of the ADM phase space. The elementary variable we will work with is not the triad itself but its densitized version, which is an su(2)-valued vector density of weight one, denoted by E_j^a and defined as
$$E_j^a := \sqrt{\det(q)} e_j^a, \tag{5}$$
where $\det(q) = det(q^{-1})^{-1} = \det(e)^{-2}$ is understood as a function of the triads. The densitized triads will be the momentum variables in the new phase space. As mentioned before in the ADM phase space the canonically conjugate momenta to q_{ab} are related to the extrinsic curvature, which we will denote by K_{ab}. The canonically conjugate configuration variable to the densitized triad is given by
$$K_a^j := K_{ab} e_k^b \eta^{jk}, \tag{6}$$
which is, like e_a^j, a su(2)-valued one-form.

For the reason that we have extended the ADM phase space by additional rotational degrees of freedom encoded in the (co)-frames we obtain the so called rotational constraints given by
$$G_j = \epsilon_{jk\ell} K_a^k E_\ell^a, \tag{7}$$
which ensure, that on shell we obtain again the ADM phase space. Given the canonical pair (K_a^j, E_j^a) we obtain the Ashtekar variables by applying two canonical transformations. The first one is a rescaling of the elementary variables, which introduces the so called Barbero–Immirzi parameter $\gamma \neq 0 \in \mathbb{C}$ into the classical theory
$$K_a^j \to {}^{(\gamma)}K_a^j := \gamma K_a^j \qquad E_j^a \to {}^{(\gamma)}E_j^a := \frac{1}{\gamma} E_j^a. \tag{8}$$
The second canonical transformation involves the spin connection, which we briefly discuss before describing the canonical transformation. Given the metric q_{ab} on M there exists a unique Levi-Civita connection ∇, also called covariant derivative, which is metric compatible, that is $\nabla q_{ab} = 0$ and torsion-free, that is $\Gamma_{bc}^a = \Gamma_{cb}^a$, where Γ_{bc}^a are the Christoffel symbols associated with q_{ab}. Once we introduce triads we have to consider tensors having spatial as well as su(2) indices and therefore we extend the covariant derivative onto tensors with mixed indices by defining
$$\nabla_a t_j^b := \partial_a t_j^b + \Gamma_{ac}^b t_j^c + \Gamma_{a\ j}^{\ k} t_k^b, \tag{9}$$

with $\Gamma_{ajk} = -\Gamma_{akj}$ so that Γ_a is an antisymmetric matrix and takes values in so(3). The extension to arbitrary tensors is obtained by linearity, the Leibniz rule and the requirements that ∇_a commutes with contractions. If we extend the metric compatibility $\nabla_a q_{bc} = 0$ to $\nabla_a e_j^b = 0$ we can express $\Gamma_{a\ j}^{\ k}$ in terms of the (co)-triads and the Christoffel-symbols Γ_{bc}^a given by

$$\Gamma_{a\ k}^{\ j} = -e_k^b \left(\partial_a e_b^j - \Gamma_{ab}^c e_c^j \right).$$

Since Γ_a takes values in so(3), we can use a basis of so(3) denoted by $\{T_1, T_2, T_3\}$ with $(T_i)_{jk} = \epsilon_{ikj}$ to expand Γ_a as $\Gamma_a^j T_j$ with Γ_a^j being the spin connection. Note that we can also consider T_i as the generators of su(2) in the adjoint representation since there exists an isomorphism between su(2) in the adjoint and so(3) in the defining representation. Using the spin connection we can now perform the second canonical transformation, which is an affine transformation, and finally leads to the connection or Ashtekar variables

$$^{(\gamma)}K_a^j \to {}^{(\gamma)}A_a^j := \Gamma_a^j + {}^{(\gamma)}K_a^j \qquad {}^{(\gamma)}E_j^a \to {}^{(\gamma)}E_j^a. \qquad (10)$$

Although Γ_a^j has, as a function of E_j^a, a complicated form it was proven [7, 8], that $({}^{(\gamma)}A_a^j, {}^{(\gamma)}E_j^a)$ build indeed a canonical pair and satisfy the following Poisson algebra

$$\{{}^{(\gamma)}A_a^j(x), {}^{(\gamma)}A_b^k(y)\} = \{{}^{(\gamma)}E_j^a(x), {}^{(\gamma)}E_k^b(y)\} = 0 \qquad (11)$$

$$\{{}^{(\gamma)}A_a^j(x), {}^{(\gamma)}E_k^b(y)\} = k \delta_k^j \delta_a^b \delta^3(x, y), \qquad (12)$$

where $k = 8\pi G_N$ with G_N being Newton's constant. In order to absorb the factor k occurring above in the definition of the elementary variables we use

$$^{(\gamma)}P_j^a := \frac{1}{k} {}^{(\gamma)}E_k^b(y) \qquad (13)$$

as the canonically conjugate momenta to ${}^{(\gamma)}A_a^j(x)$ in the following.

Let us briefly comment on the role of the Barbero–Immirzi-parameter. For each choice of γ we obtain a different set of canonical variables to coordinatize the phase space of general relativity. At this point the choice is arbitrary but might be determined from other physical situations like for instance the computation of the black hole entropy (*see Chapter 7*). In the literature different choices of γ have been discussed, as for example $\gamma = \pm i$ [7] and $\gamma \in \mathbb{R}$ [9] and $\gamma \in \mathbb{C}$ [10]. The choice $\gamma = \pm i$ is special in the sense that
(i) The Hamiltonian constraint — and consequently its later quantization — simplifies, and
(ii) on classical solutions ${}^{(\gamma)}A_a^j$ has the natural geometric meaning of the restriction to M of the self-dual part of the space-time Lorentz connection.
However, in this case the connection ${}^{(\gamma)}A_a^j$ is complex (in the Lorentzian signature) leading to an additional reality condition for ${}^{(\gamma)}A_a^j$ whose implementation on the quantum level is highly non-trivial. Therefore currently, one mainly works with real γ and real connection variables. From now on we will drop the label ${}^{(\gamma)}$ and just use

(A_a^j, P_j^a) in order to keep our notation more clearly and always keep in mind that the construction of the Ashtekar variables involves the Barbero–Immirzi parameter.

As mentioned in the last section the introduction of the Ashtekar variables allows to describe general relativity very close to the language of other gauge theories used in quantum field theory and this point will become clear when we discuss the form of the constraints in terms of Ashtekar variables. We saw that with the extension of the ADM phase space we obtained the rotational constraint in (7). Expressed in terms of (A_a^j, P_j^a) it has the form

$$G_j = \partial_a E_j^a + \epsilon_{jk}{}^\ell A_a^k P_\ell^a =: \mathcal{D}_a P_j^a, \tag{14}$$

where we introduced a new covariant derivative $\mathcal{D}$, that involves instead of the spin connection the SU(2) connection A_a^j. In terms of these new variables the rotational constraints have the form of an SU(2) Gauss law known from Yang–Mills gauge theory. Hence, in terms of the connection variables general relativity can be understood as a SU(2) gauge theory. The remaining constraint, that were already present in the ADM-formalism, are the (spatial) diffeomorphism constraints $\mathcal{C}_a$ and the Hamiltonian constraint $\mathcal{C}$. Using the connection variables and considering the $G_j = 0$ constraint hypersurface, these are given by

$$\mathcal{C}_a = F_{ab}^j P_j^b \quad \mathcal{C} = \frac{k\gamma^2}{2} \frac{\epsilon_j{}^{mn} P_m^a P_n^b}{\sqrt{\det(q)}} \left(F_{ab}^j - (1+\gamma^2)\epsilon^{jk\ell} K_a^k K_b^m \right), \tag{15}$$

where we dropped the term proportional to the Gauss constraint in $\mathcal{C}_a$ and F_{ab}^j is the curvature associated with the connection A_a^j

$$F_{ab}^j = \partial_a A_b^j - \partial_b A_a^j + \epsilon^j{}_{k\ell} A_a^k A_b^\ell \tag{16}$$

and $K_a^j = A_a^j - \Gamma_a^j$ is considered as a function of (A, P) and $\det(q)$ as a function of P. Let us introduce the smeared version of the above constraints

$$\mathcal{C}_G(\Lambda) := \int_M d^3x (\Lambda^j G_j)(x), \quad \mathcal{C}_{\text{Diff}}(\vec{N}) := \int_M d^3x (N^a \mathcal{C}_a)(x) \quad \mathcal{C}(N) := \int_M d^3x (N\mathcal{C})(x). \tag{17}$$

Here Λ^j is lie-algebra-valued smearing field and N and N^a are the lapse function and the shift vector respectively, which in the ADM-formalism are related to the 00 and 0a components of the (inverse) metric $g^{\mu\nu}$ by

$$g^{00} = N^{-2}, \quad g^{0a} = N^{-2} N^a. \tag{18}$$

An aspect that will be later important when the quantization of the (smeared) constraints is discussed is that they satisfy the following constraint algebra

$$\{\mathcal{C}_G(\Lambda), \mathcal{C}_G(\Lambda)\} = \mathcal{C}_G(\Lambda), \quad \{\mathcal{C}_G(\Lambda), \mathcal{C}_{\text{Diff}}(\vec{N})\} = -\mathcal{C}_G(\mathcal{L}_{\vec{N}}\Lambda), \tag{19}$$

$$\{\mathcal{C}_{\text{Diff}}(\vec{N}), \mathcal{C}_{\text{Diff}}(\vec{N}')\} = \mathcal{C}_{\text{Diff}}(\mathcal{L}_{\vec{N}}\vec{N}'), \quad \{\mathcal{C}_G(\Lambda), \mathcal{C}(N)\} = 0 \tag{20}$$

$$\{\mathcal{C}(N), \mathcal{C}(N')\} = -\mathcal{C}_{\text{Diff}}(\vec{S}), \quad S^a := \frac{P_j^a P_k^b \eta^{jk}}{|\det(q)|} \left(N N'_{,b} - N' N_{,b} \right). \tag{21}$$

The subalgebra of $\mathcal{C}(N)$ and $\mathcal{C}_{\text{Diff}}(\vec{N})$ encodes the diffeomorphism invariance at the canonical level and can be also derived from purely geometrical considerations [11]; see also the discussion in *Chapter 2*. It will play a pivotal role in the quantization of the constraint operators because one requires that the corresponding constraint operators satisfy an analogue commutator algebra in order to carry over the classical symmetries into the quantum theory.

Let us finally summarize: We have formulated general relativity in terms of connection variables (A, P). The corresponding action in the 3+1-picture is given by

$$S = \int_{\mathbb{R}} dt \int_M d^3x \left(\dot{A}_a^j P_j^a - \left(\Lambda^j G_j + N\mathcal{C} + N^a \mathcal{C}_a \right) \right). \tag{22}$$

The 'Hamiltonian' H is given by

$$H = \mathcal{C}_G(\Lambda) + \mathcal{C}(N) + \mathcal{C}_{\text{Diff}}(\vec{N}) \tag{23}$$

and, as mentioned before, is a linear combination of constraints only. The Hamiltonian equation of motion

$$\dot{A}_a^j(x) = \{A_a^j(x), H\}, \qquad \dot{P}_j^a(x) = \{P_j^a(x), H\} \tag{24}$$

together with the constraints

$$\mathcal{C}_G(\Lambda) = 0, \qquad \mathcal{C}(N) = 0, \qquad \mathcal{C}_{\text{Diff}}(\vec{N}) = 0 \tag{25}$$

are completely equivalent to Einstein's equations in vacuum

$$R_{\mu\nu} - \frac{1}{2} g_{\mu\nu} R = 0. \tag{26}$$

For the reason that the 'Hamiltonian' H vanishes on the constraint hypersurface, the evolution generated by H is interpreted as gauge transformations and not as a physical evolution. A discussion on how physical evolution can be implemented in the context of general relativity in the framework of observables can for instance be found in Ref. [12].

We have discussed the connection formulation for space-times of dimension 4. In $D+1$ dimensions a spatial metric has $\frac{D(D+1)}{2}$ degrees of freedom, while a frame in D dimensions includes D^2 degrees of freedom. Consequently, we need $D^2 - \frac{D(D-1)}{2} = \frac{D(D+1)}{2}$ constraints in order to recover the corresponding ADM formulation in $D+1$ dimensions. Note that $\frac{D(D+1)}{2}$ is precisely the dimension of $SO(D)$ and thus it would be a natural choice for a gauge group here. However, an $SO(D)$ connection has $\frac{D^2(D-1)}{2}$ degrees of freedom and the only dimension for D in which the number of degrees of freedom of the D-bein and the $SO(D)$ connection coincide is the special case $D = 3$.

However, this does not mean, that there exists no connection variable formulation in higher dimensions. Recently, it has been shown that one can introduce a different extension of the ADM phase space and formulate general relativity in terms of SO(D+1) Yang Mills variables [13]. In order to match the degrees of freedom of the ADM phase space and thus general relativity, the formulation in Ref. [13] includes additional constraints, that have to be implemented.

Although we have restricted our discussion to the vacuum case here, the connection formulation can be generalized to gravity coupled to matter by simply performing a 3+1-split also for the matter action [14] (see also Ref. [15] for a pedagogical introduction to this topic). We then obtain further degrees of freedom in phase space describing the matter part of the theory. The constraints and hence also the 'Hamiltonian' will then include additional contributions from the matter degrees of freedom. In the next section we will also restrict the discussion for simplicity to the vacuum case and show how a quantum theory for the connection formulation can be constructed.

3. Holonomy-Flux Algebra and its Representation(s)

The connection formulation of general relativity discussed in the last section is the classical starting point for loop quantum gravity. Before we explain in detail how this works for the connection formulation of general relativity let us briefly recall how canonical quantization is used in quantum mechanics.

3.1. *Canonical Quantization in Quantum Mechanics*

In quantum mechanics we choose as the classical starting point the phase space coordinatized by (q^j, p_j), that satisfy the so called Heisenberg algebra

$$\{q^j, q^k\} = 0 \qquad \{p_j, p_k\} = 0 \qquad \{q^j, p_k\} = \delta^j_k. \tag{27}$$

To formulate the quantum theory, we introduce an abstract *-algebra[a] $\mathfrak{A}$ of operators generated by $\hat{q}$, $\hat{p}$ and $\mathbb{1}_{\mathfrak{A}}$. Since we want to replace Poisson brackets by commutators in the quantum theory we set

$$[\hat{q}^j, \hat{p}^k] =: i\hbar \widehat{\{q^j, p_k\}} \qquad (\hat{q}^j)^* = \overline{q}^j \qquad (\hat{p}_j)^* = \overline{p}_j, \tag{28}$$

where the bar denotes complex conjugation. The task is now to find a representation of this abstract *-algebra, that is a map $\pi : \mathfrak{A} \to \mathcal{L}(\mathcal{H})$ from the algebra into the subalgebra of linear operators on a Hilbert space $\mathcal{H}$, which has the following properties

$$\pi(c\hat{a} + c'\hat{a}') = c\pi(\hat{a}) + c'\pi(\hat{a}') \qquad \pi(\hat{a}\hat{a}') = \pi(\hat{a})\pi(\hat{a}') \qquad \pi(a^*) = \pi(\hat{a})^\dagger, \tag{29}$$

where a is an element of the algebra generated by $\{q^j, p_j, \mathbb{1}_{\mathfrak{A}}\}$, $\dagger$ denotes the adjoint operation and π furthermore has to satisfy

$$[\pi(\hat{q}^j), \pi(\hat{p}^k)] = i\hbar\pi(\mathbb{1}_{\mathfrak{A}}) = i\hbar\mathbb{1}_{\mathcal{H}}.$$

---

[a]A *-algebra is an algebra with an involution, that is a map $* : \mathfrak{A} \to \mathfrak{A}$ where $a \mapsto a^*$ has the following properties $(ca + c'a')^* = \overline{c}a^* + \overline{c}'a'^*$, $(aa')^* = a'^*a^*$ and $(a^*)^* = a$ for all $a, a' \in \mathfrak{A}, c, c' \in \mathbb{R}$.

In the case of quantum mechanics (QM) the representation is well known and called the Schrödinger representation. The Hilbert space $\mathcal{H}$ is $\mathcal{H} = L_2(\mathbb{R}^3, d^3x)$ and the explicit form of the representation is

$$(\pi(\hat{q}^j)\psi)(x) = x^j \psi(x) \qquad (\pi(\hat{p}_j)\psi)(x) = -i\hbar \frac{\partial \psi}{\partial x^j}(x). \qquad (30)$$

Hence, the configuration variables become multiplication and the momenta derivation operators. We realize that formulating the classical theory requires two main choices for any quantum theory. The first choice is the classical Poisson algebra, that we take as a starting point for the quantization. Different choices will in general lead to different algebras and therefore finally also to different quantum theories. Secondly, even if we restrict our discussion to one particular choice of the classical Poisson algebra, in general there exists more than one possible representation of this algebra. Any of those representations can in principle define a different quantum theory, unless they are unitary equivalent. We call two representations π_1 and π_2 unitary equivalent if there exists an unitary operator $U : \mathcal{H}_1 \to \mathcal{H}_2$ such that $U\pi_1(\hat{a})U^{-1} = \pi_2(\hat{a})$ for all $\hat{a} \in \mathfrak{A}$. In the context of QM the famous Stone-von-Neumann uniqueness theorem states that under very mild assumptions on the representation the Schrödinger representation is up to unitary equivalence the unique representation for QM. This theorem was announced by Stone in 1930 and the first complete proof was given by von Neumann [17]. The actual proof uses the Weyl — instead of the Heisenberg — algebra, whose generators are the exponentiated versions of the q^j's and p_j's discussed above. However, since one of the assumptions for the representation is that it should be weakly continuous, the operators $\hat{q}^j$ and $\hat{p}_j$ also exist in this representation and one can also recover the Heisenberg commutation relations coming from the Weyl algebra. So far we have only considered kinematical requirements for the choice of the representation. Of course the dynamics plays as an important role as it does in the classical theory. Therefore, we are only interested in those representations that allow to implement the generators of the classical dynamics as operators. In the case of standard QM, this is the Hamiltonian, which usually is a simple function on the phase space. Hence, in the Schrödinger representation the corresponding operators can be implemented. In the case of general relativity using Dirac quantization we have to find representations for which the classical constraints can be quantized on the kinematical Hilbert space. We will see in the following discussion, that this requirement forces us to introduce a different representation than the usual Fock representation used in perturbative quantum field theory in Minkowski space-time.

3.2. *The Holonomy–Flux–Algebra* $\mathfrak{A}$

Now we take the connection formulation of general relativity as our classical starting point for the quantization. The difference with classical mechanics is that general relativity is a field theory and hence the variables $(A_a^j(x), P_j^a(x))$ are too singular to be directly promoted to operators. Therefore one quantizes not $(A_a^j(x), P_j^a(x))$

themselves but particular smeared versions of these elementary variables. In the case of standard canonical quantum field theory, one uses a 3-dimensional smearing over M for the basic field variables and their conjugate momenta. However, this kind of smearing is defined with respect to a particular background metric. For general relativity we will choose a different way of smearing $(A_a^j(x), P_j^a(x))$, which in particular has the property to be independent of any background metric and leads to basic variables similar to those used in ordinary lattice gauge theory. The SU(2)-connection A_a^j is an su(2)-valued one-form and thus it is natural to integrate the connection along oriented curves $e : [0,1] \to M, s \mapsto e(s)$ in M, which we call edges. If we further take the path-ordered exponential of this integral, we obtain the holonomy associated with the connection A given by

$$A(e) := \mathcal{P}\exp(\int_e A) \tag{31}$$

$$= \mathbb{1}_2 + \sum_{n=0}^{\infty} \int_0^1 ds_1 \int_{s_1}^1 ds_2 \cdots \int_{s_{n-1}}^1 ds_n A(e(s_1)) A(e(s_2)) \cdots A(e(s_n)), \tag{32}$$

where $A(e(s_i)) := A_a^j(e(s_i))\tau_j \dot{e}^a(s_i)$ and τ_j denotes a basis of su(2). Let us consider an edge $e : [0,1] \to M, s \mapsto e(s)$ in M with beginning point $b(e) = e(0)$ and final point $f(e) = e(1)$ and let $t \in [0,1]$. Then the holonomy $A(e) = A(e,1)$ is the unique solution of the following differential equation

$$\frac{d}{dt}A(e,t) = A(e,t) A_a^j(e(t)) \tau_j \dot{e}^a(t) \quad \text{with} \quad A(e,0) = \mathbb{1}_2 \tag{33}$$

which describes the parallel transport from $b(e)$ to $f(e)$ along the edge e. In our case the holonomy is an element of the group SU(2). Under the composition of two edges $e_1 \circ e_2$, for which the final and beginning point are the same and under the inversion of edges e^{-1}, the holonomy behaves as

$$A(e_1 \circ e_2) = A(e_1) A(e_2) \qquad A(e^{-1}) = A^{-1}(e). \tag{34}$$

Note that e^{-1} is obtained from e by reversing the orientation of the edge.

Similar variables are also used in ordinary lattice gauge theory with the corresponding connections of the gauge theories of the standard model. The reason for this is that the holonomies transform very simply under gauge transformations. While the connection transforms as $A^g = gAg^{-1} - dgg^{-1}$ under SU(2) gauge transformation, the transformed holonomy is $A^g(e) = g(e(0))A(e)g(e(1))^{-1} = g(b(e))A(e)g(f(e))^{-1}$. Hence, the transformation acts only at the beginning and final points of the curve and this simple behavior is of advantage when later gauge invariant quantities in the quantum theory will be constructed. For instance, the famous Wilson-loop defined as $Tr(\mathcal{P}\exp(\oint_\beta A))$ is the holonomy of a given connection A along a closed loop β and one example of a gauge invariant observable because the trace allows to cyclic permute the matrices and $b(e) = f(e)$ for a loop so that $g(f(e))^{-1}g(b(e)) = \mathbb{1}_G$ can be used, where $\mathbb{1}_G$ denotes the unit element in the gauge group G. Considering the conjugate variable P_j^a also here exists — from the geometric perspective — a natural smearing. The densitized triad P_j^a is a su(2)-valued vector density of weight $+1$. Introducing a su(2)-valued smearing field f^j, $P^a(f) := f^j P_j^a$ is vector density and hence dual to a (pseudo-) 2-form in three dimensions, using that ϵ_{abc} carries density weight -1. Given this (pseudo-) 2-form the natural smearing is over two-dimensional surfaces, thus we define the conjugate variables, the so called (electric) fluxes[b] as

$$P(S,f) = \int_S f^j (*P)_j = \int_S f^j \epsilon_{abc} P_j^a dx^b \wedge dx^c. \tag{35}$$

If one computes the Poisson bracket between the holonomies and fluxes the result depends on the position of the edge e relative to the surface S. In order to discuss this in detail we introduce the notion of an elementary edge. We have to consider 4 different cases for the elementary edges. If $S \cap e = 0$ we call e an edge of type out. If $S \cap e = e$ and hence e lies entirely inside S we call e of type in. If e is not of type in but $S \cap e \neq 0$ we consider as elementary edges only those, which have one intersection point, denoted by p, with S in its end points. If e lies above S, we call e of type up and if e lies below S of type down. Furthermore, we distinguish the cases where p is the beginning point $b(e)$ and the final point $f(e)$ respectively. Any edge e can be written as a composition of elementary edges by introducing appropriate additional vertices. Using this classification we have

$$\{A(e), P(S,f)\} = -\frac{\kappa(S,e)}{2} \times \begin{cases} A(e)\tau_j f^j(b(e)) & \text{if } S \cap e = b(e) \\ -\tau_j f^j(f(e))A(e) & \text{if } S \cap e = f(e), \end{cases} \tag{36}$$

with

$$\kappa(S,e) = \begin{cases} +1 & \text{if } e \text{ of type up} \\ 0 & \text{if } e \text{ of type in or out} \\ -1 & \text{if } e \text{ of type down.} \end{cases} \tag{37}$$

As discussed in the context of QM, we have to find a suitable Poisson algebra, which encodes the underlying classical theory. In the case of QM this was the Heisenberg- and Weyl-algebra respectively. In both cases the Hilbert space associated with the representation is an $L_2(\mathbb{R}^3, d^3x)$-space, that is the space of square integrable function over $\mathbb{R}^3$ with the standard Lebesgue measure d^3x on $\mathbb{R}^3$. Hence, we see for QM the Hilbert space underlying the representation involves the construction of a measure on $\mathbb{R}^3$, which is the classical configuration space for classical mechanics. For general relativity we consider a classical field theory and in terms of the connection formulation the classical configuration space $\mathcal{A}$ is the space of smooth connections. As usual in canonical field theories, the quantum theory is not based on the classical configuration space, but requires the introduction of a larger space,

---

[b]The name (electric) flux comes from analogy to electrodynamics where in the Hamiltonian framework the canonical momentum is precisely the electric field and integrating it over a surface gives the electric flux.

that includes not only smooth connections but also so called generalized or distributional connections and is called the quantum configuration space denoted by $\overline{\mathcal{A}}$. Thus, for loop quantum gravity, we have to construct a measure on the quantum configuration space. For this reason we will choose our classical Poisson algebra in such a way that it can be easily extended from the classical configuration space $\mathcal{A}$ to the quantum configuration space $\overline{\mathcal{A}}$. For this purpose, we introduce so called cylindrical functions on $\mathcal{A}$. So far we have restricted our discussion to an arbitrary but single edge e. Now we generalize this picture and introduce the notion of a graph α. A graph α consists of a finite collection of edges $\{e_1,\cdots,e_n\}$ in M, whereas the edges intersect only in their beginning or final points. These intersection points are called vertices of α. For a given graph α, we denote the set of edges by $E(\alpha)$ and the set of vertices by $V(\alpha)$. In order to give the definition of a cylindrical function, we denote the subset of connections associated with a graph α by $\mathcal{A}_\alpha \subset \mathcal{A}$. $\mathcal{A}_\alpha$ contains all connections A_{e_i} associated with the edges $\{e_i\}$ of the graph α. Then there exists a map

$$I_E : \mathcal{A}_\alpha \to SU(2)^n \quad \text{with} \quad A \in \mathcal{A}_\alpha \mapsto I_E(A) := (A(e_1),\cdots,A(e_n)) \tag{38}$$

and we can use this I_E to define smooth cylindrical functions[c] defined with respect to a given graph α with edges $\{e_1,\cdots,e_n\}$ as

$$f_\alpha(A) = F_\alpha(I_E(A)) = F_\alpha(A(e_1),\cdots,A(e_n)), \tag{39}$$

where $F_\alpha : SU(2)^n \to \mathbb{C}$ is a C^∞–function on n copies of SU(2). A function f on $\mathcal{A}$ is said to be cylindrical if it can be written in the above form for some graph α. Since each f_α depends only on a finite number of holonomies, we need to consider all possible graphs α, that can be embedded into M in order to describe the Poisson algebra underlying gravity in connection variables. A graph α' is said to be larger than a given graph α, if every edge e can be written as a finite combination of edges e'_i of α', that is $e = e_1'^{s_1} \circ \cdots \circ e_\ell'^{s_\ell}$ for some set of edges $\{e'_i | i = 1,\cdots,\ell\}$ of α' where $s = \pm 1$. Note that every function f on $\mathcal{A}$, which is cylindrical with respect to a given graph α will automatically be cylindrical with respect to any larger graph α'. This allows to define an equivalence relation on $\bigcup_\alpha Cyl_\alpha$. Given $f, f' \in \bigcup_\alpha Cyl_\alpha$ we can find α, α' such that $f \in Cyl_\alpha$ and $f' \in Cyl_{\alpha'}$. We say that f and f' are equivalent, denoted by $f \sim f'$, provided that f, f' agree for all larger graphs $\alpha'' > \alpha, \alpha'$. We define the space of smooth cylindrical functions on $\mathcal{A}$ as

$$Cyl := \bigcup_\alpha Cyl_\alpha / \sim . \tag{40}$$

Thus, Cyl consists of equivalence classes of functions on the spaces Cyl_α. Cyl can be shown to be an Abelian C^*-algebra defined by point wise operations and with the supremum-norm. In order to choose the Poisson algebra underlying loop quantum gravity, we still have to discuss the conjugate momentum variables associated with the smooth cylindrical functions on $\mathcal{A}$. The latter will be the flux vector fields on

---

[c]Here a cylindrical function f is said to be smooth if any of its representatives f_α on G^n is smooth.

Cyl, which we denote by $X(f,S) \in V(Cyl)$ and which are the Hamiltonian vector fields of $P(S,f)$, where $V(Cyl)$ includes not only the Hamiltonian but all vector fields on Cyl. The action of $X(S,f)$ on f_α is given by

$$(X(S,f)f_\alpha)(A) := (\{f_\alpha, P(S,f)\})(A)$$
$$= \frac{k}{2} \sum_{e \in E(\alpha)} \frac{\kappa(e,S)}{2} \begin{cases} A(e)\tau_j f^j(b(e)) & \text{if } S \cap e = b(e) \\ -\tau_j f^j(f(e))A(e) & \text{if } S \cap e = f(e) \end{cases} \frac{\partial F_\alpha(\{A(e)_{e \in E(\alpha)}\})}{\partial A(e)_{AB}}$$
(41)

where A, B denote SU(2)-indices. Finally, we can now define the classical Poisson algebra, which will be the starting point for our quantization in the next section, and which is called the holonomy–flux algebra $\mathfrak{A}$:

- The classical Poisson algebra underlying loop quantum gravity is the Lie *-subalgebra of $Cyl \times V(Cyl)$ generated by the smooth cylindrical functions and flux vector fields on Cyl. The involution on the algebra is just complex conjugation. This algebra is called the holonomy–flux algebra and will be denoted by $\mathfrak{A}$.

We will discuss the representation of the holonomy–flux algebra in the next section.

4. The Ashtekar-Lewandowski Representation and the Kinematical Hilbert Space of LQG

So far we have discussed smooth cylindrical functions on the classical configuration space $\mathcal{A}$. For the derivation of the kinematical Hilbert space underlying the representation of the holonomy–flux algebra, we have to construct a measure on the quantum configuration space $\overline{\mathcal{A}}$. The necessity of $\overline{\mathcal{A}}$ can be also understood from the following perspective: In order to obtain a kinematical Hilbert space $\mathcal{H}$ from Cyl, we need to take the Cauchy-completion with respect to a norm defined on Cyl. This completion will include objects as limit points, which cannot be understood as functions on $\mathcal{A}$, but are more general objects such as distributions on $\mathcal{A}$. The strategy one adopts is to look for a larger quantum configuration space $\overline{\mathcal{A}}$ such that $\mathcal{H}$ is isomorphic to an L_2-space over $\overline{\mathcal{A}}$ with respect to some measure on $\overline{\mathcal{A}}$. As we will see below the action of the flux vector fields on Cyl can be easily extended from cylindrical functions on $\mathcal{A}$ to cylindrical functions on $\overline{\mathcal{A}}$ by the introduction of left- and right-invariant vector fields on SU(2). A measure on $\overline{\mathcal{A}}$ can be defined by using the fact that any cylindrical function over a graph α can be expressed via the map I_E in (38) by means of functions F on $SU(2)^n$. On $SU(2)^n$ a natural measure exists, using n copies of the Haar measure on SU(2). This allows to firstly define a measure on $\overline{\mathcal{A}}_\alpha$, which includes all, not necessarily smooth connections $\{A_{e_i}\}$ along the edges of the graph α, and thus an inner product on Cyl_α for all α given by

$$\langle f_\alpha, \tilde{f}_\alpha \rangle := \int_{SU(2)^n} \prod_{i=1}^n d\mu_H(A(e_i)) \overline{F_\alpha(A(e_1), \cdots, A(e_n))} F_\alpha(A(e_1), \cdots, A(e_n)), \quad (42)$$

where $d\mu_H(g)$ denotes the Haar measure on SU(2). Taking the closure of Cyl_α with respect to the corresponding norm of the above defined inner product, we obtain Hilbert spaces $\mathcal{H}_\alpha := L_2(\overline{A}_\alpha, d\mu_\alpha)$ for all graphs α. The kinematical Hilbert space $\mathcal{H}$ can then be constructed using projective techniques, because $\overline{\mathcal{A}}$ can be understood as the projective limit of the $\overline{A}_\alpha$'s. Given the measures μ_α on $\overline{A}_\alpha$, they can be used to construct a measure denoted by μ_{AL}, called the Ashtekar–Lewandowski measure, on $\overline{\mathcal{A}}$. For this purpose, we have to discuss how an inner product can be defined in case the functions $f_{\alpha'}, \tilde{f}_{\alpha''}$ are cylindrical with respect to two different graphs α' and α'' respectively. Given this situation, we can use that Cyl has the property, that we can always find a common graph $\alpha > \alpha', \alpha''$ with respect to which $f, \tilde{f}$ are cylindrical. Hence, we can use α to define an inner product for $f_{\alpha'}, \tilde{f}_{\alpha''}$. Here we associate trivial holonomies to $f_{\alpha'}$ and $f_{\alpha''}$ respectively to those edges in α, which are not contained in α' and α'' respectively. Cylindrical consistency ensures that the inner product on $\overline{\mathcal{A}}$ does not depend on the particular choice of the common graph α. For instance, if we take as the common graph just the union $\alpha := \alpha' \cup \alpha''$, then the inner product defined with respect to α should yield the same value as if we further unify the graph α with another graph α''' not contained in α'. Also, the inner product should be the same for two graphs α and $\tilde{\alpha}$ when $\tilde{\alpha}$ can be obtained from α just by subdividing edges of α by means of the introduction of additional vertices. Thus, we define the inner product on $\overline{\mathcal{A}}$ for $f, \tilde{f} \in Cyl$ as

$$\langle f_\alpha, \tilde{f}_\alpha \rangle := \int_{SU(2)^n} \prod_{i=1}^n d\mu_H(A(e_i)) \overline{F_\alpha(A(e_1), \cdots, A(e_n))} F_\alpha(A(e_1), \cdots, A(e_n)), \quad (43)$$

where α is a common graph with respect to which f and $\tilde{f}$ are cylindrical. Considering the closure of Cyl with respect to the corresponding norm gives the kinematical Hilbert space $\mathcal{H} = L_2(\overline{\mathcal{A}}, d\mu_{AL})$, which is the space of square integrable functions over $\overline{\mathcal{A}}$ with respect to the Ashtekar-Lewandowski measure. Now, given the kinematical Hilbert space $\mathcal{H}$ we can discuss the representation π of the holonomy–flux algebra. The space Cyl is dense in $\mathcal{H}$ and therefore we can define the action of the elementary operators in the Ashtekar–Lewandowski representation on Cyl. The holonomy operators act as multiplication operators and hence we obtain for cylindrical functions

$$(\pi(f)\psi)(A) = (\hat{f}\psi)(A) = f(A)\psi(A) \quad (44)$$

for $\psi \in \mathcal{H}$. The flux vector fields become derivation operators and their explicit action is given by

$$(\pi(P(S,f))\psi)(A) = \hat{P}(S,f)\psi(A) = (X(S,f)\psi)(A) \quad (45)$$

for $\psi \in \mathcal{H}$, that lie in the domain of $\hat{P}(S,f)$. We will express the right-hand side of the equation above now by means of the left- and right-invariant vector fields on SU(2) denoted by L_j and R_j respectively. Given a function $f : SU(2) \to \mathbb{C}$ and $g \in SU(2)$ these are defined as

$$(R_j f)(g) := \frac{d}{dt}\left(f(e^{t\tau_j}g)\right)_{t=0} \qquad (L_j f)(g) := \frac{d}{dt}\left(f(ge^{t\tau_j})\right)_{t=0}. \quad (46)$$

Thus, we can define the action of the flux operators on f_α in Cyl_α as

$$\hat{P}(S,f)f_\alpha(A) = \frac{\hbar}{2} \sum_{v \in V(\alpha)} f^j(v) \sum_{\substack{e \in E(\alpha) \\ e \cap v \neq \emptyset}} \kappa(e,S)\hat{Y}_j^{(v,e)} f_\alpha(A), \qquad (47)$$

with

$$\hat{Y}_j^{(v,e)} := \mathbb{1}_\mathcal{H} \times \mathbb{1}_\mathcal{H} \times \cdots \times \mathbb{1}_\mathcal{H} \times \begin{Bmatrix} iR_j^e \\ -iL_j^e \end{Bmatrix} \times \mathbb{1}_\mathcal{H} \times \cdots \times \mathbb{1}_\mathcal{H}, \quad \text{if} \quad \begin{Bmatrix} e \text{ outgoing at } v \\ e \text{ ingoing at } v \end{Bmatrix} \qquad (48)$$

This finishes our discussion on the kinematical representation of loop quantum gravity. The next subsection will briefly deal with the question whether there exists other than the already introduced representation for the kinematical Hilbert space of loop quantum gravity.

4.1. Other Representations of the Holonomy–flux-algebra $\mathfrak{U}$

In the last section we discussed in detail how the kinematical representation for loop quantum gravity looks like. As we have seen the algebra underlying loop quantum gravity is the holonomy–flux algebra $\mathfrak{U}$ and one possible representation of this algebra is the Ashtekar–Lewandowski representation (AL-representation) introduced above. In the context of quantum mechanics we already briefly mentioned that given a choice of a classical algebra in general more than one possible representation of the algebra exists and thus in general different quantum theories can be obtained from the same classical starting point. This is a particularly interesting aspect in the case of general relativity since it is a field theory and in contrast to quantum mechanics an analog to Stone–von Neumann theorem does not exist. As a consequence, in the context of field theories, in principle, infinitely many unitarily non-equivalent representations could exist. However, in practice finding representations of a given algebra can be a challenging task and often we are happy to have found one at all. Nevertheless it is an interesting question to ask what kind of assumptions in the AL-representation have to be required in order to make it, under those assumptions, the — up to unitary equivalence — unique representation of the holonomy–flux algebra.

An answer to this question is given by the so called LOST-theorem [18, 19] and yields progress in two directions. On the one hand, we learn what kind of characteristic properties the AL-representations has and on the other hand, we can try to look for new representations by violating one of those assumptions. What are the assumptions needed in the LOST-theorem? As required in most physical theories one of the assumptions is that the representation should be irreducible. This means that any vector in $\mathcal{H}$ is a cyclic vector. A cyclic vector Ω is a vector in $\mathcal{H}$ for which the set $\{\pi(a)\Omega \,|\, a \in \mathfrak{U}\}$ is dense in $\mathcal{H}$. Further assumptions are related to the (gauge) symmetries of general relativity formulated in connection variables. As usual for quantum theories one requires that the classical symmetries should be implemented by unitary operators. In the context of the holonomy–flux algebra $\mathfrak{U}$ the LOST-theorem includes an assumption on a positive linear functional on the holonomy–flux algebra so that this is automatically fulfilled for the spatial diffeomorphisms and the SU(2)-gauge transformations. The positive linear functional is used in the context of the Gelfand–Naimark–Segal theorem to construct a cyclic representation of $\mathfrak{U}$. Moreover, the LOST-theorem assumes that there is at least one vector $\Omega \in \mathcal{H}$ that is invariant under diffeomorphisms. These assumptions are strong enough to restrict the number of possible representations of the holonomy-flux algebra, up to unitarily equivalence, to one single representation, the AL-representation, which is summarized in the theorem below [18, 19]

Theorem 1. *There is only one cyclic representation of the holonomy–flux algebra $\mathfrak{U}$ with diffeomorphism invariant cyclic vector — the Ashtekar–Lewandowski representation.*

Characteristic properties of the AL-representation are:

- As we will see in Section 6 so called geometric operators associated with length, volume and area have purely discrete spectra, giving already an idea that quantum geometry could yield to a new fundamental picture of geometry.
- Although operators for the holonomy $A(e)$ exist, there are no operators representing the connection A_a^j directly in this representation.
- Similarly, also for the spatial diffeomorphisms the infinitesimal generators do not exist, but only finite diffeomorphisms are implemented as unitary operators.

A different representation, that is not unitary equivalent to the AL-representation was rather recently discussed in the literature and is the so called Koslowski–Sahlmann [20–22] representation (KS-representation). The way the LOST-theorem is circumvented is that in the KS-representation the spatial diffeomorphism is not implemented unitarily, as will be discussed more in detail below. In the context of the above mentioned GNS theorem, associated with the AL-representation is a so called GNS vacuum state, which in the case of AL-representation describes an extremely degenerate situation of an empty geometry. Here the smooth classical spatial geometry is expected to arise through some coarse-graining procedure that describes the transition from the deep quantum to the classical regime. Therefore an interesting question is whether the observed smoothness of classical geometry can already be described at the quantum level without applying any coarse-graining. Following this idea Koslowski [20] considered a slight modification of the AL-representation, in which he extended the representation of the flux operators. In particular, the representation of the fluxes is changed by adding a c-number term

$$\pi_{P^{(0)}}(P(S,f)) = \hat{P}(S,f) + P^{(0)}(S,f)\mathbf{1}_\mathcal{H} \quad \text{with} \quad P^{(0)}(S,f) := \int_S f^j (*P^{(0)})_j, \tag{49}$$

where $P^{(0)}(S,f)$ is the classical value of the flux with respect to a background geometry given by the densitized triad $E^{(0)} = kP^{(0)}$. For this reason we labeled the representation $\pi_{P^{(0)}}$ by $P^{(0)}$ in order to distinguish between the AL- und KS-representation. The Hilbert space associated with $\pi_{P^{(0)}}$ is the same as in the AL-representation, that is $\mathcal{H}_{P^{(0)}} = \mathcal{H}$ and the action of the holonomies and the cylindrical function respectively agrees, thus

$$\pi_{P^{(0)}}(f) = \pi(f) = \hat{f}. \tag{50}$$

Note that in the case $P^{(0)} = 0$ we recover the AL-representation. In this sense the representations $\pi_{P^{(0)}}$ can be understood as a family of representations, with one member being the AL-representation. However, for other choices than $P^{(0)} = 0$ the AL- and the KS-representation are not unitarily equivalent and therefore could in principle describe different physics. We have already mentioned above that spatial diffeomorphism are not implemented unitarily in the KS-representation, being however one of the assumptions in the LOST-theorem. By this we mean, that if $\hat{U}(\phi)$ denotes the unitary operator implementing spatial diffeomorphisms ϕ in the AL-representation, then in general we have

$$\hat{U}(\phi)\pi_{P^{(0)}}(P(S,f))\hat{U}^\dagger(\phi) \neq \pi_{P^{(0)}}(P(\phi(S), \phi^*f)). \tag{51}$$

The reason that the above equality fails is that the quantity $P^{(0)}$ is fixed and will not transform under the action of $\hat{U}(\phi)$. Since also in the context of the KS-representation spatial diffeomorphism play an important role, it was shown in Ref. [21] that by enlarging the Hilbert space $\mathcal{H}_{P^{(0)}}$ one can define unitary operators that implement spatial diffeomorphisms that also involve the background field $P^{(0)}$ and hence implement the corresponding automorphisms of the SU(2) principle fiber bundle on which the whole mathematical formulation of the theory is based. In the context of the (enlarged) Hilbert space of the KS-representation, denoted by $\mathcal{H}_{KS}$, an orthonormal basis of the form $\{|s, P^{(0)}\rangle\}$ exists where s denotes a standard spin network in the AL-representation, which are discussed in detail in the next subsection and which provide an orthonormal basis for $\mathcal{H}$, and $P^{(0)}$ denotes, as before, a background field. The inner product in $\mathcal{H}_{KS}$ is of the following form

$$\langle s', P'^{(0)} | s, P^{(0)} \rangle = \langle s' | s \rangle_{AL} \delta_{P'^{(0)}, P^{(0)}}, \tag{52}$$

where $\langle s' | s \rangle_{AL}$ denotes the inner product in the AL-representation. The action of the cylindrical functions and fluxes in $\mathcal{H}_{KS}$ is given by

$$\hat{f}|s, E\rangle = |\hat{f}s, E\rangle \qquad \hat{P}(S,f)|s, E\rangle = |\hat{P}(S,f)s, E\rangle + P^{(0)}(S,f)|s, E\rangle. \tag{53}$$

As shown in Refs. [21] and [23] the KS-representation based on this enlarged Hilbert space supports a unitary implementation of the spatial diffeomorphisms as well as the SU(2) gauge transformation (for which similar problems occur) and the diffeomorphism and SU(2) gauge invariant Hilbert space can be constructed using the technique of group averaging that is discussed in more detail in *Chapter 2*.

In Ref. [24] it was pointed out that if one considers also higher order commutators such as for instance the element of $\mathfrak{U}$ given by $(0, [\hat{P}(S, f_1), [\hat{P}(S, f_2), \hat{P}(S, f_3))]])$ then one can derive the following identity for the double commutator

$$[\hat{P}(S, f_1), [\hat{P}(S, f_2), \hat{P}(S, f_3))]] = \frac{1}{4}\hat{P}(S, [f_1, [f_2, f_3]]) \tag{54}$$

and thus using the AL-representation of the holonomy–flux algebra $\mathfrak{U}$ the elements $(0, [\hat{P}(S, f_1), [\hat{P}(S, f_2), \hat{P}(S, f_3))]])$ and $(0, \frac{1}{4}\hat{P}(S, [f_1, [f_2, f_3]]))$ need to be identified. Let us now consider the situation in the KS-representation. There we have

$$[\pi_{P^{(0)}}(P(S, f_1)), [\pi_{P^{(0)}}(P(S, f_2)), \pi_{P^{(0)}}(P)(S, f_3))]] \tag{55}$$
$$= [\hat{P}(S, f_1), [\hat{P}(S, f_2), \hat{P}(S, f_3))]],$$

where the equality above is true because the constant contributions of the background fields $P^{(0)}$ cancel in the double commutator. As a consequence we obtain

$$[\pi_{P^{(0)}}(P(S, f_1)), [\pi_{P^{(0)}}(P(S, f_2)), \pi_{P^{(0)}}(P)(S, f_3))]] = \frac{1}{4}\hat{P}(S, [f_1, [f_2, f_3]]). \tag{56}$$

However, due to the part coming from the background field in $\pi_{P^{(0)}}(P(S, f))$ we have

$$\pi_{P^{(0)}}(P(S, [f_1, [f_2, f_3]])) \neq \hat{P}(S, [f_1, [f_2, f_3]]). \tag{57}$$

The suggestion in [24] to cure this problem is the modification of commutation relations of the standard holonomy–flux algebra by an appropriate central term. As also discussed in Ref. [24] it is still an open question whether the introduction of such a central term is sufficient in the context of further higher order commutators, that could yield additional relations among the algebra elements.

A different point of view is taken in Ref. [25] where the holonomy–flux algebra is extended by the so called background exponentials denoted by $\beta_{P^{(0)}}(A)$ whose explicit form is given by

$$\beta_{P^{(0)}}(A) := e^{i\int_S P^{(0)} \cdot A} \quad \text{with} \quad P^{(0)} \cdot A := (P^{(0)})_i^a A_a^i. \tag{58}$$

Next to the holonomy and flux action in $\mathcal{H}_{KS}$ given above these background exponentials act as

$$\hat{\beta}_{P'^{(0)}}|s, P^{(0)}\rangle = |s, P'^{(0)} + P^{(0)}\rangle. \tag{59}$$

We have discussed in the last section that the AL-representation is a representation of the holonomy–flux algebra $\mathfrak{U}$. If we instead consider the holonomy–flux algebra enlarged by these background exponentials, called the holonomy-background-exponential-flux algebra in Ref. [25], then it was shown [25] that the KS-representation can be also understood as a representation of the holonomy-background-exponential-flux algebra.

4.2. Spin Networks as an Orthonormal Basis of the Kinematical Hilbert Space

A useful orthonormal basis of the kinematical Hilbert space $\mathcal{H}$ is given by so called spin network basis. Also here we will take advantage of an already existing natural orthonormal basis in the Hilbert space $L_2(SU(2), d\mu_H)$. Let us consider the equivalence classes of finite dimensional, unitary, irreducible representations of SU(2) on a representation space V_j and take one representative of it denoted by π^j. We denote the dimension of π^j by $\dim(\pi^j)$. We define the following functions on SU(2)

$$b^j_{mn} : SU(2) \to \mathbb{C}, \quad g \mapsto \langle g \,|\, b^j_{mn}\rangle := \sqrt{\dim(\pi^j)}\pi^j_{mn}(g), \quad m,n = 1,\cdots,\dim(\pi^j). \tag{60}$$

Using the Haar measure μ_H on SU(2), we can define an inner product for b^j_{mn} as

$$\langle b^j_{mn}, b^{j'}_{m'n'}\rangle := \int_{SU(2)} d\mu_H(g)\sqrt{2j+1}\pi^j_{mn}(g)\sqrt{2j'+1}\pi^{j'}_{m'n'}(g), \tag{61}$$

where we used $\dim(\pi^j) = 2j+1$ in the case of SU(2). The Peter–Weyl theorem proves that the set of functions $\{b^j_{mn}\}$ build an orthonormal basis of $L_2(SU(2), d\mu_H)$. In particular the proof is true for any compact Lie group G. Hence, in our case $G = $SU(2) we have

$$\langle b^j_{mn}, b^{j'}_{m'n'}\rangle = \delta^{j,j'}\delta_{mm'}\delta_{nn'}. \tag{62}$$

The Hilbert space $L_2(SU(2), d\mu_H)$ decomposes into a direct sum over all inequivalent irreducible representations labeled by j

$$L_2(SU(2), d\mu_H) = \bigoplus_j \mathcal{H}_j \quad \text{with} \quad \mathcal{H}_j := V_j \otimes V_j^*, \tag{63}$$

where V_j^* denotes the dual space of V_j. A basis in $\mathcal{H}_j$ is given by $\{b^j_{mn} \,|\, m,n \in -j, -j+1, \cdots, j-1, j\}$. Now, we will use this fact to construct the spin network basis of $\mathcal{H} = L_2(\overline{\mathcal{A}}, d\mu_{AL})$. For this purpose we first consider the Hilbert spaces $\mathcal{H}_\alpha$ associated with a fixed graph α, which can be identified with $L_2(SU(2)^n, d^n\mu_H)$. For this reason we can construct an orthonormal basis of $\mathcal{H}_\alpha$ simply by introducing the so called spin network functions (SNF)

$$|s^{\vec{j}}_{\alpha,\vec{n},\vec{m}}\rangle \;:\; \overline{\mathcal{A}}_\alpha \to \mathbb{C}, \quad A \mapsto \langle A \,|\, s^{\vec{j}}_{\alpha,\vec{n},\vec{m}}\rangle \tag{64}$$

$$\langle A \,|\, s^{\vec{j}}_{\alpha,\vec{n},\vec{m}}\rangle := \sqrt{2j_{e_1}+1}\cdots\sqrt{2j_{e_n}+1}\pi^{j_{e_1}}_{m_{e_1}n_{e_1}}(A(e_1))\cdots\pi^{j_{e_n}}_{m_{e_n}n_{e_n}}(A(e_n)),$$

with

$$\vec{j} := \{j_{e_1},\cdots,j_{e_n}\}, \quad \vec{m} := \{m_{e_1},\cdots,m_{e_n}\}, \quad \vec{n} := \{n_{e_1},\cdots,n_{e_n}\}. \tag{65}$$

A decomposition in terms of irreducible representations of SU(2) associated with each edge of the graph α is given by

$$\mathcal{H}_\alpha = \bigoplus_{\vec{j}} \mathcal{H}_{\alpha,\vec{j}} \quad \text{with} \quad \mathcal{H}_{\alpha,\vec{j}} := \bigotimes_{i=1}^n \mathcal{H}_{j_{e_i}}, \tag{66}$$

with $\mathcal{H}_{j_{e_i}}$ defined as in equation (63). We choose a fixed set of representations $\vec{j}$ and will discuss how $\mathcal{H}_{\alpha,\vec{j}}$ can be further decomposed, which will be of advantage when we discuss the solutions to the Gauss constraint later on. Let us choose an arbitrary vertex $v_i \in V(\alpha)$ and consider all edges $\{e_i\}$ intersecting at v_i. Let us assume that α has $m = |V(\alpha)|$ vertices. Then we can rewrite $\mathcal{H}_{\alpha,\vec{j}}$ as

$$\mathcal{H}_{\alpha,\vec{j}} = \bigotimes_{i=1}^{m} \mathcal{H}_{v_i} \quad \text{with} \quad \mathcal{H}_{v_i} := \bigotimes_{\substack{e \in E(\alpha) \\ e_i \cap v_i \neq \emptyset}} \mathcal{H}_{j_{e_i}}. \quad (67)$$

The operators $\hat{Y}_j^{(v_i,e_i)}$ satisfy $[\hat{Y}_j^{(v_i,e_i)}, \hat{Y}_k^{(v_i,e_i)}] = i\epsilon_{jk}{}^\ell \hat{Y}_\ell^{(v_i,e_i)}$, where we have chosen the basis $\{\tau_j\}$ in such a way that $[\tau_j, \tau_k] = \epsilon_{jk}{}^\ell \tau_\ell$. They can be interpreted as components of angular momentum operators. For different edges $e_i \neq e_j$ these operators commute. A natural basis in the context of angular momentum operators is the eigenbasis $\{|jm\rangle\}$, that is labelled by the angular momentum j and the magnetic quantum number m. Let us restrict our discussion to the case of one edge first and denote the abstract angular momentum Hilbert space by $\mathcal{H}^{jm}$ and the associated spin network Hilbert space for this edge e by $\mathcal{H}_{jm}$. Then the corresponding SNF are

$$\langle A \mid j_e m_e \rangle_{n_e} := \sqrt{2j_e + 1} \pi^{j_e}_{m_e n_e}(A(e)). \quad (68)$$

For fixed n these states are orthogonal likewise to the angular momentum eigenstates $|j_e m_e\rangle$. Using the definitions of the operators $\hat{Y}_j^{(v,e)}$ in terms of left- and right-invariant vector fields their action on $|jm\rangle_n$ is given by

$$\hat{Y}_k^{(v,e)} |j_e m_e\rangle_n = \sum_{\tilde{m}_e} \left\{ \begin{array}{c} i\pi^{j_e}_{m_e \tilde{m}}(\tau_k) \\ -i\pi^{j_e}_{m_e \tilde{m}_e}(\tau_k) \end{array} \right\} |j_e \tilde{m}_e\rangle_{n_e}. \quad (69)$$

In order to rewrite this in terms of standard angular momentum operators $\hat{J}_j^{(v,e)}$ and their eigenbasis $|jm\rangle$ we construct for fixed n a unitary map $W : \mathcal{H}^{jm} \to \mathcal{H}_{jm}$ that satisfies $W \hat{J}_k^{(v,e)} W^{-1} = \hat{Y}_k^{(v,e)}$ and is explicitly given by

$$W : \mathcal{H}^{jm} \to \mathcal{H}_{jm} \quad |jm;n\rangle \mapsto W|j_e m_e; n_e\rangle = \sum_{\tilde{m}_e} \pi^{j}_{m_e \tilde{m}_e}(\epsilon) |j_e \tilde{m}_e\rangle_{n_e}, \quad (70)$$

with $\epsilon := i\sigma_2 = \begin{pmatrix} 0 & 1 \\ -1 & 0 \end{pmatrix}$. The inverse map W^{-1} is then just given by

$$W^{-1} : \mathcal{H}_{jm} \to \mathcal{H}^{jm} \quad |j_e m_e\rangle_n \mapsto W^{-1}|j_e m_e\rangle_n = \sum_{\tilde{m}_e} \pi^{j_e}_{m_e \tilde{m}_e}(\epsilon^{-1}) |j_e \tilde{m}_e; n_e\rangle. \quad (71)$$

Now we go back to SNF associated with a graph α. The discussion above shows that we can apply the unitary map W edgewise and have

$$W^{-1} \pi^{j_e}_{m_e n_e}(A(e)) = \pi^{j_e}_{m_e \tilde{m}_e}(\epsilon^{-1}) \frac{\langle A \mid j_e \tilde{m}_e; n_e \rangle}{\sqrt{2j_e + 1}}, \quad (72)$$

here summation over repeated indices is assumed. Hence, the spin network function $|s^{\vec{j}}_{\alpha,\vec{m},\vec{n}}\rangle$ can be rewritten in the abstract angular momentum basis as

$$\langle A|s^{\vec{j}}_{\alpha,\vec{m},\vec{n}}\rangle = \pi^{j_{e_1}}_{m_{e_1}\tilde{m}_{e_1}}(\epsilon^{-1})\langle A\,|\,j_{e_1}\tilde{m}_{e_1};n_{e_1}\rangle\cdots\pi^{j_{e_n}}_{m_{e_n}\tilde{m}_{e_n}}(\epsilon^{-1})\langle A\,|\,j_{e_n}\tilde{m}_{e_n};n_{e_n}\rangle. \tag{73}$$

By means of the unitary map W we identify $|s^{\vec{j}}_{\alpha,\vec{m},\vec{n}}\rangle$ with abstract angular momentum states and the operators $\hat{Y}^{(v,e)}_j$ with angular momentum operators $\hat{J}^{(v,e)}_j$ and thus we can also discuss the further decomposition of $\mathcal{H}_{\alpha,\vec{j}}$ in the context of angular momentum coupling theory. Let introduce the following operator associated with the vertex v_i

$$(\hat{J}^{(v_i)})^2 := \eta^{jk}\hat{J}^{(v_i)}_j\hat{J}^{(v_i)}_k \quad \text{with} \quad \hat{J}^{(v_i)}_j := \sum_{\substack{e\in E(\alpha)\\ e\cap v_i\neq\emptyset}}\hat{J}^{(v_i,e)}_j, \tag{74}$$

where η^{jk} denotes again the Cartan–Killing metric for su(2). For each v_i the operator $(\hat{J}^{(v_i)})^2$ acts only on $\mathcal{H}_{v_i}$ non trivially and has the eigenvalues $l_{v_i}(l_{v_i}+1)$, where the particular value of l_{v_i} are determined by the values $\{j_{e_i}\}$ associated to the edges, that intersect in v_i. l_{v_i} can be interpreted as the total angular momentum to which the individual angular momenta associated to the edges couple to. Hence, given the operators $(\hat{J}^{(v_i)})^2$ at each vertex v_i we can label their associated eigenspaces by l_{v_i} and denote them by $\mathcal{H}_{\alpha,\vec{j},l_{v_i}}$. Likewise to the decomposition in terms of irreducible representations $\vec{j}$ associated to the edges the Hilbert space $\mathcal{H}_{\alpha,\vec{j}}$ further decomposes into the following direct sum

$$\mathcal{H}_{\alpha,\vec{j}} = \bigoplus_{\vec{l}}\mathcal{H}_{\alpha,\vec{j},\vec{l}} \quad \text{with} \quad \vec{l} = (l_{v_1},\cdots,l_{v_m}),\ \mathcal{H}_{\alpha,\vec{j},\vec{l}} := \bigotimes_{i=1}^{m}\mathcal{H}_{\alpha,\vec{j},l_{v_i}}. \tag{75}$$

Thus, the Hilbert space associated with a given graph α can be rewritten as

$$\mathcal{H}_\alpha = \bigoplus_{\vec{j},\vec{l}}\mathcal{H}_{\alpha,\vec{j},\vec{l}} \tag{76}$$

and states in this Hilbert space are characterized by the irreducible representations, that are associated to the edges and vertices of the graph. For this reason we can label the SNF also by this data yielding $|s_{\alpha,\vec{j},\vec{l}}\rangle$. The difference on the form in (64) is that here the coupling basis for angular momenta has been used for the Hilbert spaces $\mathcal{H}_{l_{v_i}}$ whereas in (64) the product basis was used. In the following sections we will use both notations depending on which one is more suitable in the given situation. Now let us focus our discussion again on the kinematical Hilbert space $\mathcal{H} = L_2(\overline{\mathcal{A}},d\mu_{AL})$. We would like to rewrite $\mathcal{H}$ as a direct sum of the individual $\mathcal{H}_\alpha$s. However, here we are faced with the following problem. Given a graph α and a cylindrical function f_α that does not depend on the holonomies of at least one of the edges of α, then this function would also be an element of $\mathcal{H}_{\tilde{\alpha}}$ for some $\tilde{\alpha}$, that has less edges and vertices. Hence, $\mathcal{H}_\alpha \cap \mathcal{H}_{\tilde{\alpha}} \neq 0$ and therefore the two spaces are not orthogonal. A similar situation occurs when a function depends on the holonomies of two adjacent edges e_1, e_2 in α such that for an edge $\tilde{e}$ in $\tilde{\alpha}$ we have $\tilde{e} = e_1 \circ e_2$. As a consequence, we have to introduce some further rules on how the irreducible representations are associated to the edges of the graph in order to write $\mathcal{H}$ as an orthogonal decomposition of the $\mathcal{H}_\alpha$s. For this purpose we introduce the notion of an admissible labeling of edges and vertices. Given a graph α we call a labeling of the edges and vertices of α by irreducible representations admissible if none of the edges carries a trivial representation and furthermore no two-valent vertex carries a trivial representation. We denote graph Hilbert spaces with admissible labelings by $\mathcal{H}'_\alpha$. Then we can rewrite the kinematical Hilbert space for LQG as

$$\mathcal{H} = \bigoplus_\alpha \mathcal{H}'_\alpha = \bigoplus_\alpha \bigoplus_{\substack{\vec{j},\vec{l} \\ \text{admissible}}} \mathcal{H}_{\alpha,\vec{j}\vec{l}}. \tag{77}$$

This decomposition will be important in the following section when we discuss the dynamics of loop quantum gravity, that is encoded in the quantum Einstein's equations of loop quantum gravity.

5. The Quantum Einstein's Equations of Loop Quantum Gravity

Following the Dirac quantization program requires in the case of loop quantum gravity to implement the Gauss, diffeomorphism and Hamiltonian constraint as operators on the kinematical Hilbert space $\mathcal{H}$ introduced in the last section. Let us denote these operators by $\widehat{\mathcal{C}}_G(\vec{\Lambda})$, $\widehat{\vec{\mathcal{C}}}(\vec{N})$ and $\widehat{\mathcal{C}}(N)$, the quantum analog of the classical Einstein's equations, the so called quantum Einstein's equation of loop quantum gravity are given by

$$\widehat{\mathcal{C}}_G(\vec{\Lambda})\psi_{\text{phys}}(A) = 0, \quad \widehat{\vec{\mathcal{C}}}(\vec{N})\psi_{\text{phys}}(A) = 0, \quad \widehat{\mathcal{C}}(N)\psi_{\text{phys}}(A) = 0, \tag{78}$$

where $\psi_{\text{phy}}(A)$ denotes the physical states, which live in the physical Hilbert space $\mathcal{H}_{\text{phys}}$. The construction of the latter requires apart from finding the (general) solution to the quantum Einstein's equations also to define an inner product on the set of physical states. In this Chapter we will restrict our discussion on the definition and solutions of the Gauss constraints. The remaining diffeomorphism and Hamiltonian constraint is discussed in detail in *Chapter 2*.

5.1. *Solutions to the Gauss Constraint: Gauge-invariant Spinnetwork Functions*

The Gauss constraint is solved using techniques from ordinary lattice gauge theory, where a similar constraint is involved in the theory. Technically, we have two possibilities to construct the solution space, which we will denoted by $\mathcal{H}^{\mathcal{G}}$. Either we can define an operator $\widehat{\mathcal{C}}_G(\vec{\Lambda})$ generating infinitesimal gauge transformation or we can consider the exponentiated version $\widehat{U}(\mathcal{C}_G)$, that generates finite gauge transformations. The solution space will be the same in both cases. How the infinitesimal gauge transformations can be implemented in the quantum theory is explained in detail for instance in Ref. [15]. Here we will consider finite gauge transformation, implemented by unitary operators. As discussed before the holonomy $A(e)$ transforms under gauge transformation as $A(e) \to A^g(e) = g(b(e))A(e)g^{-1}(e)$. The matrix elements of representations of $A(e)$ have thus the following transformation behavior

$$\pi^j_{m_e n_e}(A(e)) \to \pi^j_{m_e n_e}(A^g(e)) = \pi^j_{m_e n_e}(g(b(e))A(e)g^{-1}(f(e))) \tag{79}$$
$$= \pi^j_{m_e \alpha_e}(g(b(e)))\pi^j_{\alpha_e \beta_e}(A(e))\pi^j_{\beta_e n_e}(g^{-1}(f(e))).$$

In order to construct gauge invariant SNF, first we write the SNF in (64) in more compact form as

$$|s^{\vec{j}}_{\alpha,\vec{n},\vec{m}}\rangle \;:\; \overline{\mathcal{A}}_\alpha \to \mathbb{C}, \quad A \mapsto \langle A | s^{\vec{j}}_{\alpha,\vec{n},\vec{m}}\rangle \tag{80}$$

$$\langle A | s^{\vec{j}}_{\alpha,\vec{n},\vec{m}}\rangle := \prod_{k=1}^n \sqrt{2j_{e_k}+1}\, \pi^{j_{e_k}}_{m_{e_k} n_{e_k}}(A(e_k)).$$

Secondly, for the reason that the gauge transformation act on the beginning and final points only, which are precisely the vertices of the graph, we rewrite the product of edges occurring above as

$$\langle A | s^{\vec{j}}_{\alpha,\vec{n},\vec{m}}\rangle := \prod_{v \in v(\alpha)} \prod_{\substack{e \in E(\alpha) \\ e \cap v \neq \emptyset}} \sqrt{2j_e+1}\, \pi^{j_e}_{m_e n_e}(A(e)). \tag{81}$$

Let us consider one individual vertex, at which we have n outgoing edges. For simplicity we will consider only outgoing edges first, but will discuss the more general case below. At the vertex v the SNF transforms under gauge transformation as

$$\langle A^g | s^{\vec{j}}_{\alpha,\vec{n},\vec{m}}\rangle \Big|_v = \prod_{\substack{e \in E(\alpha) \\ e \cap v \neq \emptyset}} \sqrt{2j_e+1}\, \pi^{j_e}_{m_e \alpha_e}(g(b(e)))\pi^{j_e}_{\alpha_e n_e}(A(e)). \tag{82}$$

Let us denote the tensor product of the Hilbert spaces associated with each edge at v as before by $\mathcal{H}_v = \otimes_{\substack{e \in E(\alpha) \\ e \cap v \neq \emptyset}} \mathcal{H}_{j_e}$. We can define a basis of $\mathcal{H}_v$ in terms of tensors of type $(0,n)$, denoted by $\{t_i\}$ with components $t_i^{\alpha_1 \cdots \alpha_n}$, one index for each representation j_e. We can define a dual basis, denoted by $\{\tilde{t}^i\}$ with components $\tilde{t}^i_{\alpha_1 \cdots \alpha_n}$ associated with $\mathcal{H}_v^* = \otimes_{\substack{e \in E(\alpha) \\ e \cap v \neq \emptyset}} \mathcal{H}_{j_e}^*$, where each $\mathcal{H}_{j_e}^*$ carries the dual representation $\overline{\pi}^{j_e}$, by requiring

$$\tilde{t}^j(t_i) = \tilde{t}^j_{\alpha_1 \cdots \alpha_n} t_i^{\alpha_1 \cdots \alpha_n} = \delta^i_j. \tag{83}$$

The gauge transformation acts on these tensors and its duals by

$$t_i^{\alpha_1 \cdots \alpha_n} \to (t')_i^{\alpha_1 \cdots \alpha_n} = \pi^{j_{e_1}}(g(v))^{\alpha_1}{}_{\beta_1} \cdots \pi^{j_{e_n}}(g(v))^{\alpha_n}{}_{\beta_n} t_i^{\beta_1 \cdots \beta_n} \tag{84}$$

$$\tilde{t}^i_{\alpha_1 \cdots \alpha_n} \to (\tilde{t}')^i_{\alpha_1 \cdots \alpha_n} = \overline{\pi}^{j_{e_1}}(g(v))^{\beta_1}{}_{\alpha_1} \cdots \overline{\pi}^{j_{e_n}}(g(v))^{\beta_n}{}_{\alpha_n} \tilde{t}^i_{\beta_1 \cdots \beta_n} \tag{85}$$

$$= \overline{\pi}^{j_{e_1}}(g^{-1}(v))_{\alpha_1}{}^{\beta_1} \cdots \overline{\pi}^{j_{e_n}}(g^{-1}(v))_{\alpha_n}{}^{\beta_n} \tilde{t}^i_{\beta_1 \cdots \beta_n}, \tag{86}$$

where we have used that the dual representation $\overline{\pi}(g(v)) = \pi(g^{-1}(v))^T$ and used the notation $\pi^j_{mn}(g(v)) = \pi^j(g(v))^m{}_n$. Now we are interested in those tensors which are invariant under gauge transformations, which will be denoted by $\{i^k\}$. In terms of their components gauge invariance means

$$\pi^{je_1}(g(v))^{\alpha_1}{}_{\beta_1} \cdots \pi^{je_n}(g(v))^{\alpha_n}{}_{\beta_n} i_k^{\beta_1\cdots\beta_n} = i_k^{\alpha_1\cdots\alpha_n} \tag{87}$$

and likewise for their corresponding dual tensors. An intertwiner i between m dual representations $\overline{\pi}^{j_1},\cdots,\overline{\pi}^{j_m}$ and n representations $\pi^{j_1},\cdots\pi^{j_n}$ is a covariant map

$$i : \bigotimes_{k=1}^{m} \mathcal{H}_{je_k} \to \bigotimes_{\ell=1}^{n} \mathcal{H}_{je_\ell} \tag{88}$$

and can also be understood as an invariant tensor in $\bigotimes_{k=1}^{m} \mathcal{H}^*_{je_k} \otimes \bigotimes_{\ell=1}^{n} \mathcal{H}_{je_\ell}$. We will use this fact to construct gauge invariant SNF. In our example we have a vertex v with n outgoing edges. We achieve that the spin network is invariant under gauge transformation at v when we contract the SNF with the corresponding intertwiner i_v at v, in our example this leads to

$$\left[\langle A | s^{\vec{j}}_{\alpha,\vec{n},\vec{m}} \rangle \Big|_v \right]_{\text{inv}} = i_{v\vec{m}} \prod_{\substack{e \in E(\alpha) \\ e \cap v \neq \emptyset}} \sqrt{2j_e + 1} \pi^{j_e}_{\vec{m}n_e}(A(e)) \tag{89}$$

$$= \sqrt{2j_{e_1}+1} \cdots \sqrt{2j_{e_n}+1} i_v^{m_1\cdots m_n} \pi^{j_{e_1}}_{m_1 n_{e_1}}(A(e_1)) \cdots \pi^{j_{e_n}}_{m_n n_{e_n}}(A(e)).$$

We generalize our discussion to a vertex, that has v_m ingoing edges and v_n outgoing edges. Again we can construct the gauge invariant part of the SNF at this vertex by contracting with an intertwiner i_v, which has components of the form $i_v^{m_1\cdots m_{v_n}}{}_{n_1\cdots n_{v_m}}$. Thus, we can construct an invariant SNF by contracting the gauge variant SNF in (64) at each vertex with a corresponding intertwiner. We will denote the gauge invariant SNF $|s^{\vec{j}}_{\alpha,\vec{i}}\rangle$, where $\vec{i} \in \{i_v \,|\, v \in V(\alpha)\}$ is the set of intertwiners associated with the graph. The gauge invariant SNF is then given by

$$|s^{\vec{j}}_{\alpha,\vec{i}}\rangle \,:\, \overline{\mathcal{A}}_\alpha \to \mathbb{C}, \quad A \mapsto \langle A | s^{\vec{j}}_{\alpha,\vec{i}} \rangle$$

$$\langle A | s^{\vec{j}}_{\alpha,\vec{i}} \rangle := \prod_{v \in V(\alpha)} i_v \prod_{\substack{e \in E(\alpha) \\ e \cap v \neq \emptyset}} \sqrt{2j_e+1}\pi^{j_e}(A(e))$$

$$= \prod_{v \in V(\alpha)} i_v^{m_1\cdots m_{v_n}}{}_{n_1\cdots n_{v_m}} \sqrt{2j_{e_1}+1} \cdots \sqrt{2j_{e_{v_n+v_m}}+1} \pi^{j_{e_1}}_{m_1 n_{e_1}}(A(e_1)) \tag{90}$$

$$\cdots \pi^{j_{e_{v_n}}}_{m_{v_n} n_{e_{v_n}}}(A(e_{v_n})) \pi^{j_{e_{v_n+1}}}_{m_{e_{v_n+1}} n_1}(A(e_{v_n+1})) \cdots \pi^{j_{e_{v_n+v_m}}}_{m_{e_{v_n+v_m}} n_{v_m}}(A(e_{v_n+v_m})).$$

Here each vertex has v_n outgoing and v_m ingoing edges and we have labeled set of edges $\{e_1,\cdots,e_{v_n+v_m}\}$ in such a way, that $e_1,\cdots,e_{v_n}$ are the outgoing edges and $e_{v_n+1},\cdots e_{v_n+v_m}$ are the ingoing edges.

Going back to the decomposition of $\mathcal{H}$ in (77), the gauge invariant Hilbert space corresponds to the case where the edges at all vertices couple to a total angular momentum of zero. Thus, we have for the gauge invariant Hilbert space denoted by $\mathcal{H}_{\text{inv}}^G$

$$\mathcal{H}_{\text{inv}}^G = \bigoplus_\alpha \bigoplus_{\substack{\vec{j},\vec{l} \\ \text{admissible}}} \mathcal{H}_{\alpha,\vec{j}\vec{l}=0}. \tag{91}$$

The Hilbert space $\mathcal{H}_{\text{inv}}^G$ and therefore the solution space of the Gauss constraint is a subspace of the kinematical Hilbert space $\mathcal{H}$. For the remaining constraints of the quantum Einstein's equations, this will be no longer the case and the construction of their corresponding solution spaces is more complicated and will be discussed in Chapter 2.

In the discussion above we have derived $\mathcal{H}_{\text{inv}}^G$ by starting with the configuration space $\overline{\mathcal{A}}$ and implemented the finite gauge transformations on $\mathcal{H}$. Afterwards the solution space $\mathcal{H}_{\text{inv}}^G$ was constructed as a subspace of $\mathcal{H}$. Alternatively, one can also obtain $\mathcal{H}_{\text{inv}}^G$ by considering the reduced quantum configuration space $\overline{\mathcal{A}}/\overline{\mathcal{G}}$, which consists of all generalized connections modulo (generalized) gauge transformations $\overline{\mathcal{G}}$. The latter are the extension of the gauge transformations $\mathcal{G}$ from the classical configuration space $\mathcal{A}$ to the quantum configurations space $\overline{\mathcal{A}}$. In this case only gauge invariant cylindrical functions are considered from the beginning and the final Hilbert space one obtains is also $\mathcal{H}_{\text{inv}}^G$.

6. Geometric Operators and Their Properties

One of the special properties of the AL-representation used in loop quantum gravity introduced in the last section is that one can define operators corresponding to geometrical objects such as volume, area and length. For the KS-representation it has been shown that geometric operators can be implemented using similar techniques as for the AL-representation [21]. This is a consequence of the choice of the particular smearing of the elementary variables discussed above yielding to the holonomy and flux variables. If we had for instance chosen a 3-dimensional smearing like for the standard Fock quantization, the implementation of these geometrical operators in the quantum theory would not be possible.

Among those geometrical operators the most simple one is the area operator from the point of view of its quantization as well as with regards to the spectrum of these operators, therefore we will discuss this operator first.

6.1. The Area Operator

The area operator was first introduced by Smolin [26] and then further analyzed by Rovelli and Smolin in the loop representation [27], which is a representation based on loops instead of graphs and that was used in the early days of loop quantum gravity. Ashtekar and Lewandowski [28] discussed the spectrum of the area operator in the connection representation. In this section we want to discuss the implementation of the area operator as well as its spectrum in detail. At the end of the section we will briefly comment on the volume and length operator.

The strategy one adopts to quantize is the following: As a first step we have to express the classical expression, such as the area, in terms of Ashtekar variables (A, P). Afterwards we need do find a regularization of it, meaning that in our case the area needs to be written as a function of holonomies and fluxes. The guiding principle for the regularization is, that in the limit where the regulator is removed, the classical area in terms of (A, P) should be recovered. Since corresponding operators for holonomies and fluxes exist, the regularized area can then be promoted to a (regularized) operator on the kinematical Hilbert space $\mathcal{H}$, whose detailed properties usually still depends on the chosen regularization. In a final step, one has to show that in the limit where the chosen regulator tends to zero a well defined operator is obtained. The classical area functional associated to a surface S is given by the following expression

$$A_S = \int_U d^2u \sqrt{\det(X^*q)}(u), \tag{92}$$

where q denotes the ADM 3-metric and $X : U \to S$ is an embedding of the surface. Here $U \subset \mathbb{R}^2$ and X^* denotes the pull back of X. The coordinates on the embedded surface S are given by the embedding functions X^a with $a = 1, 2, 3$ and let us denote the two coordinates parametrizing the surface by u_1 and u_2. Given the embedding we can construct two tangent vector fields on S

$$X^a_{,u_1} := \frac{\partial X^a}{\partial u_1}, \quad X^a_{,u_2} := \frac{\partial X^a}{\partial u_2} \tag{93}$$

and also a co-normal vector field n_a that is determined from the condition

$$n_a X^a_{,u_i} = 0 \quad \text{for} \quad i = 1, 2. \tag{94}$$

The determinant in the area functional can be expressed as

$$\det(X^*q) = q_{u_1 u_1} q_{u_2 u_2} - q_{u_1 u_2} q_{u_2 u_1} = \left(X^a_{,u_1} X^b_{,u_1} X^c_{,u_2} X^d_{,u_2} - X^a_{,u_1} X^b_{,u_2} X^c_{,u_2} X^d_{,u_1} \right) q_{ab} q_{cd}. \tag{95}$$

In order to quantize the area functional we need to express it in terms of Ashtekar variables. For this purpose we consider the expression $\det(q) n_a n_b q^{ab}$ and we can express the inverse metric as

$$q^{ab} = \frac{1}{2} \frac{1}{\det(q)} \epsilon^{acd} \epsilon^{bef} q_{ce} q_{df}. \tag{96}$$

Furthermore, we see from (94) that $n_a = \epsilon_{abc} X^c_{,u_1} X^d_{,u_2}$ yielding

$$\det(q) n_a n_b q^{ab} = \det(q) n_a n_b \frac{1}{2} \frac{1}{\det(q)} \epsilon^{acd} \epsilon^{bef} q_{ce} q_{df}$$

$$= \epsilon_{ak\ell} X^k_{,u_1} X^\ell_{,u_2} \epsilon_{bmn} X^m_{,u_1} X^n_{,u_2} \frac{1}{2} \epsilon^{acd} \epsilon^{bef} q_{ce} q_{df}$$

$$= q_{u_1 u_1} q_{u_2 u_2} - q_{u_1 u_2} q_{u_2 u_1}. \tag{97}$$

The inverse metric has a simple form in Ashtekar variables given by $q^{ab} = \frac{1}{k\gamma} P_j^a P_k^b \delta^{jk}/\det(P)$ and depends only on the densitized triad. From $P_j^a = k\gamma\sqrt{\det(q)}e_j^a$ we get $\det(q) = k^3\gamma^3 \det(P)$ yielding

$$\det(q)q^{ab} = k^2\gamma^2 P_j^a P_k^b \delta^{jk} \tag{98}$$

from which we can conclude using (97) that

$$\sqrt{\det(X^*q)} = k\gamma\sqrt{n_a n_b P_j^a P_k^b \delta^{jk}} = k\gamma\sqrt{P_j^\perp P_k^\perp \delta^{jk}}, \tag{99}$$

where $P_j^\perp$ denotes the projection of P_j^a in normal direction with respect to the surface. Note that often one chooses the basis $\tau_j := -i\sigma_j/2$ in su(2) with σ_j being the Pauli matrices for which the Cartan-Killing metric on su(2) η_{jk} becomes $\eta_{jk} := Tr(ad(\tau_j)ad(\tau_k)) = -2\delta_{jk}$ and then one uses the Killing metric in the expression above and adjusts the pre-factors accordingly.

In order to quantize the area functional we need to choose a regularization of the classical expression. For this purpose, we choose a family of non-negative densities $f_u^\epsilon(u')$ on the surface S as regulators, which tend to $\delta_u(u')$ in the limit $\epsilon \to 0$, that is

$$\lim_{\epsilon \to 0} f_u^\epsilon(u') = \delta_x(y), \tag{100}$$

where $\delta_u(u')$ is the delta-function on S peaked at u. Given $f_u^\epsilon(u')$ we can define a regularized version of $P_j^\perp(u)$ denoted by $[P_j^\perp]^\epsilon$ and defined as

$$[P_j^\perp]^\epsilon(u) := \int_S d^2u' f_u^\epsilon(u') P_j^\perp(u'). \tag{101}$$

In the limit where the regulator is removed we have

$$\lim_{\epsilon \to 0}[P_j^\perp]^\epsilon(u) = P_j^\perp(u). \tag{102}$$

Using $[P_j^\perp]^\epsilon$ and a point-splitting, a common technique used in quantum field theory, we can define a regularized expression for the area functional as

$$[A_S]^\epsilon := k\gamma \int_S d^2u \left(\int_S d^2u' \int_S d^2u'' f_u^\epsilon(u') P_j^\perp(u') f_u^\epsilon(u'') P_k^\perp(u''') \delta^{jk} \right)^{\frac{1}{2}}$$

$$= k\gamma \int_S d^3u \left| [P_j^\perp]^\epsilon(u)[P_k^\perp]^\epsilon(u)\delta^{jk} \right|^{\frac{1}{2}}. \tag{103}$$

Obviously, we have $\lim_{\epsilon\to 0}[A_S]^\epsilon = A_S$ in the classical theory. To define a regularized area operator $[\hat{A}_S]^\epsilon$ we use the following strategy: We replace $P_j^\perp(u')$ in (101) by the operator $\hat{P}_j^\perp := -i\hbar \frac{\delta}{\delta A_\perp^j}$ yielding a regularized operator $[\hat{P}_j^\perp]^\epsilon$. Afterwards we have to compute the action of $[\hat{P}_j^\perp]^\epsilon$ on SNF and check whether $[\hat{P}_j^\perp]^\epsilon$ yields a well defined operator. This is indeed the case and one obtains

$$[\hat{P}_j^\perp]^\epsilon(u)|s_{\alpha,\vec{m},\vec{n}}^{\vec{j}}\rangle = \frac{\hbar}{2} \sum_{v \in V(\alpha)} f_u^\epsilon(v) \sum_{\substack{e \in E(\alpha) \\ e \cap v \neq \emptyset}} \kappa(S,e) \hat{Y}_j^{(v,e)} |s_{\alpha,\vec{m},\vec{n}}^{\vec{j}}\rangle, \tag{104}$$

where the operators $\hat{Y}_j^{(v,e)}$ have been defined in (48). Hence, we can rewrite the regularized area operator in the following form

$$[A_S]^\epsilon |s^{\vec{j}}_{\alpha,\vec{m},\vec{n}}\rangle = 4\pi\gamma\ell_p^2 \int_S d^2u \left| \left(\sum_{v\in V(\alpha)} f_u^\epsilon(v) \sum_{\substack{e\in E(\alpha)\\e\cap v\neq\emptyset}} \kappa(S,e)\hat{Y}_j^{(v,e)} \right)^2 \right|^{\frac{1}{2}} |s^{\vec{j}}_{\alpha,\vec{m},\vec{n}}\rangle, \tag{105}$$

where we have used the definition of the Planck length $\ell_p = \hbar G_N = \frac{8\pi}{\hbar}k$. Next we choose ϵ sufficiently small such that for a given $u \in S$ $f_u^\epsilon(v)$ is non-vanishing only for at most one vertex v. Thus, we have $f_u^\epsilon(v)f_u^\epsilon(v') = \delta_{v,v'}(f_u^\epsilon(v))^2$ and we obtain

$$[A_S]^\epsilon |s^{\vec{j}}_{\alpha,\vec{m},\vec{n}}\rangle = 4\pi\gamma\ell_p^2 \int_S d^2u \sum_{v\in V(\alpha)} f_u^\epsilon(v) \left| \left(\sum_{\substack{e\in E(\alpha)\\e\cap v\neq\emptyset}} \kappa(S,e)\hat{Y}_j^{(v,e)} \right)^2 \right|^{\frac{1}{2}} |s^{\vec{j}}_{\alpha,\vec{m},\vec{n}}\rangle. \tag{106}$$

As a final step we have to remove the regulator yielding to a well defined area operator $\hat{A}_S$ on the kinematical Hilbert space $\mathcal{H}$ of the form

$$[A_S]|s^{\vec{j}}_{\alpha,\vec{m},\vec{n}}\rangle := \lim_{\epsilon\to 0}[A_S]^\epsilon |s^{\vec{j}}_{\alpha,\vec{m},\vec{n}}\rangle \tag{107}$$

$$= 4\pi\gamma\ell_p^2 \int_S d^2u \sum_{v\in V(\alpha)} \delta_u(v) \left| \left(\sum_{\substack{e\in E(\alpha)\\e\cap v\neq\emptyset}} \kappa(S,e)\hat{Y}_j^{(v,e)} \right)^2 \right|^{\frac{1}{2}} |s^{\vec{j}}_{\alpha,\vec{m},\vec{n}}\rangle.$$

From the above expression for the area operator we realize that in the sum over all vertices of the graph α only those vertices will contribute which are intersection points of the surface S as otherwise $\kappa(S,e) = 0$. For this reason we can write the area operator in more compact form by introducing the set $I(S)$ of intersection points of edges of type up and type down, that is given by

$$I(S) = \{v \in e \cap S | \kappa(S,e) \neq 0,\ e \in E(\alpha)\}. \tag{108}$$

This yields to the final form of the area operator that we will use in the following

$$\hat{A}_S |s^{\vec{j}}_{\alpha,\vec{m},\vec{n}}\rangle = 4\pi\gamma\ell_p^2 \sum_{v\in I(S)} \left| \left(\sum_{e\ \text{at}\ v} \kappa(S,e)\hat{Y}^{(e,v)} \right)^2 \right|^{\frac{1}{2}} |s^{\vec{j}}_{\alpha,\vec{m},\vec{n}}\rangle. \tag{109}$$

Let us now discuss the spectrum of the area operator. At each intersection point $v \in I(S)$ we have edges of type up, edges of type down and edges of type in that will not contribute to the spectrum. In order to write the expression under the square root in (109) in compact form we introduce the following operators:

$$\hat{Y}_j^{v,u} := \sum_{e\in E(v,u)} \hat{Y}_j^{(v,e)} \qquad \hat{Y}_j^{v,d} := \sum_{e\in E(v,d)} \hat{Y}_j^{(v,e)}. \tag{110}$$

Here $E(v,u), E(v,d)$ denote all edges of type up and down respectively that intersect each other in the point v. Then we have for each intersection point v

$$\left(\sum_{\substack{e\in E(\gamma)\\e\cap v\neq\emptyset}} \kappa(S,e)\hat{Y}_j^{(v,e)} \right)^2 = \left(\hat{Y}^{v,u} - \hat{Y}^{v,d} \right)^2$$

$$= (\hat{Y}^{v,u})^2 + (\hat{Y}^{v,d})^2 - 2\hat{Y}^{v,u}\hat{Y}^{v,d}$$
$$= 2(\hat{Y}^{v,u})^2 + 2(\hat{Y}^{v,d})^2 - (\hat{Y}^{v,u} + \hat{Y}^{v,d})^2. \tag{111}$$

We used in the second line that $[\hat{Y}_j^{v,u}, \hat{Y}_k^{v,d}] = 0$. Furthermore, the operators $(\hat{Y}^{v,u})^2$, $(\hat{Y}^{v,d})^2$ and $(\hat{Y}^{v,u} + \hat{Y}^{v,d})^2$ mutually commute. Moreover, we choose an explicit basis $\tau_j = -i\sigma_j/2$ for which the operators $\hat{Y}^{(v,e)}$ satisfy the usual angular momentum algebra given by $[\hat{Y}_i^{(v,e)}, \hat{Y}_j^{(v,e)}] = \epsilon_{ijk}\hat{Y}_k^{(v,e)}$. Then we have that the operators $(\hat{Y}^{(v,e)})^2 \equiv \delta^{jk}\hat{Y}_j^{(e,v)}\hat{Y}_k^{(v,e)}$ locally act as

$$-\delta^{ij}R_iR_j = -\langle R,R\rangle \equiv -\Delta_{SU(2)}, \text{ or } -\delta^{ij}L_iL_j = -\langle L,L\rangle \equiv -\Delta_{SU(2)}, \quad (112)$$

where $-\Delta_{SU(2)}$ is the positive definite SU(2) Laplacian with spectrum $j(j+1)$, due to our choice of basis for su(2). Hence, the same holds for the operators $(\hat{Y}^{v,u})^2$, $(\hat{Y}^{v,d})^2$, and $(\hat{Y}^{v,u} + \hat{Y}^{v,d})^2$, they act as Laplacians in the respective direct sum of representations. Therefore the spectrum of the operators involved in (111) can be easily computed and we obtain

$$\text{Spec}(\hat{A}_S) = 4\pi\gamma\ell_p^2 \sum_{v \in I(S)} \sqrt{2j_{u,v}(j_{u,v}+1) + 2j_{d,v}(j_{d,v}+1) - j_{u+d,v}(j_{u+d,v}+1)}.$$

$$(113)$$

Here $j_{u,v}, j_{d,v}$ denote the total angular momentum of the edges of type up (down respectively) at the intersection point v and $j_{u+d,v}$ total coupled angular momentum of the up and down edges whose values range between $|j_{u,v} - j_{d,v}| \leq j_{u+d,v} \leq j_{u,v} + j_{d,v}$. Let us consider the eigenvalue at one intersection point v. The smallest possible eigenvalue that we can get occurs when either $j_{u,v} = 0$ and $j_{d,v} = \frac{1}{2}$ or vice versa. The eigenvalue denoted by λ_0 is non vanishing and given by

$$\lambda_0 = 2\pi\gamma\ell_p^2\sqrt{3} \quad (114)$$

and is known as the area gap in loop quantum gravity. The area gap plays an important role in the description of black hole physics within loop quantum gravity and black hole entropy calculations can be used the fix the value of the Immirzi parameter γ as discussed in *Chapter 7*.

6.2. The Volume Operator

The volume operator enters crucially into the construction of the dynamics of the quantum Einstein's equations for the reason that the classical co-triad is expressed as the Poisson bracket between the connection and the classical volume functional using the Thiemann identity (see *Chapter 2*). In the case of the area operator, the area functional depends on the momenta P_j^a only, which is also true for the classical volume functional, that for a given region R in the spatial manifold Σ reads

$$V_R = \int_R d^3x \sqrt{\det(q)} = (k\gamma)^{\frac{3}{2}} \int_R d^3x \sqrt{|\det(P_j^a)|}. \quad (115)$$

Likewise to the case of the area operator we need to choose a regularization of V_R in order to write the volume functional in terms of fluxes $P(S, f)$ for which well defined operators exist yielding as a first step a regularized expression of the classical volume functional V_R. For the latter it is natural to choose a partition $\mathcal{P}^\epsilon$ of the spatial region R in terms of cubic cells C^ϵ and adapted 2-surfaces for each cubic cell. For this purpose we introduce a coordinate system (x^a) and assume that each C^ϵ has a volume of less than ϵ in the chosen coordinate system and that two different cells share only points on their boundaries. For each cubic cell C^ϵ we introduce three 2-surfaces $\{S_a \,|\, a = 1, 2, 3\}$ chosen in a way such that the coordinates components x^a are constant along S_a for $a = 1, 2, 3$ following the notation in Ref. [30]. Furthermore, each S_a has the property that it divides C^ϵ into two disjoint parts. We can now use the surfaces $\{S_a \,|\, a = 1, 2, 3\}$ to formulate a regularized volume functional denoted by V_R^ϵ as a function of fluxes over the surfaces $\{S_a \,|\, a = 1, 2, 3\}$. Going back to the definition of the flux $P(S, f)$ in (35) we will choose as the smearing functions f^j the su(2) basis elements τ_j and define $P_j(S) := P(S, \tau_j)$. Given this, we have

$$V_R^\epsilon = (k\gamma)^{\frac{3}{2}} \sum_{C^\epsilon \in \mathcal{P}^\epsilon} \sqrt{|Q_{C^\epsilon}|} \quad (116)$$

with

$$Q_{C^\epsilon} := \frac{1}{3!} \epsilon^{jk\ell} \epsilon^{abc} P_j(S_a) P_k(S_b) P_\ell(S_c). \quad (117)$$

In the classical theory we have $\lim_{\epsilon \to 0} V_R^\epsilon = V_R$, however in the quantum theory the removal of the regular has to be taken with more care. While in the case of the area operator after the regulator has been removed the final operator does not depend on the chosen background structure of the regularization, a different situation occurs for the volume operator. In the case of the volume operator once the regulator is removed, the resulting operator still depends on the chosen partition and thus carries a memory of the chosen regularization. As a consequence this operator depends on the chosen background structure during the regularization procedure and therefore the limit does not yield an appropriate candidate for a volume operator because it fails to be covariant under spatial diffeomorphisms. This problem can be circumvented by first averaging over the possible background structures, whose dependence enters into the volume operator in a rather simple way, before removing the regulator. The requirements that we obtain a well defined operator when the regulator is removed as well as that the final operator is covariant under spatial diffeomorphisms, are restrictive enough to uniquely determine the final form of the operator up to a global constant, that we will denote regularization constant c_{reg} in the following.

In the literature two different volume operators exist, one introduced by Rovelli and Smolin (RS) [27] and one introduced by Ashtekar and Lewandowski (AL) [29], which come out of a priori equally justified but different regularization techniques. In the classical theory both regularized versions, the RS- as well as the AL-volume — although being of different kind — yield the classical volume functional once the regulator is removed. However, in the quantum theory the removal of the regulator is more subtle and this is the reason why one ends up with two different quantum operators. Both volume operators act non-trivially only on vertices where at least three edges intersect. At a given vertex the operators have the following form

$$\hat{V}_{v,\text{RS}} = c_{\text{RS}} \sum_{e_I \cap e_J \cap e_K = v} \left| \hat{Q}_{IJK} \right|^{\frac{1}{2}}$$

$$\hat{V}_{v,\text{AL}} = c_{\text{AL}} \left| \sum_{e_I \cap e_J \cap e_K = v} \epsilon(e_I, e_J, e_K) \hat{Q}_{IJK} \right|^{\frac{1}{2}}. \quad (118)$$

Here $\hat{Q}_{IJK} := \epsilon^{ijk} \hat{Y}_i^{(v,e_I)} \hat{Y}_j^{(v,e_J)} \hat{Y}_k^{(v,e_K)}$ is an operator involving only flux operators and thus right and left invariant vector fields and $c_{\text{RS}}, c_{\text{AL}}$ are regularization constants. The sum runs over all ordered triples of edges intersecting at the vertex v. A detailed discussion about the regularization of the volume operator can for instance be found in Refs. [15] and [30]. The main differences between these two operators is that the RS-operator is not sensitive to the orientation of the triples of edges and is therefore covariant under homeomorphisms. The AL-operator has likewise to the $\kappa(S, e)$ in the area operator a similar sign factor $\epsilon(e_I, e_J, e_K)$ that can take the values $\{+1, 0, -1\}$ and is the sign of the cross product of the tangent vectors at v of the triple of edges e_i, e_j, e_k that intersect at this vertex v. Furthermore, the sum over triples of edges involved in both operators occurs outside the square root in case of the RS and inside the square root in case of the AL-operator. Due to the sign factor $\epsilon(e_I, e_J, e_K)$ the operator $\hat{V}_{\text{AL}}$ is covariant only under diffeomorphisms.

The spectral analysis of the volume operator is more complicated than for the area operator and can in general not be computed analytically. A general formula for the computation of matrix elements of the AL-volume operator has been derived in Ref. [31]. Those techniques have been used to analyze the spectrum of the volume operator numerically up to a vertex valence of 7 in a series of papers [32]. Their work showed that the spectral properties of the volume operator depend on the embedding of the vertex that enters via the sign factors $\epsilon(e_i, e_j, e_k)$ into the construction of the AL-operator. Particularly, the presence of a volume gap, that is a smallest non-vanishing eigenvalue, depends on the geometry of the vertex. A consistency check for both volume operators has been discussed in Ref. [33] where the Thiemann identity, discussed in detail in *Chapter 2*, has been used to define an alternative flux operator. The alternative flux operator is then compared to the usual flux operator and consistency of both operators could for instance fix the undetermined regularization constant $c_{\text{AL}} = \ell_p^3 / \sqrt{48}$ in the volume operator. Furthermore, the RS-operator did not pass this consistency check and the reason that it worked for the AL-operator is exactly the presence of those sign factors $\epsilon(e_I, e_J, e_K)$ in the AL-operator.

A technique to compute matrix elements of the volume operator with respect to semiclassical states analytically was developed in Ref. [34]. This method relies on the idea of an expansion of the matrix elements of the volume operator in a power series of matrix elements of operators, that can be computed analytically. These operators in the expansion are chosen in such a way that the error caused by this expansion can be estimated and can be well controlled.

6.3. *The Length Operator*

A length operator for LQG was introduced in Ref. [35]. The length operator is in some sense the most complicated one among the kinematical geometrical operators. Let us recall the length of a curve $c : [0,1] \to \Sigma$ classically is given by

$$\ell(c) = \int_0^1 \sqrt{q_{ab}(c(t))\dot{c}^a(t)\dot{c}^b(t)}dt = \int_0^1 \sqrt{e_a^i(c(t))e_b^j(c(t))\dot{c}^a(t)\dot{c}^b(t)\delta_{ij}}dt, \qquad (119)$$

here $\dot{c}^a$ denotes the components of the tangent vector associated to the curve. When we express the metric q_{ab} in terms of Ashtekar variables we obtain

$$q_{ab} = \frac{k}{4}\epsilon_{acd}\epsilon_{bef}\epsilon_{ijk}\epsilon^{imn}\frac{P_j^c P_k^d P_e^m P_f^n}{\det(P)}, \qquad (120)$$

which is a non-polynomial function in terms of the electric fields and therefore a regularization in terms of flux operators similar to the area and volume operator does not exist. Furthermore, the denominator being the square of the volume density cannot be defined on a dense set in $\mathcal{H}$ because it has a huge kernel. One possibility to quantize the length used in Ref. [35] is to use for the co-triads that occur in (119) the Thiemann identity and replace them by a Poisson bracket between the connection and the volume functional. This yields a length operator that involves a square root of two commutators between holonomy operators along the curve c and the volume operator. In this way the inverse volume density can be avoided and the volume occurs only linearly in the commutator. Also, the length operator does not change the graph or the spin labels of the edges likewise to the area and volume operator. However, since the length operator becomes even a function of the volume operator its spectral analysis becomes even more complicated than for the volume operator itself and very little about the spectrum of the length operator is known except for low valence vertices.

Another length operator was introduced in Ref. [36], where the Thiemann identity was not used for the quantization. The regularization adapted in Ref. [36] is motivated from the dual picture of quantum geometry and uses that the curve can be expressed as an intersection of two surfaces. This allows to express the tangent vector of the curve in terms of the normals of the surfaces. The inverse volume issue discussed above is circumvented by using a Tikhonov regularization for the inverse RS-volume-operator. For this length operator the spectral properties have only been analyzed for a vertex of valence 4, which is monochromatic, that is all spins are identical. Another alternative length operator for LQG has been discussed in [37] where a different regularization has been chosen such that the final length operator can be expressed in terms of other geometrical objects — the area, volume and flux operators. In this work the AL-operator is used and the inverse volume operator is also defined using a Tikhonov regularization similar to the one in Ref. [36].

7. Summary

In this Chapter we presented a brief introduction to the kinematical setup of loop quantum gravity [15, 16]. Loop quantum gravity can be understood as a framework for canonically quantizing general relativity. This approach leads to a quantum theory based on quantum geometry for the reason that not only the matter part of the theory but also the geometry itself is quantized. In Section 1 we briefly mentioned earlier attempts to canonically quantize general relativity using ADM-variables. However, these approaches could only provide a quantum theory that was constructed at a rather formal level since neither the functional analytical details about the kinematical Hilbert space had been worked out nor could the dynamics of the quantum theory be implemented rigorously. But precisely the quantization of the constraints that encode the dynamics of the quantum theory needs to be understood in great detail if one wants to analyze characteristic properties and consequences of quantum geometry. Particularly, the Hamiltonian constraint is a non-polynomial function of the elementary phase space variables in contrast to the Hamiltonian used in other gauge theories in the context of the standard model of particle physics. Nevertheless, these earlier results were important because they already showed what kind of complications occur if one tries to carry over the standard quantization used in ordinary quantum mechanics to general relativity. Progress regarding this aspect was made when the connection variables were introduced by Ashtekar [7] leading to a reformulation of general relativity in terms of an SU(2) gauge theory as discussed in Section 2. As a consequence it involves next to the spatial diffeomorphism and the Hamiltonian constraint also known from the ADM-formalism an additional SU(2) Gauss constraint. Although, the Hamiltonian constraint keeps its non-polynomial form also with respect to these new variables the advantage of the connection formulation is, that general relativity can be formulated in the language of ordinary gauge theories. This leads to a form of the constraint in the new variables that looks much closer to what we are familiar with from other gauge theories. Therefore, techniques developed in those fields could be taken as a point of reference for constructing the quantum theory underlying loop quantum gravity. Taking this into account the choice of holonomies and fluxes as presented in Section 3 is a very natural choice as elementary phase space variables for the theory. We introduced the notion of cylindrical functions and flux vector fields acting on them in order to give a precise definition of the holonomy–flux algebra used in loop quantum gravity. The choice of the classical algebra and its related properties are important in the sense that the corresponding quantum theory will of course depend on the particular choice because we obtain the quantum theory by finding representations of the underlying classical algebra. In the case of the holonomy–flux algebra the first representation that was found is the Ashtekar–

Lewandowski [38, 39] representation discussed in Section 4.[d] Interestingly, later the LOST-theorem [18] proved that this is the only representation of the holonomy–flux algebra if the symmetries of the theory, particularly the spatial diffeomorphisms are taken very seriously. Other representations that violate one of the assumptions used in the LOST-theorem were found by Sahlmann and Koslowski [22]. We finished Section 4 by introducing spin networks which provide an orthonormal basis for the kinematical Hilbert space. Beside being a very useful tool as far as computations in loop quantum gravity are concerned they also deliver insight into the question how quantum states look like in loop quantum gravity. Each spin network is defined on a graph that consists of a finite number of edges that are one-dimensional objects embedded into the spatial manifold we obtained from 3+1 split. These edges are labeled with so called spin quantum numbers and the vertices of the graph carry intertwiners. These data can be understood as describing a particular state of quantum geometry at the kinematical level and by varying these data we would obtain different states of quantum geometry. To go beyond the kinematical level we have to consider the dynamics of quantum geometry that is described by the quantum Einstein's equations. These are the classical analogue of Einstein's equations in general relativity. In the context of Dirac quantization for constrained systems the formulation of the quantum Einstein's equations requires to implement the classical constraints as operators on the kinematical Hilbert space. If one considers a reduced phase space quantization approach for loop quantum gravity [12], then formulating the dynamics requires to define a (physical) Hamiltonian on the physical Hilbert space. The latter is obtained by quantizing directly the reduced phase space. A more detailed presentation of the quantum dynamics can be found in *Chapter 2*. In Section 5 we only start to introduce the topic of quantum dynamics and we restrict our discussion to the construction of solutions to the Gauss constraint only. The corresponding solutions are gauge invariant spin network functions and the remaining dynamical operators associated with finite spatial diffeomorphisms and the infinitesimal Hamiltonian constraint are well defined on the gauge invariant Hilbert space. We finished this Chapter with a brief review on geometrical operators. These are operators associated with geometrical quantities like length, area and volume. That these operators can be implemented is a special property of the kinematical representation used in loop quantum gravity and related to the fact that holonomies as well as fluxes are used as the elementary variables. In a Fock representation, used in ordinary quantum field theory, those operators are not well defined. At the kinematical level the spectrum of the area operator can be computed analytically and interestingly it turns out to be discrete and a smallest non-vanishing eigenvalue exists in a so called area gap. For the volume and length operator the complete

---

[d]Since this Chapter was submitted, Dittrich, Geiller and others have opened up the possibility of constructing a representation that is complementary to the Ashtekar–Lewandowski representation discussed here. See, e.g., B. Dittrich and M.Geiller, arXiv:1604.05195, and *Chapter 5* in this volume.

spectrum is still unknown but one has analyzed the volume operator for special spin networks states with low valence [32].

The kinematical setup introduced in this Chapter is the mathematical foundation for most of the research done in loop quantum gravity. In the context of loop quantum cosmology, that is a symmetry reduced model for loop quantum gravity and introduced in *Chapter 6*, the kinematical representation discussed here is adopted and specialized to the context of cosmological models. Also the particular implementation of the quantum Einstein's equations discussed in the *Chapter 2* is closely related to the choice of the kinematical representation. In the context of black hole physics the area operator plays an important role and provides new insights on a quantum mechanical description of the black hole entropy as discussed in *Chapter 7*. Furthermore, a motivation for spin foam models, which aims to provide the corresponding covariant formulation of loop quantum gravity in the context of path integral quantization, is again the kinematical framework presented in this Chapter. Therefore, also in the covariant approach the kinematical Hilbert space plays an important role. More details on the covariant approach can be found in *Chapters 3, 4 and 5*.

References

[1] S. Deser, R. Arnowitt and C. W. Misner, "Consistency of Canonical Reduction of General Relativity," New York: Belfer Graduate School of Science, Yeshiva University, 1964.
[2] R. P. Geroch, "The domain of dependence," *J. Math. Phys.* **11**, 437 (1970).
[3] C. Rovelli, "Quantum reference systems," *Class. Quant. Grav.* **8**, 317 (1991).
C. Rovelli, "Time in quantum gravity: Physics beyond the Schrödinger regime," *Phys. Rev. D* **43**, 442 (1991).
C. Rovelli, "Quantum mechanics without time: A model," *Phys. Rev. D* **42**, 2638 (1990).
[4] B. Dittrich, "Partial and complete observables for Hamiltonian constrained systems," *Gen. Rel. Grav.* **39**, 1891 (2007) [arXiv:gr-qc/0411013]. B. Dittrich, "Partial and complete observables for canonical general relativity," *Class. Quant. Grav.* **23**, 6155 (2006) [arXiv:gr-qc/0507106].
[5] P. A. M. Dirac, "Lectures on quantum mechanics," *J. Math. Phys.* **1**, 434 (1960).
[6] B. S. DeWitt, "Quantum theory of gravity. 1. The canonical theory," *Phys. Rev.* **160**, 1113 (1967).
[7] A. Ashtekar, "New variables for classical and quantum gravity," *Phys. Rev. Lett.* **57**, 2244 (1986).
[8] M. Henneaux, J. E. Nelson and C. Schomblond, "Derivation of Ashtekar variables from tetrad gravity," *Phys. Rev. D* **39** (1989) 434-7.
[9] J. F. Barbero G., "Real Ashtekar variables for Lorentzian signature space times," *Phys. Rev. D* **51**, 5507 (1995) [arXiv:gr-qc/9410014].
[10] G. Immirzi, "Real and complex connections for canonical gravity," *Class. Quant. Grav.* **14**, L177 (1997) [arXiv:gr-qc/9612030].
[11] S. A. Hojman, K. Kuchar and C. Teitelboim, "Geometrodynamics regained," *Annals Phys.* **96**, 88 (1976).
[12] K. Giesel and T. Thiemann, "Algebraic quantum gravity (AQG). IV. Reduced phase

space quantisation of loop quantum gravity," *Class. Quant. Grav.* **27**, 175009 (2010) [arXiv:0711.0119 [gr-qc]].

[13] N. Bodendorfer, T. Thiemann and A. Thurn, "New variables for classical and quantum gravity in all dimensions I. Hamiltonian analysis," *Class. Quant. Grav.* **30**, 045001 (2013) [arXiv:1105.3703 [gr-qc]].
N. Bodendorfer, T. Thiemann and A. Thurn, "New variables for classical and quantum gravity in all dimensions II. Lagrangian analysis," *Class. Quant. Grav.* **30**, 045002 (2013) [arXiv:1105.3704 [gr-qc]].
N. Bodendorfer, T. Thiemann and A. Thurn, "New variables for classical and quantum gravity in all dimensions III. Quantum theory," *Class. Quant. Grav.* **30**, 045003 (2013) [arXiv:1105.3705 [gr-qc]].
N. Bodendorfer, T. Thiemann and A. Thurn, "New variables for classical and quantum gravity in all dimensions IV. Matter coupling," *Class. Quant. Grav.* **30**, 045004 (2013) [arXiv:1105.3706 [gr-qc]].

[14] A. Ashtekar, J. D. Romano, R. S. Tate, "New variables for gravity: Inclusion of matter," *Phys. Rev. D* **40**, 2572 (1989).

[15] T. Thiemann, *Modern Canonical Quantum General Relativity*, Cambridge, UK: Cambridge Univ. Pr. (2007) 819 p [gr-qc/0110034].

[16] C. Rovelli, *Quantum Gravity*, Cambridge, UK: Cambridge Univ. Pr. (2007) 488 p.

[17] J. von Neumann, "On the uniqueness of the Schrödinger operators," *Math. Ann.* **104** (1931), 570-578 (in German).

[18] J. Lewandowski, A. Okolow, H. Sahlmann and T. Thiemann, "Uniqueness of diffeomorphism invariant states on holonomy–flux algebras," *Commun. Math. Phys.* **267**, 703 (2006) [arXiv:gr-qc/0504147].

[19] C. Fleischhack, "Representations of the Weyl algebra in quantum geometry," *Commun. Math. Phys.* **285**, 67 (2009) [arXiv:math-ph/0407006].

[20] T. A. Koslowski, "Dynamical Quantum Geometry (DQG Programme)," arXiv:0709.3465 [gr-qc].

[21] H. Sahlmann, "On loop quantum gravity kinematics with non-degenerate spatial *Class. Quant. Grav.* **27**, 225007 (2010) [arXiv:1006.0388 [gr-qc]].

[22] T. Koslowski and H. Sahlmann, "Loop quantum gravity vacuum with non-degenerate geometry," *SIGMA* **8**, 026 (2012) [arXiv:1109.4688 [gr-qc]].

[23] M. Campiglia and M. Varadarajan, "The Koslowski–Sahlmann representation: gauge and diffeomorphism invariance," *Class. Quant. Grav.* **31**, 075002 (2014) [arXiv:1311.6117 [gr-qc]].

[24] A. Stottmeister and T. Thiemann, "Structural aspects of loop quantum gravity and loop quantum cosmology from an algebraic perspective," arXiv:1312.3657 [gr-qc].

[25] M. Campiglia and M. Varadarajan, "The Koslowski-Sahlmann representation: Quantum configuration space," *Class. Quant. Grav.* **31**, 175009 (2014) [arXiv:1406.0579 [gr-qc]].

[26] L. Smolin, "Recent developments in nonperturbative quantum gravity," arXiv:hep-th/9202022.

[27] C. Rovelli and L. Smolin, "Discreteness of area and volume in quantum gravity," *Nucl. Phys. B* **442**, 593 (1995) [Erratum-ibid. B **456**, 753 (1995)] [arXiv:gr-qc/9411005].

[28] A. Ashtekar and J. Lewandowski, "Quantum theory of geometry. 1: Area operators," *Class. Quant. Grav.* **14**, A55 (1997) [arXiv:gr-qc/9602046].

[29] A. Ashtekar and J. Lewandowski, "Quantum theory of geometry. 2. Volume operators," *Adv. Theor. Math. Phys.* **1**, 388 (1998) [arXiv:gr-qc/9711031].

[30] A. Ashtekar and J. Lewandowski, "Background independent quantum gravity: A Status report," *Class. Quant. Grav.* **21**, R53 (2004) [arXiv:gr-qc/0404018].

[31] J. Brunnemann and T. Thiemann, "Simplification of the spectral analysis of the volume operator in loop quantum gravity," *Class. Quant. Grav.* **23**, 1289 (2006) [arXiv:gr-qc/0405060].

[32] J. Brunnemann and D. Rideout, "Properties of the volume operator in loop quantum gravity. I. Results," *Class. Quant. Grav.* **25**, 065001 (2008) [arXiv:0706.0469 [gr-qc]].
J. Brunneman and D. Rideout, "Properties of the volume operator in loop quantum gravity. II. Detailed presentation," *Class. Quant. Grav.* **25**, 065002 (2008) [arXiv:0706.0382 [gr-qc]].
J. Brunnemann and D. Rideout, "Oriented matroids – Combinatorial structures underlying loop quantum gravity," *Class. Quant. Grav.* **27**, 205008 (2010) [arXiv:1003.2348 [gr-qc]].

[33] K. Giesel and T. Thiemann, "Consistency check on volume and triad operator quantisation in loop quantum gravity. I," *Class. Quant. Grav.* **23**, 5667 (2006) [arXiv:gr-qc/0507036].
K. Giesel and T. Thiemann, "Consistency check on volume and triad operator quantisation in loop quantum gravity. II," *Class. Quant. Grav.* **23**, 5693 (2006) [arXiv:gr-qc/0507037].

[34] K. Giesel and T. Thiemann, 'Algebraic quantum gravity (AQG). III. Semiclassical perturbation theory," *Class. Quant. Grav.* **24**, 2565 (2007) [gr-qc/0607101].

[35] T. Thiemann, "A Length operator for canonical quantum gravity," *J. Math. Phys.* **39**, 3372 (1998) [arXiv:gr-qc/9606092].

[36] E. Bianchi, "The length operator in loop quantum gravity," *Nucl. Phys. B* **807**, 591 (2009) [arXiv:0806.4710 [gr-qc]].

[37] Y. Ma, C. Soo and J. Yang, "New length operator for loop quantum gravity," *Phys. Rev. D* **81**, 124026 (2010) [arXiv:1004.1063 [gr-qc]].

[38] A. Ashtekar and C. J. Isham, "Representations of the holonomy algebras of gravity and nonAbelian gauge theories," *Class. Quant. Grav.* **9**, 1433 (1992) [hep-th/9202053].

[39] A. Ashtekar and J. Lewandowski, "Projective techniques and functional integration for gauge theories," *J. Math. Phys.* **36**, 2170 (1995) [gr-qc/9411046].

Chapter 2

Quantum Dynamics

Alok Laddha

Chennai Mathematical Institute, Siruseri, Chennai, 6000103, India

Madhavan Varadarajan

Raman Research Institute, Sadashivanagar, Bengaluru 560080, India

1. Introduction

As discussed in *Chapter 1*, by virtue of its diffeomorphism invariance, the classical Hamiltonian dynamics of gravity is generated by constraints. The definition of its quantum dynamics then necessitates the construction of quantum operator versions of these constraints. A non-trivial requirement on these constraint operators is that they provide an anomaly free representation of their classical Poisson bracket algebra. As for any gauge theory, the anomaly free requirement ensures that the quantum theory has the correct number of degrees of freedom. In addition, the work of Hojman, Kuchař and Teitelboim (HKT) [1] implies that such a requirement is closely connected with the emergence of spacetime covariance in the quantum theory. The transformations generated by the diffeomorphism and Hamiltonian constraints can be interpreted in terms of deformations of the Cauchy slice within the dynamically generated spacetime. HKT show that the algebra of these deformations obtains a characteristic structure which depends only on the spacetime covariance of the gravitational dynamics and is independent of its finer details. Hence, it is believed that a notion of spacetime covariance can emerge in quantum theory only if this algebra is represented without anomalies.

Since the basic variables of LQG are an $SU(2)$ connection and its conjugate triad, the classical theory also has a 'Gauss Law' constraint which generates triad rotations and imposes $SU(2)$ invariance. It follows that we may ignore this constraint as long as we work with (linear combinations of) the $SU(2)$ gauge invariant spin network states discussed in *Chapter 1*. Accordingly in canonical LQG, our aim is to construct the action of the diffeomorphism and Hamiltonian constraint operators on such states in such a way as to ensure an anomaly free representation of their classical Poisson brackets, and, in accordance with the tenets of Dirac quantization, identify physical states with the kernel of these constraints.

The most direct way to construct the constraint operators is to express the classical constraints in terms of the basic phase space functions of the theory, namely the holonomies and fluxes, and replace these functions by their quantum operator correspondents in the resulting expressions. Unfortunately, this strategy cannot be employed due to the following reason. The expressions for the classical constraints involve the connection and its curvature. These local fields must then be re-expressed in terms of non-local holonomies. While in classical theory these local fields at a point can be obtained through the limiting behavior of a set of holonomies around loops which shrink to the point, in quantum theory the limit of the corresponding operators does not exist. The limit of this shrinking procedure is ill defined because the background independent Hilbert space is unable to distinguish a 'large loop' from a 'small' one if they are diffeomorphic images of each other.

The above obstruction can be bypassed for the diffeomorphism constraints by shifting attention to the *finite* transformations which they generate. These finite transformations correspond to the group of finite spatial diffeomorphisms. The anomaly free requirement translates to the requirement that the group structure of these transformations be correctly represented in quantum theory. Since LQG provides a unitary representation of the group of spatial diffeomorphisms, this requirement is satisfied. The identification of diffeomorphism invariant states then corresponds to the identification of states which are invariant under these unitary transformations. The Hilbert space of such states, $\mathcal{H}_{\text{Diff}}$, can be constructed by an application of *group averaging* techniques [2]. We describe this, by now standard material, in the first part of Section 2. The above considerations bypass the need for operators corresponding to the *generators* of finite diffeomorphisms. Nevertheless, it is of interest to enquire if the action of such operators can be defined as a suitable limit of a family of operators corresponding to finite diffeomorphisms in the neighborhood of identity. For the same reason which precludes the definition of a connection operator as a limit of holonomies, this limit does not exist on the LQG Hilbert space. However, Lewandowski and Marolf showed that it *is* possible to define such a limit on an appropriate deformation of diffeomorphism invariant states [3]. In the second part of Section 2 we exhibit the space of these deformed states known as a *habitat* and show that it supports a representation of the Poisson bracket algebra of the diffeomorphism constraints.

The Poisson bracket of the Hamiltonian constraint with itself involves *structure functions*. As a result, standard group averaging techniques, which are predicated on a Lie group structure, cannot be applied to the finite transformations generated by the Hamiltonian constraint. Indeed, a satisfactory treatment of the Hamiltonian constraint in quantum theory constitutes one of the key open problems in canonical LQG and we shall devote the rest of this Chapter to a discussion of this problem. In the absence of group averaging techniques, one attempts a direct construction of the Hamiltonian constraint operator by replacing the classical fields which comprise it with quantum operators. As mentioned above, due to the conflict between the

non-locality of the holonomies and the local nature of the classical fields, this cannot be done *exactly*. Hence, Thiemann adopted the following step by step procedure:
(1) Fix a one parameter family of triangulations, T_ϵ of the manifold where ϵ is a positive parameter whose vanishing signals the 'continuum limit' of infinitely fine triangulation.
(2) *Approximate* the local fields comprising the constraint by classical holonomies around small loops and classical fluxes through small areas, the smallness being characterized by ϵ, so that the approximations become exact in the continuum limit.
(3) Replace the holonomy-flux functions by the corresponding operators and attempt to define the continuum limit of the action of the resulting finite triangulation Hamiltonian constraint operator $\hat{C}_\epsilon$, the hope being that while the action of the individual approximants to the local fields admit no such limit, the conglomeration of these approximants which form $\hat{C}_\epsilon$ together, does admit such a limit.

Remarkably, Thiemann was able to implement these steps successfully. Unfortunately, this implementation suffers from the following shortcomings. First, in the continuum limit, the operator action depends on the choice of the holonomy-flux approximants (associated with the finite triangulation) from an infinite number of candidates. Second, the anomaly free requirement cannot provide a discriminatory tool for the correct set of choices. The reason is that in this implementation, the continuum limit of the commutator of a pair of Hamiltonian constraints as well as the operator version of the classical Poisson bracket can be seen to vanish for a very large class of choices of approximants. Finally, direct analogs of these choices are found to be physically inappropriate in toy model and minisuperspace situations [4, 5]. Hence, current research focuses on removing these shortcomings from Thiemann's remarkable work. Accordingly the layout of the rest of this Chapter is as follows. In Section 3, we review Thiemann's work, discuss its inadequacies and suggest a strategy to overcome them, based in part on the study of toy models [4]. The habitat construction of Section 2 will be seen to play a key role in this discussion. In Section 4 we describe recent work on the quantum dynamics of a weak coupling limit of Euclidean gravity. Section 5 is devoted to work on Euclidean gravity. Section 6 contains a discussion of directions for further research. The exciting line of research which emerges from the considerations of Sections 3-6 may be summarized as follows: Attempt to code the action of the Hamiltonian constraint in terms of *phase space dependent* spatial diffeomorphisms, and, use the key feature of LQG — namely, the unitary action of (phase space independent) diffeomorphisms — to implement this operator action.

2. Spatial Diffeomorphism Invariance

We refer to the set of states generated by the action of all diffeomorphisms on a given state as the orbit of the state. Clearly, this orbit is invariant under diffeomorphisms. Intuitively, if we construct a state which is the sum over all states in an orbit, such a state would be diffeomorphism invariant. We shall see that this intuitive idea can be rigorously implemented by interpreting the summing procedure as a *group averaging map* from kinematic to diffeomorphism invariant states [7]. Accordingly, Section 2.1 describes the defining properties of a group averaging map and their role in the construction of the corresponding group invariant Hilbert space. Starting from an intuitive 'sum over states in an orbit', Section 2.2 constructs an essentially *unique* group averaging map for the group of spatial diffeomorphisms in LQG as well as the corresponding Hilbert space of diffeomorphism invariant states.[a] Section 2.3 constructs the Lewandowski-Marolf habitat and shows that it supports a representation of the Poisson bracket algebra of the diffeomorphism constraints.

2.1. *Group Averaging Maps*

Let the group G be represented unitarily on the Hilbert space $\mathcal{H}$. A Group Averaging map η is an anti linear map $\eta : \mathcal{D} \to \mathcal{D}'$ from a dense domain $\mathcal{D} \subset \mathcal{H}$ that is preserved under the unitary action of G, to the space $\mathcal{D}'$ of complex linear mappings on $\mathcal{D}$ ($\mathcal{D}'$ is called the *algebraic dual* of $\mathcal{D}$), satisfying the following three properties [2, 7]:

(1) $\forall \psi_1 \in \mathcal{D}$, $\eta(\psi_1) \in \mathcal{D}'$ is G-invariant:

$$\eta(\psi_1)[\hat{U}(a)\psi_2] = \eta(\psi_1)[\psi_2] \quad \forall a \in \text{G}, \ \psi_2 \in \mathcal{D} \tag{1}$$

(2) η is real and positive:

$$\eta(\psi_1)[\psi_2] = \overline{\eta(\psi_2)[\psi_1]}, \quad \eta(\psi_1)[\psi_1] \geq 0 \quad \forall \psi_1, \psi_2 \in \mathcal{D} \tag{2}$$

(3) η commutes with the observables:

$$\eta(\psi_1)[\hat{O}\psi_2] = \eta(\hat{O}^\dagger \psi_1)[\psi_2] \quad \forall \psi_1, \psi_2 \in \mathcal{D}, \ \forall \hat{O} \in \mathcal{O}. \tag{3}$$

where $\mathcal{O}$ is the set of 'strong observables' which, together with their adjoints preserve $\mathcal{D}$ so that:

$$\hat{O} \in \mathcal{O} \iff \hat{O}, \hat{O}^\dagger : \mathcal{D} \to \mathcal{D}, \ \hat{U}(a)\hat{O} = \hat{O}\hat{U}(a) \quad \forall a \in \text{G}. \tag{4}$$

Note that the algebraic dual space supports an anti-representation of $\mathcal{O}$ through the dual action of operators on this space: The action of $\hat{O} \in \mathcal{O}$ on $\Psi \in \mathcal{D}'$ yields $\hat{O}\Psi \in \mathcal{D}'$ where $\hat{O}\Psi$ is defined through its action on any $\psi \in \mathcal{D}$ as $\hat{O}\Psi(\psi) := \Psi(\hat{O}^\dagger \psi)$.

Given a group averaging map η, the G-invariant Hilbert space $\mathcal{H}_\text{G}$ is obtained as follows [2]: Let $\mathcal{V}_\text{G} \subset \mathcal{D}'$ be the span of dual vectors of the form $\eta(\psi)$. The sesquilinear form $\langle \eta(\psi_1), \eta(\psi_2) \rangle_\text{G} := \eta(\psi_2)[\psi_1]$ provides an inner product on $\mathcal{V}_\text{G}/\sim$ where the quotient is over zero-norm states. Property 2 implies it is an inner product, and $\mathcal{H}_\text{G}$ is defined as the completion of $\mathcal{V}_\text{G}/\sim$ under this inner product. Property 3 ensures that strong observables satisfy the correct adjointness relations on $\mathcal{H}_\text{G}$ if they do so on $\mathcal{D}$ [8].

---
[a] A detailed version of the material in Section 2.2 can be found in Ref. [6].

2.2. The Diffeomorphism Invariant Hilbert Space

Note that if there exists a left and right invariant measure da on G, the formal expression $\eta(|\psi\rangle) := (\int_G da\, \hat{U}(a)|\psi\rangle)^\dagger := \int_G da\, \langle\psi|\hat{U}(a^{-1})$ serves as a candidate definition of a group averaging map. Indeed if this integral expression can be given meaning as an element of $\mathcal{D}'$, the reader may verify that this expression satisfies the defining properties of a group averaging map. Since, to the best of our knowledge, no such measure is available for the diffeomorphism group G := Diff, we replace the integral over the group by a *sum* as follows.

Let $|s\rangle$ be an $SU(2)$ gauge invariant spin net. Let the coarsest graph underlying it be $\gamma(s)$ with edges $\{e\}$, vertices $\{v\}$, edge colors $\{j_e, e \in \gamma(s)\}$ and intertwiners $\{\mathcal{I}_v, v \in \gamma\}$. The unitary action $\hat{U}(\phi)$ of a diffeomorphism ϕ on $|s\rangle$ yields the spin net $\hat{U}(\phi)|s\rangle =: |\phi(s)\rangle$ which is based on the graph $\phi(\gamma(s))$ with edge colorings $\{j_{\phi(e)} = j_e\}$ and intertwiners $\{\mathcal{I}_{\phi(v)} = \mathcal{I}_v\}$. Let $[s]$ denote the set of distinct (and hence mutually orthornormal) diffeomorphic images of $|s\rangle$ under Diff.

We define the action of the putative group averaging map η on the spin net state $|s\rangle$ as

$$\eta(|s\rangle) = \eta_{[s]} \sum_{|\bar{s}\rangle \in [s]} \langle\bar{s}|, \qquad (5)$$

where $\eta_{[s]}$ is a positive parameter which we shall determine below. Since the elements of $[s]$ are orthonormal, it follows that the natural action of $\eta(|s\rangle)$ on any spin net either vanishes or equals $\eta_{[s]}$ from which it follows that η maps $\mathcal{D}$ to $\mathcal{D}'$ where $\mathcal{D}$ is the finite span of spin net states and $\mathcal{D}'$ is its algebraic dual. Since the sum is over all distinct diffeomorphic images of $|s\rangle$, it follows (as can be explicitly checked) that property (1) holds. It is useful to rewrite this sum as follows. Let Sym_s be the subset of Diff which leaves s invariant. It follows that elements of the orbit $[s]$ are in correspondence with *right cosets* of Diff by Sym_s so that we may write:

$$\eta(|s\rangle) := \eta_{[s]} \sum_{c \in \text{Diff}/\text{Sym}_s} (\hat{U}(\phi_c)|s\rangle)^\dagger \qquad (6)$$

where ϕ_c is any element of the coset c.

Next, we show that $\mathcal{D}$ and its image by η split into *superselection* sectors i.e. sectors which cannot be mapped to each other by any element of $\mathcal{O}$. Let s_1, s_2 be spin nets with graph labels $\gamma(s_1), \gamma(s_2)$ such that $\gamma(s_1) \neq \gamma(s_2)$. It is straightforward to see that there are infinitely many diffeomorphisms ϕ_{s_1} which move $|s_2\rangle$ but keep $|s_1\rangle$ invariant.[b] From the diffeomorphism invariance of elements of $\mathcal{O}$ we have that

$$\langle s_2|\hat{O}|s_1\rangle = \langle s_2|\hat{O}\hat{U}^\dagger(\phi_{s_1})|s_1\rangle = \langle s_2|\hat{U}^\dagger(\phi_{s_1})\hat{O}|s_1\rangle \qquad (7)$$

so that the state $\hat{O}|s_1\rangle$ has the same component along spin networks of the form $|\phi_{s_1}(s_2)\rangle$. Since there are infinitely many of the latter, it follows from $\hat{O}: \mathcal{D} \to \mathcal{D}$

---
[b]These are generated by any vector field which vanishes everywhere on $\gamma(s_1)$ but is transverse to an open subset of $\gamma(s_2)$.

that $\langle s_2|\hat{O}|s_1\rangle = 0$. This shows that at the kinematic level two spin nets lie in different superselection sectors unless their underlying graphs coincide. It is then straightforward to see that a similar argument implies that their group averaged images lie in different superselection sectors unless their underlying graphs are diffeomorphic.

Adopting the attitude that each such superselection sector provides a different physical realization, an unambiguous specification of the putative group averaging map only requires a determination of the parameters $\eta_{[s]}$ within each such sector. We show that the imposition of requirement (3) uniquely determines these coefficients within such a sector (up to an irrelevant overall constant which can be absorbed into the Hilbert space inner product for the sector). We define the subgroup, Sym_s^0 of Sym_s to be the set of edge preserving elements of Diff:

$$\text{Sym}_s^0 := \{\phi \in \text{Diff} : \phi(e) = e \quad \forall e \in \gamma(s)\}. \tag{8}$$

It is straightforward to check that Sym_s^0 is a normal subgroup of Sym_s and that the quotient group $D_s := \text{Sym}_s/\text{Sym}_s^0$ is the finite group of allowed edge permutations[c] which leave s invariant. We denote the cardinality of D_s by $|D_s|$. Clearly $|D_s|$ is a diffeomorphism invariant number i.e. $|D_s| = |D_{\bar{s}}|$ for all $|\bar{s}\rangle \in [s]$.

Let s_1 and s_2 be two spin networks based on diffeomorphic graphs (for otherwise property (3) trivializes). We want to impose the condition

$$\eta(\hat{O}|s_1\rangle)[|s_2\rangle] = \eta(|s_1\rangle)[\hat{O}^\dagger|s_2\rangle], \tag{9}$$

for all $\hat{O} \in \mathcal{O}$. Since $\hat{O} : \mathcal{D} \to \mathcal{D}$, the vector $\hat{O}|s_1\rangle$ admits an expansion of the form,

$$\hat{O}|s_1\rangle = \sum_{i=1}^{n} \lambda_i \hat{U}(\phi_i)|s_2\rangle + |\chi\rangle \quad \text{with} \quad \langle s_2|\hat{U}(\phi)|\chi\rangle = 0 \,\forall\, \phi \in \text{Diff}. \tag{10}$$

The vectors $\lambda_i \hat{U}(\phi_i)|s_2\rangle$ represent the components of $\hat{O}|s_1\rangle$ along the orbit of $|s_2\rangle$ and are taken to be orthogonal; $|\chi\rangle$ encodes the remaining vectors orthogonal to the span of the orbit of $|s_2\rangle$. It is straightforward to check that the left-hand side of (9) then evaluates to $\eta(\hat{O}|s_1\rangle)[|s_2\rangle] = \eta_{[s_2]} \sum_i \bar{\lambda}_i$.

To evaluate the right-hand side, we first rewrite $\eta(|s_1\rangle)$ as a sum over $\text{Sym}_{s_1}^0$ cosets as follows. Consider the auxiliary map defined by

$$\eta^0(|s\rangle) := \eta_{[s]} \sum_{c \in \text{Diff}/\text{Sym}_s^0} (\hat{U}(\phi_c)|s\rangle)^\dagger. \tag{11}$$

where, similar to (6), $\text{Diff}/\text{Sym}_s^0$ is the set of right cosets of Diff by Sym_s^0 and ϕ_c is a choice of diffeomorphism in each such coset c. It then follows that:

$$\eta^0(|s\rangle) = |D_s|\eta(|s\rangle). \tag{12}$$

Using (11) and (12) and the commutativity of Diff with $\mathcal{O}$, it is straightforward to check that:

$$\eta(|s_1\rangle)[\hat{O}^\dagger|s_2\rangle] = \eta_{[s_1]} \sum_i \bar{\lambda}_i x_i \tag{13}$$

---

[c]By 'allowed' we mean that there exists a diffeomorphism which implements each such permutation.

where

$$x_i := |D_{s_1}|^{-1} \sum_{c \in \text{Diff}/\text{Sym}^0_{s_1}} \overline{\langle s_2|\hat{U}(\phi_c)|\phi_i(s_2)\rangle}, \qquad (14)$$

Since $\langle \phi_i(s_2)|\hat{O}|s_1\rangle = \lambda_i \neq 0$ it follows from the discussion centered around (7) that $\gamma(\phi_i(s_2)) = \gamma(s_1)$. In particular $\text{Sym}^0_{\phi_i(s_2)} = \text{Sym}^0_{s_1}$ and the sum is independent of the representative choices $c \to \phi_c$. Let $c_i := \phi_i^{-1}\text{Sym}^0_{s_1}$ be the right coset of Diff by $\text{Sym}^0_{s_1}$ which contains ϕ_i^{-1}. All other cosets can be obtained as $\phi_i^{-1}\phi\text{Sym}^0_{s_1}$ for appropriate $\phi \in \text{Diff}$. Nonzero contributions to (14) come from elements $\phi \in \text{Sym}_{s_2^{(i)}}$. It follows that there are $|D_{\phi_i(s_2)}|$ such terms and so we obtain:

$$x_i = |D_{s_1}|^{-1}|D_{s_2}| \qquad (15)$$

where we have used the diffeomorphism invariance of $|D_{\phi_i(s_2)}|$ to replace it with $|D_{s_2}|$. Assuming $\hat{O}$ is such that $\sum_i \lambda_i \neq 0$ (if no such observable exist, the states would be superselected) we conclude that in order to satisfy (9) we must have that $\eta_{[s_1]}/\eta_{[s_2]} = |D_{s_1}|/|D_{s_2}|$. This implies that $\eta_{[s]} = C|D_s|$ for some $C > 0$, where C is constant within the superselection sector containing $\eta(|s\rangle)$. With this choice of parameters $\eta_{[s]}$ it is straightforward to verify that equation (6) satisfies all the defining properties of a group averaging map so that the space of diffeomorphism invariant states can then be converted to a Hilbert space $\mathcal{H}_{\text{Diff}}$ following the steps sketched in Section 2.1.

It is also straightforward to verify that with the same choice of parameters $\eta_{[s]}$, the group averaging map can be extended to act on the dense domain of *cylindrical functions* and that the properties (1)-(3) hold with $\mathcal{O}$ extended to include Diff invariant observables which preserve this larger domain. Thus, our treatment here rigorously and *uniquely* derives the group averaging maps of Refs. [8, 9].

2.3. *The Lewandowski–Marolf habitat*

Consider, for simplicity, a spin net state $|s\rangle$ which has no non-trivial discrete symmetries (so that D_s consists only of the identity element) and whose underlying graph $\gamma(s)$ has n vertices. Clearly, these properties are diffeomorphism invariant and, hence, shared by any spinnet in $[s]$. Next, consider any complex valued function f on n copies of the Cauchy slice i.e. $f : \Sigma^n \to \mathbf{C}$. Let the vertices of $|\bar{s}\rangle \in [s]$ be located at the points $\bar{v}_1, .., \bar{v}_n$. The n-tuple of points $(\bar{v}_1, .., \bar{v}_n)$ define a point in Σ^n. Denote the evaluation of f on this point by $f(\bar{v}_1, .., \bar{v}_n)$ and define the *habitat* state $\Psi_{[s],f} \in \mathcal{D}'$ as:

$$\Psi_{[s],f} = \sum_{|\bar{s}\rangle \in [s]} \langle \bar{s}|f(\bar{v}_1,..,\bar{v}_n). \qquad (16)$$

Comparing this expression with that of equation (5), we see that habitat states are 'deformations' of diffeomorphism invariant ones, the deformation being introduced by the weight function f. The finite span of such states for all choices of $[s], f$ is called the Lewandowski-Marolf (LM) habitat and the weight functions are called *vertex smooth functions* by virtue of their arguments.[d]

Next, consider a one parameter family of diffeomorphisms $\phi_\xi(\epsilon), \epsilon \in \mathbf{R}$ generated by the vector field ξ. From the dual action of $\hat{U}(\phi_\xi(\epsilon))$ on $\mathcal{D}'$, it follows that

$$\hat{U}(\phi_\xi(\epsilon))\Psi_{[s],f} = \sum_{|\bar{s}\rangle \in [s]} \langle \phi_\xi(-\epsilon)(\bar{s})|f(\bar{v}_1,..,\bar{v}_n) = \sum_{|\bar{s}\rangle \in [s]} \langle \bar{s}|f(\phi_\xi(\epsilon)(\bar{v}_1),..,\phi_\xi(\epsilon)(\bar{v}_n))$$
(17)

which in turn implies that

$$\lim_{\epsilon \to 0} \frac{\hat{U}(\phi_\xi(\epsilon)) - \mathbf{1}}{i\epsilon} \Psi_{[s],f} = -i\Psi_{[s],\mathcal{L}_\xi f}.$$
(18)

where $\mathcal{L}_\xi$ denotes the Lie derivative with respect to ξ so that the right-hand side is a new habitat state with associated weight function $\mathcal{L}_\xi f$. The above equations are equalities between elements of $\mathcal{D}'$ and can be easily checked, for example, by evaluating their action on $\mathcal{D}$. Equivalently, equation (17) can also be checked directly by evaluating the (dual) action of $\hat{U}(\phi_\xi(t))$ on each bra in the sum (16) and re-labeling the summation index. Equation (18) then also follows straightforwardly from equation (17) by evaluating the limit on each such bra. Equation (18) shows that the habitat supports the action of the *generator* of spatial diffeomorphisms $\hat{D}(\vec{\xi})$ which is just the diffeomorphism constraint $D_a(x)$ smeared with a shift vector field ξ^a. Clearly the commutator of two such generators yields the commutator of the associate shift vector fields and the kernel of the diffeomorphism constraint (for any choice of shift) consists of habitat states with f constant, which, as expected are just the diffeomorphism invariant states of Section 2.2.

3. The Thiemann Construction

Section 3.1 reviews the key features of Thiemann's seminal work [8, 10]. Section 3.2 suggests improvements to Thiemann's detailed choices while retaining his basic framework. Section 3.3 discusses the role of toy models in arriving at these choices. Before we start, we list some useful classical identities and expressions.

The unit density weight Hamiltonian constraint for Lorentzian gravity, C_L can be written as $C_L = C_E + (C_L - C_E)$ where C_E is the Hamiltonian constraint for Euclidean gravity,

$$C_E = \frac{\epsilon^{ijk} E_i^a E_j^b}{2\sqrt{q}} F_{abk}.$$
(19)

On the Gauss Law constraint surface the Poisson bracket between a pair of Hamiltonian constraints smeared with lapses N, M is proportional to the diffeomorphism constraint:

$$\{C_L(N), C_L(M)\} = -\{C_E(N), C_E(M)\} = \int_\Sigma (N\nabla_a M - M\nabla_a N) q^{ab} D_b(x)$$
(20)

---

[d]If D_s is non-trivial, f is required to be symmetric under the action of D_s on its arguments.

Let $V_\mathcal{R}$ be the spatial volume of a region $\mathcal{R}$. Then, a key identity discovered by Thiemann [10] is:

$$\eta_{abc}\frac{E_i^a E_j^b \epsilon^{ijk}}{\sqrt{q}}(x) = \{A_c^k(x), V_\mathcal{R}\}, \quad x, \mathcal{R} \text{ s.t. } x \in \mathcal{R} \tag{21}$$

The following holonomy expansions for an edge e of coordinate size ϵ and loop α_{ab} of coordinate area ϵ^2 hold with $\hat{e}^a$ being the unit coordinate tangent to e:

$$h_e = 1 - \frac{i}{2}A_a^i \hat{e}^a \sigma_i + O(\epsilon^2) \quad h_{\alpha_{ab}} = 1 - \frac{i}{2}\epsilon^2 F_{ab}^i \sigma_i + O(\epsilon^3) \tag{22}$$

3.1. Thiemann's Hamiltonian Constraint and its Commutator

Thiemann discovered a remarkable rewriting of $C_L - C_E$ in terms of Poisson brackets of $V_\mathcal{R}$ with C_E which he then used to construct $\widehat{C_L - C_E}$ in terms of the commutator between $\hat{V}_\mathcal{R}, \hat{C}_E$ [10]. For simplicity we restrict attention in this Chapter to the construction of $\hat{C}_E$.

Let $T(\epsilon)$ be a one parameter family of triangulations of Σ with cells of coordinate size $O(\epsilon^3)$ (in any fixed coordinate atlas) so that in the $\epsilon \to 0$ 'continuum' limit, this triangulation becomes infinitely fine. To each such cell $\triangle$, associate the edges $\{s_i, i = 1, .., e_\triangle\}$ and the loops $\{\alpha_J, J = 1, .., L_\triangle\}$. Each such edge is of coordinate length $O(\epsilon)$ and each such loop of coordinate area $O(\epsilon^2)$. Fix one point $x_\triangle$ in each cell $\triangle$. A finite triangulation approximant $C_{E,T(\epsilon)}(N)$ which agrees with $C_E(N)$ in the continuum limit is constructed as follows. Use equation (21) to re-express the triad part of equation (19). Write the resulting expression for $C_E(N)$ as a sum of integrals, one for each cell. Approximate each cell integral by approximating its integrand and multiplying the result by the coordinate measure ϵ^3. Use holonomies along the edges and loops associated with each cell in equation (22) to construct the desired approximant to the integrand for each cell. It is straightforward to see that such a procedure yields an approximant $C_{E,T(\epsilon)}(N)$ of the form:

$$C_{E,T(\epsilon)}[N] = \sum_{\triangle \in T(\epsilon)} N(x_\triangle) \sum_{i=1}^{e_\triangle} \sum_{J=1}^{L_\triangle} \mathcal{C}^{iJ} Tr(\, h_{\alpha_J(\triangle)} h_{s_i(\triangle)} \{h_{s_i(\triangle)}^{-1}, V_\mathcal{R}\}). \tag{23}$$

Here $\mathcal{C}^{iJ}$ are numerical constants which depend on $e_\triangle, L_\triangle$ but are otherwise independent of $T(\epsilon)$. Note that all factors of ϵ have disappeared in this expression. In the next section, we shall see that this remarkable property follows from the unit density nature of C_E. The action of the corresponding operator on some spin network state $|s\rangle$ is obtained by replacing the classical functions in the above expression by their quantum correspondents and the Poisson brackets by commutators:

$$\hat{C}_{E,T(\epsilon)}[N]|s\rangle = -i\hbar \sum_{\triangle \in T(\epsilon)} N(x_\triangle) \sum_{i=1}^{e_\triangle} \sum_{J=1}^{L_\triangle} \mathcal{C}^{iJ} Tr(\, \hat{h}_{\alpha_J(\triangle)} \hat{h}_{s_i(\triangle)} [\, \hat{h}_{s_i(\triangle)}^{-1}, \hat{V}_\mathcal{R} \,])|s\rangle \tag{24}$$

Next, note that since $\hat{V}_\mathcal{R}$ only acts at vertices of spin nets, it follows that this action is trivial unless some of the edges s_i and loops α_J intersect some vertex of $\gamma(s)$.

To ensure this, the triangulation $T(\epsilon)$ and the choices of these edges and loops are *tailored* to $\gamma(s)$ so that each vertex v of $\gamma(s)$ is contained in one (and only one) cell, $\triangle(v)$, and in this cell the edges and loops are based at this vertex. Denoting this choice of 'graph adapted triangulation' by $T(\epsilon, \gamma(s))$ and the vertex set of $\gamma(s)$ by $V(\gamma(s))$, we obtain:

$$\hat{C}_{E,T(\epsilon,\gamma(s))}[N]|s\rangle = -i\hbar \sum_{\substack{\triangle(v) \in T(\epsilon,\gamma(s)) \\ v \in V(\gamma(s))}} N(v) \sum_{i=1}^{e_\triangle} \sum_{J=1}^{L_\triangle} \mathcal{C}^{iJ} Tr(\hat{h}_{\alpha_J(\triangle)} \hat{h}_{s_i(\triangle)} [\hat{h}_{s_i(\triangle)}^{-1}, \hat{V}_\mathcal{R}]) |s\rangle$$
(25)

where we have chosen $x_\triangle = v$ for the cell $\triangle(v)$. Equation (25) has the structure of a sum over vertex contributions, each such contribution being multiplied by the evaluation of the lapse at the vertex in question. Motivated by the diffeomorphism covariance of C_E (19), Thiemann imposed certain restrictions on the choice of regulating loops and edges which give rise to each of these vertex contributions. For our purposes the main consequence of these restrictions is as follows (for details see [8, 9]). For any vertex of $\gamma(s)$, the vertex contributions at two different values of ϵ are diffeomorphic. We shall refer to this property as *regulator covariance*. Next, we focus on the question of how small ϵ should be for the action (25) to be defined. One may think that this action is defined for a 'small enough ϵ' which necessarily depends on the graph $\gamma(s)$.[e] However, Thiemann was able to parameterize the one parameter family of triangulations $T(\epsilon, \gamma(s))$ in such a way that the regulator covariant action (25) acquires a dependence which is *uniform* in ϵ. For our purposes here this uniformity property may be stated as follows. There exists $\bar{\epsilon}$ such that the action (25) is well defined for all $\epsilon < \bar{\epsilon}$ and such that the value of $\bar{\epsilon}$ is the *same* for all spin nets $|s\rangle$. We shall refer to this uniformity property of such covariant regulators as *uniform regulator covariance*. It follows from uniform regulator covariance that the action of an 'ϵ-approximant' Hamiltonian constraint on the finite span of spinnets, $\mathcal{D}$, can be defined by linear extension of (25) as:

$$\hat{C}_{E,\epsilon}[N] \sum_{I=1}^{M} a_I |s_I\rangle := \sum_{I=1}^{M} a_I \hat{C}_{E,T(\epsilon,\gamma(s_I))}[N]|s_I\rangle$$
(26)

Next note that, as $\epsilon \to 0$, the action (26) does not converge within the kinematic Hilbert space because the deformed spinnets generated in equation (25) at different values of ϵ live on different graphs and are, therefore, mutually orthonormal. However, uniform regulator covariance ensures that the evaluation of diffeomorphism invariant states on the right-hand side of (26) does converge in the continuum limit: for any diffeomorphism invariant state $\Phi \in \mathcal{H}_{\text{Diff}}$ and any $|\psi\rangle \in \mathcal{D}$, we have that $\Phi(\hat{C}_{T(\epsilon)}[N]|\psi\rangle)$ is *independent* of ϵ because uniform regulator covariance implies

---
[e]For example, let this action be defined for all $\epsilon < \epsilon_1$ for $|s_1\rangle$. One may think that for a spin net $|s_2\rangle$ such that $\gamma(s_2)$ has a 'high enough density' of vertices in the neighborhood of a given vertex of $\gamma(s_1)$, the action (25) on $|s_2\rangle$ must hold for a smaller range of ϵ. The loophole in this way of thinking is that the notion of 'high enough density' is a background dependent one.

that the actions (26) at two different values of ϵ are diffeomorphic. It follows that:

$$\Phi(\lim_{\epsilon\to 0}\hat{C}_{E,\epsilon}(N)|\psi\rangle) = \Phi(\hat{C}_{E,\epsilon_0}(N)|\psi\rangle) \qquad (27)$$

where ϵ_0 is *any* value of $\epsilon < \bar{\epsilon}$. An interpretation of the limit (27) was provided by Thiemann in terms of continuity with respect to a topology on the space of operators which preserve $\mathcal{D}$ as follows. The pair $\Phi \in \mathcal{H}_{\text{Diff}}, |\psi\rangle \in \mathcal{D}$ defines a *seminorm* $||\hat{O}||_{\Phi,\psi} := |\Phi(\hat{O}|\psi\rangle)|$ for any operator $\hat{O}: \mathcal{D} \to \mathcal{D}$. Since such operators form a vector space, the family of seminorms $||\ ||_{\Phi,\psi}\ \forall \Phi, |\psi\rangle$ define a topology on this space. It follows from (27) that in this operator topology by virtue of uniform regulator covariance, the one parameter family of ϵ-approximants $\hat{C}_{E,\epsilon}(N)$ converge uniformly so that:

$$\lim_{\epsilon\to 0}\hat{C}_{E,\epsilon}(N) = \hat{C}_{E,\epsilon_0}(N). \qquad (28)$$

The above operator topology is called the Uniform Rovelli Smolin (URS) topology. It follows that the URS topology continuum limit constraint operator (28) is well defined on the kinematic Hilbert space.[f] It turns out that the notion of convergence in this topology is powerful enough to enable a computation of the continuum limit of the commutator between a pair of finite triangulation constraints [10].

We may also attempt a definition of a continuum limit operator $\hat{C}_E(N)$ through the dual action of kinematic operators on $\mathcal{H}_{\text{Diff}}$ as follows:

$$((\hat{C}_E(N))^\dagger \Phi)(|\psi\rangle) := \lim_{\epsilon\to 0}(\hat{C}^\dagger_{E,\epsilon}(N)\Phi)(|\psi\rangle) := \Phi(\lim_{\epsilon\to 0}\hat{C}_{E,\epsilon}(N)|\psi\rangle) = \Phi(\hat{C}_{E,\epsilon_0}(N)|\psi\rangle) \qquad (29)$$

However the continuum limit of the dual action (29) does not preserve $\mathcal{H}_{diff}$ due to the presence of the c-number lapse N. It is then not clear how to compute the continuum limit of the action of a second constraint on that of the first. This precludes an analysis of their commutator. As discussed in Section 2.3, Lewandowski and Marolf deformed the space of diffeomorphism invariant states into their habitat to remedy this situation. This was done in a remarkably delicate manner to obtain the following properties: (a) the dual action of Thiemann's finite triangulation Hamiltonian constraint admits a well defined $\epsilon \to 0$ limit on this habitat, and, (b) the dual action maps the habitat into itself so that the commutator can be computed.

We now discuss the choice dependence of the continuum limit of the action of Thiemann's constraint (the discussion is applicable to the URS as well as the habitat continuum limit). The deceptively simple treatment of the constraint, hides Thiemann's ingenious uniform regulator covariant assignment of regulating edges and loops. Clearly, the final operator action on a spin net depends on the diffeomorphism invariant aspects of the placement of these loops and edges relative to the spin net graph as well as diffeomorphism invariant properties of the loops and edges themselves. A study of Thiemann's work then reveals that, as may already

---

[f]Note however that the one parameter choice available for the value of ϵ_0 implies that this limit is not unique; related to this is that the URS topology is not Hausdorff.

be obvious to the reader, there are, in general, infinitely many diffeomorphically distinct ways in which to choose these edges and loops and their placements so that this definition of quantum dynamics is plagued by an infinite ambiguity [9]. In addition, while equations (22) use the defining $j = \frac{1}{2}$ representation, in principle higher spin representations may also be used to obtain curvature and connection approximants [11]. It turns out that the LM habitat based continuum limit operator also suffers from exactly the same infinite choice problems which occur for the limits (29) and (28) essentially due to the (partial) labeling of habitat states by diffeomorphism classes of spin nets (16).

To summarize: While uniform regulator covariance seeks to promote the background independence of LQG kinematics to its dynamics by making the regulating procedure as independent of background structures as possible, the continuum limit of the action still depends on an infinitely manifold choice of background structures.

3.2. *An Improved Strategy*

We refer to the commutator between two Hamiltonian constraints (or a finite triangulation approximant to this commutator) as the LHS and the operator version of their Poisson bracket (20) as the RHS. As mentioned in Section 1, a strategy to reduce the vast choice in the definition of the Hamiltonian constraint is to demand that the continuum limit action of the commutator between two finite triangulation constraints is non-anomalous i.e. the (continuum limit of) LHS and RHS agree. Unfortunately, it turns out that for *all* the choices discussed in Section 3.1, the continuum limit of the LHS and as well as the RHS vanish [9]. This 'consistent trivialization' manifests in the URS as well as the habitat continuum limit [3, 9, 10, 12]. Classically, the RHS is proportional to the diffeomorphism constraint, so that its vanishing in the URS continuum limit is expected. However, its vanishing on the habitat runs counter to intuition because the habitat states are not diffeomorphism invariant.

For this reason, as well as for concreteness, let us restrict attention to the habitat case. As we show now, the vanishing of the RHS is a direct consequence of working with unit density Hamiltonian constraints. Finite triangulation approximants to various local fields can be constructed from holonomies around ϵ length loops, fluxes through ϵ^2 size surfaces and volumes of ϵ^3 size regions so that the following factors of ϵ are associated with the continuum objects below:

$$\hat{E}_i^a \sim \frac{\widehat{\text{flux}}}{\epsilon^2} \quad \widehat{\sqrt{q}} \sim \frac{\widehat{\text{Volume}}}{\epsilon^3} \quad \hat{F}_{ab}^i \sim \frac{\widehat{\text{holonomy}} - 1}{\epsilon^2}. \tag{30}$$

Due to background independence, the action of the flux, volume and holonomy operators on any state yields states which have Hilbert norm of order unity. We shall refer to such operators as *finite operators* to underline this order unity behavior. Equation (30) can be used to write any finite triangulation approximant in terms of powers of ϵ and finite operators. The reader may check that the above relations together with the replacement $d^3x \sim \epsilon^3$ imply the expected cancellation of factors of ϵ for the unit density Hamiltonian constraint (19). It is straightforward to check that the RHS then has an overall factor of ϵ. The RHS can be ordered [8, 10] so as to also reduce to a (finite) sum of contributions, one for each vertex of the spin net. As a result the RHS can be expressed in the form $\epsilon\,(\widehat{\text{Finite}})$ and the RHS vanishes in the continuum limit *irrespective of the finer details of the finite operator*.

The trivialization of the RHS implies that the only reduction of ambiguity available through the anomaly-free requirement is that the LHS also vanish. For the simplest set of choices of finite triangulation approximants, the new vertices created by the Hamiltonian constraint are such that the action of the second Hamiltonian constraint on them is trivial. As a consequence, the antisymmetric combination of the actions of two such constraints results in the vanishing of the LHS. This indicates that if there is a way to remove extra factors of ϵ in the RHS, then depending on the nature of the operator $\widehat{\text{Finite}}$, one may be able to get a nontrivial RHS but a vanishing LHS, thus ruling out such a choice. There are also other classes of choices [3] for which the LHS vanishes in the habitat calculation and, one may hope that a computation resulting in a rescaled RHS could render many of these choices anomalous and rule them out.[g] It turns out [8, 10] that the finite operator $\widehat{\text{Finite}}$ acts as the difference between an ϵ size diffeomorphism and the identity, i.e. $\widehat{\text{Finite}} \sim (\hat{U}(\phi_\epsilon) - 1)$. A rescaling of the RHS by ϵ^{-1} would then result in the operator $(\hat{U}(\phi_\epsilon) - 1)$. This operator maps the habitat weight function f (16) to the new function $(\phi_{\epsilon*}f) - f$ which vanishes in the continuum limit. Comparing with the non-trivial operator action in equation (18), it follows that this vanishing difference can be converted to a *non-vanishing derivative* with another factor of ϵ^{-1}. This motivates us to consider a rescaling of the RHS by ϵ^{-2} which in turn corresponds to one additional factor of ϵ^{-1} in the Hamiltonian constraint. From equation (30), this may be achieved by multiplying the constraint (19) by $q^{\frac{1}{6}}$. Thus, the above discussion motivates the consideration of a density weight $\frac{4}{3}$ Hamiltonian constraint. The extra factor of ϵ^{-1} implies that at finite triangulation the constraint takes the form of a finite operator *divided* by ϵ, which would blow up in the URS topology continuum limit. However, if it were possible to code the action of the Hamiltonian constraint in terms of differences of small diffeomorphisms and the identity, then similar to equation (18) one may hope that the continuum limit action on some appropriate 'habitat' also yields a derivative of a weight function. This suggests that one should seek an interpretation of the transformations generated by the Hamiltonian constraint in terms of *diffeomorphisms*. The existence of such an interpretation is hinted at by the following remarkable classical identity [13]:

$$\sum_{i=1}^{3}\{D[\vec{N}_i], D[\vec{M}_i]\} = (2\alpha - 1)\{C_{E,\alpha}(N), C_{E,\alpha}(M)\} = -(2\alpha - 1)\{C_{L,\alpha}(N), C_{L,\alpha}(M)\}. \tag{31}$$

Here $N_i^a := q^{-\alpha} N E_i^a$, the lapse N has density weight $(2\alpha - 1)$ and the α subscripts indicate rescaling of the Euclidean and Lorentzian constraints by $\sqrt{q}^{1-2\alpha}$

---
[g]Indeed, an ad hoc 'rescaling' of the LHS and RHS by such factors points to such a conclusion [12].

i.e. $C_{E,\alpha} = \sqrt{q}^{1-2\alpha} C_E, C_{L,\alpha} = \sqrt{q}^{1-2\alpha} C_L$. This identity suggests that important aspects of the action of the Hamiltonian constraint could perhaps be expressed in terms of *phase space dependent* diffeomorphisms.[h] To summarize: Our considerations above suggest that Thiemann's choices could be improved by (a) the consideration of higher density weight constraints and their confrontation with the anomaly free requirement, and, (b) an attempt to express the action of the constraints in terms of (phase space dependent) diffeomorphisms.

3.3. *Toy Models and the Form of Holonomy Approximants*

Two useful toy models are that of Parameterized field theory (PFT) [19] and the Husain-Kuchař (HK) model [20]. PFT is free scalar field theory on a fixed flat spacetime in diffeomorphism invariant disguise. In 1+1 dimensions, the free scalar field splits into left and right moving modes. The PFT phase space also splits into left and right sets of variables and its dynamics is describable in terms of the action of two commuting sets of spatial diffeomorphisms, one for each set of variables. PFT dynamics is solvable in an LQG representation because, similar to Section 2, physical states can be constructed by group averaging over these two sets of transformations [4]. The question is whether it is possible to implement a Thiemann like construction of the PFT Hamiltonian constraint in such a way that its action annihilates the physical states constructed by group averaging methods. The answer is in the affirmative. By choosing appropriate holonomy approximants, the finite triangulation constraint is constructed in terms of "small diffeomorphisms minus the identity" so that its action kills physical states. In order to obtain this form of the constraint, the representation labels of the holonomy approximants for PFT need to be adapted to the edge labels of the state being acted upon; this is in contrast to the standard choices in LQG wherein the representation of regulating holonomies is chosen once and for all [10, 11]. Some of the state labels in PFT are eigenvalues of 'triad-like' PFT operators thus indicating that PFT holonomy approximants depend on these PFT 'triad' operators! As discussed in *Chapter 6*, this general feature of the dependence of connection/curvature approximants on the triad was anticipated already in the so-called Improved LQC Dynamics [5], wherein such a dependence is *crucial* for obtaining the correct long distance physics. Evidence for this general feature is also provided by a quantization of the HK model using Thiemann's finite triangulation framework. The HK model is just gravity without the Hamiltonian constraint so that its solutions reside in $\mathcal{H}_{\text{Diff}}$ of Section 2.1. The question, similar to the PFT case, is whether the diffeomorphism constraint itself (rather than the unitaries for finite diffeomorphisms which figure in the group averaging map of Section 3.1) can be constructed as a continuum limit of finite triangulation approximants in such a way that the constraint operator annihilates states in $\mathcal{H}_{\text{Diff}}$. As shown in Ref. [21], the answer is in the affirmative

---

[h]Note that $\alpha = \frac{1}{2}$ corresponds to density weight one constraints and that the identity trivializes for this choice. We take this as further motivation to consider higher than unit density constraints.

provided that the curvature approximants, once again, show (in this case, a very complicated) dependence on the triad.

In PFT one can also analyze the constraint algebra. Similar to the LQG Hamiltonian constraint, the operator correspondent of the RHS for PFT trivializes for density weight one constraints due to the 'overall factor of ϵ'. A similar counting to that of the previous section indicates the use of density weight two constraints as a nontrivial probe of the PFT constraint algebra. Similar to the LQG case, these constraints are kinematically singular due to an overall factor of ϵ^{-1}. Nevertheless, their continuum limit action is well defined on an appropriate habitat of states corresponding to the product of a pair of 'left moving' and 'right moving' LM like habitats. The constraint algebra is represented non-trivially, without anomaly on this product habitat. Similar to the conversion of a trivial contribution into a derivative (18), the non-triviality of this representation is due to the factor of ϵ^{-1} i.e. due to the kinematically singular property. Similar support for the consideration of such singular operators is exhibited in the finite triangulation construction of the diffeomorphism constraint [21] and a non-trivial (LM) habitat based representation of its constraint algebra.

In summary, the study of PFT, the HK model and LQC has been very useful both in providing support, as well as detailed inputs, to the strategy (a)-(b) outlined at the end of Section 3.2.

4. Euclidean Gravity: A Weak Coupling Limit

While useful, the toy models of the previous section suffer from the simplification that their constraint algebras are Lie algebras. It is desirable to hone our ideas on a model whose constraint algebra has structure functions similar to the gravitational case. Just such a model was introduced by Laddha and its quantum constraint algebra studied in Refs. [14, 15]. Its 3+1 version is exactly Smolin's novel weak coupling limit of Euclidean gravity [16]. This section is devoted to an application of the strategy suggested in Section 3.2 to a spin network based quantization of this model. We choose to focus on the 3+1 case below. Our treatment follows that of Refs. [13, 17].

4.1. Classical Theory

The model is described in terms of a triplet of $U(1)$ connections $A_a^i, i = 1, 2, 3$ and conjugate electric fields E_i^a and we shall refer to it as the $U(1)^3$ model. We shall think of E_i^a as defining, as in gravity, a doubly densitized contravariant metric $\sum_{i=1}^{3} E_i^a E_i^b$. The dynamics of the model is driven by constraints which are $U(1)^3$ counterparts of the ones for Euclidean gravity:

$$G[\Lambda] = \int \Lambda^i \partial_a E_i^a \qquad D[\vec{N}] = \int E_i^a \mathcal{L}_{\vec{N}} A_a^i \qquad C[N] = \tfrac{1}{2} \int N q^{-\tfrac{1}{3}} \epsilon^{ijk} E_i^a E_j^b F_{ab}^k$$

(32)

where $F_{ab}^i := \partial_b A_a^i - \partial_a A_b^i$ is the curvature of the Abelian connection A_a^i and we have used a density weight $\frac{4}{3}$ Hamiltonian constraint in accordance with the suggestion of Section 3.2. The constraint algebra is isomorphic to that of gravity. The structure functions appear as in gravity in the Poisson bracket between a pair of Hamiltonian constraints. When the Gauss Law is satisfied this Poisson bracket evaluates to:

$$\{C[N], C[M]\} = D[\vec{\omega}], \qquad \omega^a := q^{-\frac{2}{3}} E_i^a E_i^b (M\partial_b N - N\partial_b M), \qquad (33)$$

where the electric field dependence of the structure function ω^a is exactly that of gravity.

Motivated by the fact that the identity (31) holds for the $U(1)^3$ theory as well, we seek to code the action of the Hamiltonian constraint in terms of phase space dependent diffeomorphisms. Additional motivation is provided by the following classical identity which holds for any vector field N^a and any $U(1)^3$ connection A_a^i:

$$N^a F_{ab}^i = \mathcal{L}_{\vec{N}} A_b^i + \partial_b \lambda^i, \qquad \lambda^i := A_a^i N^a. \qquad (34)$$

Applied to the Hamiltonian constraint $C(N)$ together with a by-parts integration, it is easy to check that when the Gauss Law is satisfied, we obtain:

$$C(N) = \tfrac{1}{2} \int \epsilon^{ijk} N_i^a F_{ab}^k E_j^b = \tfrac{1}{2} \int \epsilon^{ijk} (\mathcal{L}_{\vec{N}_i} A_b^k) E_j^b, \qquad (35)$$

where the triple of 'electric shift vector fields' $N_i^a, i = 1, 2, 3$ is defined as

$$N_i^a := N q^{-\frac{1}{3}} E_i^a, \qquad (36)$$

4.2. Quantum Kinematics

The Hilbert space is spanned by orthonormal $U(1)^3$ 'charge' network states. Similar to the $SU(2)$ case of LQG, each such state $|c\rangle \equiv |\gamma, \{\vec{q}_I\}\rangle$ is labelled by a closed graph γ whose oriented edges e_I are colored by representation labels of $U(1)^3$ i.e. by a triplet of integer valued *charges* $(q_I^i, i = 1, 2, 3) \equiv \vec{q}_I$. Charge net states can be thought of as wave functions in a connection representation so that $|c\rangle \equiv c(A)$,

$$c(A) := \prod_{k=1}^3 \exp i \sum_I q_I^k \int_{e_I} A_a^k dx^a. \qquad (37)$$

$U(1)^3$ holonomy operators act on $c(A)$ by multiplication and electric fields by differentiation, $\hat{E}_i^a \sim -i\hbar \frac{\delta}{\delta A_a^i}$. It is then easy to check that (a) charge network states are invariant under $U(1)^3$ gauge transformations only if the sum of charges at each vertex vanishes, (b) that charge net states are eigenvectors of the electric field operator valued distribution, and, (c) that (b) implies that the operator $\hat{E}_i(S)$ which measures the electric flux through the surface S is also diagonalized, its eigenvalues being computed by visualizing each such state as electric lines of force along the edges of the graph, each edge carrying flux $\vec{q}_I$ so that the integer valued eigenvalue is just the total flux through the surface S.

4.3. Finite Triangulation Constraint Operators: Motivational Heuristics and Final Expressions

We shall not worry about overall factors of constants and $\hbar$ below. Equation (35) suggests that the action of the Hamiltonian constraint operator on a gauge invariant state charge net state $c(A)$ is

$$\hat{C}(N)\, c(A) \sim \int \epsilon^{ijk} (\mathcal{L}_{\vec{N}_i} A_b^k) \frac{\delta}{\delta A_b^j}\, c(A) \tag{38}$$

Note that since the electric shift depends on the electric field through (36) it becomes operator valued in quantum theory and on first sight it is not clear what to make of the symbol $\mathcal{L}_{\vec{N}_i}$ in quantum theory. Fortunately, the fact that charge net states diagonalize the electric field operator implies that with a suitable choice of operator ordering, we may replace the electric shift operator $\hat{\vec{N}}_i(x)$ by its eigenvalue:

$$\hat{N}_i^a(x)\, c(A) := N\hat{E}_i^a \widehat{q^{-\frac{1}{3}}}(x)\, c(A) =: N_i^a(x)\, c(A). \tag{39}$$

The eigenvalue $N_i^a(x)$ is referred to as the *quantum shift*. It can be non-vanishing only when x lies on a vertex v of the charge net because, as in LQG, inverse powers of $\hat{q}$ when regularized can act non-trivially only at vertices. The detailed computation [13] of the quantum shift involves the use of a regulating coordinate patch in the neighborhood of v. The final result is that $N_i^a(x=v)$ takes the form

$$N_i^a(v) \sim N(v) \sum_{I_v} \hat{e}_{I_v}^a q_I^i \lambda_v. \tag{40}$$

Here λ_v comes from the action of $\widehat{q^{-\frac{1}{3}}}$ and is constructed out of the $\{\vec{q}_{I_v}\}$ [13]. Vertices for which $\lambda_v \neq 0$ are called *non-degenerate* vertices and those for which $\lambda_v = 0$ are called *degenerate*. Clearly, the quantum shift vanishes on degenerate vertices. The $\hat{e}_{I_v}^a q_{I_v}^i$ terms in (40) come from the action of the electric field operator through functional differentiation of $c(A)$ at the vertex v, with $\vec{e}_{I_v}$ denoting the I_vth edge tangent at v.

We note that the quantum shift inherits a coordinate dependence in two ways. First, the lapse, being a density weighted object, requires a coordinate patch for its evaluation; consequently $N(v)$ is evaluated in the regulating coordinate patch. Second the edge tangents are of unit coordinate norm in the regulating coordinates. We shall return to a discussion of the coordinate dependence of $N_i^a(v)$ when we discuss the issue of diffeomorphism covariance in Section 4.5. Let us return to equation (38) and use (40) therein to obtain:

$$\hat{C}(N)\, c(A) \sim \int_\Sigma \epsilon_i^{jk} \sum_v \lambda_v \sum_{I_v} N(v) q_{I_v}^i (\mathcal{L}_{\vec{e}_{I_v}} A_b^k) \frac{\delta}{\delta A_b^j}\, c(A) \tag{41}$$

Similar arguments may be applied to the $U(1)^3$ electric diffeomorphism constraint $D(\vec{N}_i) := \int_\Sigma N_i^a F_{ab}^j E_j^b$ which, modulo the Gauss Law reduces to $D[\vec{N}_i] =$

$\int_\Sigma d^3x (\mathcal{L}_{\vec{N}_i} A_b^j) E_j^b$ and suggests the following action of $\widehat{D[\vec{N}_i]}$ on $c(A)$:

$$\widehat{D[\vec{N}_i]}\, c(A) \sim \int_\Sigma \sum_v \lambda_v \sum_{I_v} N(v) q_{I_v}^i (\mathcal{L}_{\vec{\tilde{e}}_{I_v}} A_b^j) \frac{\delta}{\delta A_b^j} c(A) \qquad (42)$$

The difference between the two expressions (41) and (42) is that the latter looks like the action of an infinitesimal diffeomorphism whereas in the former expression, the ϵ^{ijk} shuffles the indices around so that the result is to augment the jth component of the connection with the diffeomorphic image (generated by the ith quantum shift) of its kth component. Next, note that for reasons discussed in Section 1, we need to replace the above expressions with finite triangulation approximants. The final expressions (43), (44) are complicated and instead of deriving them in detail, we motivate their form through the argumentation below. Unfortunately, we are obliged to omit some important steps in our arguments below due to space limitations and we urge the reader to consult [13] for a comprehensive account.

We replace the action of the Lie derivative with respect to the quantum shift with the difference of a δ-sized 'diffeomorphism' minus the identity. If the quantum shift was a smooth vector field like its classical counterpart, there would be no need to use the inverted commas around 'diffeomorphism'. However, since the quantum shift abruptly vanishes except at v, it is not a smooth vector field and one needs to endow the δ sized finite triangulation approximant associated with the symbol $\mathcal{L}_{\vec{\tilde{e}}_{I_v}}$ in equations (41), (42) with some operational meaning in quantum theory. This is where we make a jump from (heuristic) logic to intuition. We visualize the deformation generated by each edge tangent $\vec{\tilde{e}}_{I_v}$ as pulling the vertex v and its immediate neighborhood in the direction $\vec{\tilde{e}}_{I_v}$ to leading order in δ. In the case of equation (42), the deformation is akin to a 'singular' diffeomorphism. The vertex v is displaced by a coordinate distance $\sim \delta$ to its deformed image v'_I and all the edges $e_{J_v}, J_v \neq I_v$ are scrunched close together in a direction opposite to that of the I_vth one to leading order in δ. Thus, in the deformed charge net, the old vertex v is replaced by v'_I, and the charges on the deformed edges are the same as those on their undeformed counterparts.

In the case of equation (41), since the triplet of $U(1)$ connections all live on the same graph, the deformed structure of the graph is the same. However due to the factor of ϵ^{ijk}, two of the three charges on each edge are 'flipped' [13]. Charge conservation due to gauge invariance then requires that the deformed graph also contain segments of the undeformed edges as well as the original vertex v. Deformations generated by the actions (41), (42) are depicted in Figure 1. Note that these actions (41), (42) do not obtain a contribution from degenerate vertices so that the sum over vertices therein is restricted to non-degenerate vertices. Due to the particular nature of the flipping, it turns out that when the Hamiltonian constraint acts at a non-degenerate vertex v of the undeformed charge net $|c\rangle$, it converts v to a degenerate vertex of the deformed charge network; in addition, the deformed vertex is expected to be non-degenerate (see footnote 17 of [13] for more discussion on this).

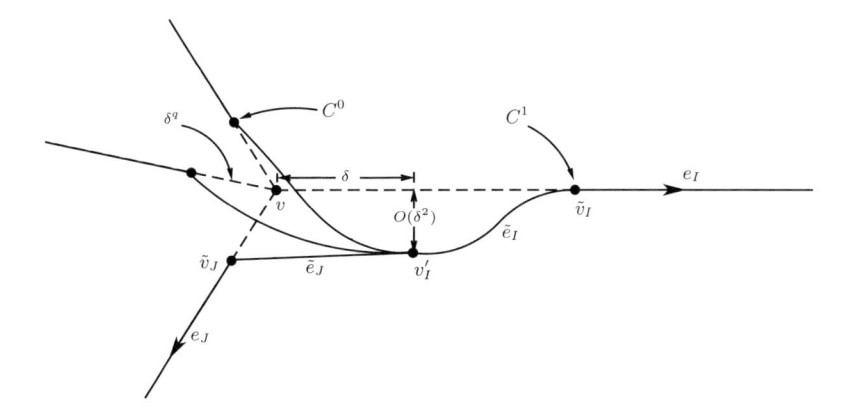

Fig. 1. A sample deformation produced at a non-degenerate vertex v along the edge e_I. In the case of $\hat{D}(\vec{N}_i)$ the dashed edges are absent and charges on the deformed edges unchanged. In the case of $\hat{C}(N)$, the dashed edges are only charged in two of the three $U(1)$ factors, but v'_I is expected to be, generically, non-degenerate. With respect to the coordinate system fixed at v, v'_I is located a distance δ from v along e_I and displaced off of e_I at a distance $O(\delta^2)$. All of the $\tilde{e}_J | J \neq I$ have tangents at v'_I which are bunched. The C^0, C^1 labels refer to the differentiability degree at which the $\tilde{e}_K$ meet e_K at $\tilde{v}_K$. [13].

To summarize: Our arguments indicate that the action of the Hamiltonian constraint *can* be coded in terms of *phase space dependent* diffeomorphisms through the use of equation (35). Its action on a charge network state is represented as a combination of 'singular' diffeomorphisms and 'charge flips'.

In order to give a flavor of the detailed expressions, we display the action of the finite triangulation Hamiltonian constraint $\hat{C}_\delta(N)$ and the electric diffeomorphism constraint $\widehat{D_\delta(\vec{N}_i)}$:

$$\hat{C}_\delta[N]\, c(A) = \frac{\hbar}{2\mathrm{i}} \sum_{v \in V(c)} \frac{3}{4\pi} N(x(v)) \lambda_v \sum_{I_v, i} q^i_{I_v} \frac{1}{\delta} \left(c(i, v'_{I_v, \delta}) - c \right) \qquad (43)$$

The various quantifiers $\{I_v, i, \delta\}$ in the argument of c specify the particular edge e_{I_v} emanating from v along which the deformation (of magnitude $\sim \delta$) was performed, and the particular flipping of the charges via internal rotation about the ith axis.

$$\hat{D}_\delta[\vec{N}_i]\, c = \frac{\hbar}{\mathrm{i}} \frac{3}{4\pi} \sum_v N(x(v)) \lambda_v \sum_{I_v} q^i_{I_v} \frac{1}{\delta} \left(c(v'_{I_v, \delta}) - c \right) \qquad (44)$$

Since there is no charge flipping in the action of $\hat{D}_\delta[\vec{N}_i]$, the deformed charge net is specified only by I_v, δ.

4.4. *Anomaly Free Continuum Limit*

Recall from Section 3.3 that the quantizations of the HK model and PFT carry a *non-trivial* representation of the constraint algebra. Accordingly, our considerations below derive their motivation from the treatment of these toy models [4, 21].

Ideally, we would like to now define continuum limits of operators on a suitable habitat similar to the case of the LM habitat for the HK model. This requires that the continuum limit Hamiltonian constraint maps the habitat to itself through its dual continuum action (see equation (29) and the related discussion). In the case of the HK model habitat states are slight deformations of the solution space of diffeomorphism states, the deformations being parameterized by vertex smooth functions f (see equation (16) and the related discussion). This structure of the habitat as a deformation of the solution space by vertex smooth functions persists in the case of PFT as well. In the $U(1)^3$ model we cannot build a putative habitat in this way because we do not have access to the space of solutions to the constraints.[i] Our task is, therefore, to construct *ab initio* a habitat with respect to which the continuum limit action of the commutator is non-trivial and anomaly free, given the finite triangulation operators (43), (44) and the identity (31). This is a very difficult object to achieve and as we shall see shortly, while we do obtain an anomaly free continuum limit using a certain set of states in the algebraic dual which are parameterized by vertex smooth functions, the continuum limit is not one of dual action. Instead it is an operator topology based limit similar to the URS continuum limit except that the role of diffeomorphism invariant states there (see discussion after equation (27)) is played by the states in the algebraic dual mentioned in the previous sentence. Since these states lie in the algebraic dual and are parameterized by vertex smooth functions, we shall call these states as *Vertex Smooth Algebraic* states (VSA states) and the finite span of these states as $\mathcal{D}_{VSA}$. The VSA states are constructed as (uncountable) sums of charge nets weighted by vertex smooth functions. Just as in (16), each LM habitat state is obtained by summing over all diffeomorphic images of some charge net $|c\rangle$, each VSA state is obtained by summing over certain deformations, to be specified below, of a suitably chosen 'primordial' charge net. We shall refer to the elements of the set of such deformed charge nets associated with the primordial charge net $|c_0\rangle$ by $B_{VSA}^{c_0}$ so that $B_{VSA}^{c_0}$ is the VSA analog of $[c]$ in the LM case.[j]

$$\Psi^f_{B_{VSA}^{c_0}} := \sum_{\bar{c} \in B_{VSA}^{c_0}} \kappa_{\bar{c}} f(V(\bar{c})) \langle \bar{c}| \qquad (45)$$

Here $V(\bar{c})$ is the set of non-degenerate vertices of $\bar{c}$[k] and $\kappa_{\bar{c}}$ is specified below. The VSA topology is specified by the seminorms defined by the pairs of elements $(\Psi^f_{B_{VSA}^{c_0}}, |c\rangle)$ similar to the URS case with $\mathcal{D}_{VSA}$ playing the role which $\mathcal{D}_{diff}$ plays

---

[i] Specifically, similar to the case of LQG (see Section 3.2), it is clear that we can construct solutions to the diffeomorphism constraints via group averaging; however we do not have solutions to the Hamiltonian constraint (which is, after all, the operator we seek to construct!). Therefore the best we can do is ensure that the space $\mathcal{D}_{VSA}$ (to be defined shortly) does contain 'vertex smooth' deformations of diffeomorphism invariant states.

[j] Reference [13] inadvertently omitted the superscript label c_0.

[k] Strictly speaking the vertices in question are not necessarily non-degenerate; their exact specification is given in Footnote 17 of Ref. [13], the important point being that Hamiltonian and electric diffeomorphism deformations conserve the number of vertices so specified.

in the URS case (see Section 3.1). The continuum limit of the commutator of two Hamiltonian constraints (i.e. the LHS of Section 3.2) is specified by the evaluation of the complex numbers:

$$\lim_{\delta \to 0} \lim_{\delta' \to 0} \Psi^f_{B^{c_0}_{VSA}} ((\hat{C}_{\delta'}[M]\hat{C}_\delta[N] - \hat{C}_{\delta'}[N]\hat{C}_\delta[M])c), \quad \forall \Psi^f_{B^{c_0}_{VSA}}, \ |c\rangle. \tag{46}$$

Similarly, the RHS continuum limit is specified by the complex numbers:

$$\lim_{\delta \to 0} \lim_{\delta' \to 0} \Psi^f_{B^{c_0}_{VSA}} ((\hat{D}_{\delta'}[\vec{M}_i]\hat{D}_\delta[\vec{N}_i] - \hat{D}_{\delta'}[\vec{N}_i]\hat{D}_\delta[\vec{M}_i])c), \quad \forall \Psi^f_{B^{c_0}_{VSA}}, \ |c\rangle. \tag{47}$$

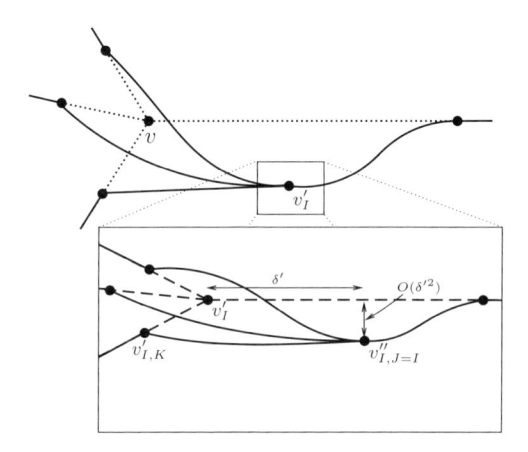

Fig. 2. Detail of the deformation generated by two successive Hamiltonian actions, in this case along the same edge $J = I$.

Our task is then to choose $B^{c_0}_{VSA}$ such that we obtain LHS and RHS continuum limits which agree with each other i.e. such that the evaluations (46) and (47) agree for all pairs $\Psi^f_{B^{c_0}_{VSA}}, \ |c\rangle$. If these limits vanish for every such pair, we would have no manifest inconsistency. However, the underlying reason for such a putative consistent trivialization would presumably have more to do with an inappropriate choice of $\mathcal{D}_{VSA}$; inappropriate in the sense that this choice would not probe features of the action of the Hamiltonian constraint in enough detail to obtain non-trivial results. This, in turn, would dilute the restrictiveness of the anomaly free requirement in the choice of finite triangulation approximants to the constraints. Thus we aim for a construction which displays non-trivial agreement of (46) and (47) for at least *some* family of pairs $(\Psi^f_{B^{c_0}_{VSA}}, \ |c\rangle)$.[1] Note that the action of two Hamiltonian constraints yield doubly deformed, charge flipped states as shown in Figure 2 and the action of two electric diffeomorphisms yield the doubly deformed states of Figure 3.

---

[1] To see the analogous statement for the diffeomorphism constraint in LQG, note that the action of the commutator between a pair of diffeomorphism constraints, $\hat{D}(\vec{\xi}_1), \hat{D}(\vec{\xi}_2)$ on the LM habitat is obtained from equation (18) with $\xi^a := \mathcal{L}_{\vec{\xi}_1} \xi_2^a$ and that $\Psi_{[s], \mathcal{L}_\xi f}(|s'\rangle)$ vanishes unless $|s'\rangle \in [s]$.

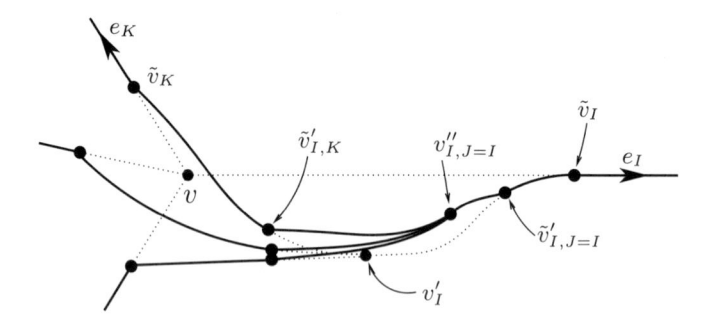

Fig. 3. Detail of the deformation generated by two successive electric diffeomorphisms, in this case along the same edge $J = I$.

Consequently, the evaluations of (46) and (47) for $|c\rangle = |c'\rangle$ and generic f will be non-trivial if we ensure that B^{co}_{VSA} is such that:

(a) if B^{co}_{VSA} contains any charge net arising from the action of two Hamiltonian constraints on $|c'\rangle$, it must contain all deformed charge nets generated by this action as well all deformed charge nets obtained by the action of two electric diffeomorphisms on $|c'\rangle$,

(b) if it contains any charge net arising from the action of two electric diffeomorphism constraints on $|c'\rangle$, it must contain all deformed charge nets generated by this action as well all deformed charge nets obtained by the action of two Hamiltonian constraints on $|c'\rangle$.

Let us call the deformed charge nets 'children' and the undeformed one the 'parent'. To ensure (a) and (b), it is clear that we need to be able to infer all the possible 'parents' of any double Hamiltonian and any double electric diffeomorphism child; if we can do this, the set of all double Hamiltonian and double electric diffeomorphism children of these parents will comprise B^{co}_{VSA}. As shown in Ref. [13] this can indeed be done (with the parents themselves arising from certain operations on the primordial state $|c_0\rangle$) by a careful study of a certain notion of 'causal structure' associated with the deformed offspring and a careful choice of the primordial charge net $|c_0\rangle$. In [13], B^{co}_{VSA} is chosen such that its elements have a single relevant vertex so that the vertex smooth functions have a single argument. With this choice of B^{co}_{VSA} the limits (46) and (47) can be evaluated. Whenever non-trivial, they agree,[m] in which case their evaluation is:

$$2\left(\frac{\hbar}{2i}\frac{3}{4\pi}\right)^2 \sum_{v\in V(c)} \sum_{I_v,i} (q^i_{I_v})^2 \lambda_v \lambda_{vI_v} \hat{e}^a_{I_v}\hat{e}^b_{I_v} \left(N(x(v))\partial_a M(x(v)) - M \leftrightarrow N\right) \partial_b f(v), \tag{48}$$

where λ_{vI_v} is evaluated at the I_vth deformed vertex (see [13]). This demonstrates the existence of a non-trivial anomaly free continuum limit action of the Hamilto-

---

[m]As shown in [13] $\kappa_{\bar{c}}$ in equation (45) needs to be chosen to be unity for those charge net labels $\bar{c}$ which describe double Hamiltonian offspring and $-\frac{1}{12}$ for double electric diffeomorphism offspring.

nian constraint i.e. one which implements equation (33) in quantum theory. From Ref. [13], one expects the existence of an infinite family of possible 'single relevant vertex' primordial states $|c_0\rangle$ and thereby a large family of VSA states. Moreover, while multi-vertex primordial states have not been explicitly constructed in Ref. [13], we do not anticipate any obstructions to their construction.

4.5. Diffeomorphism Covariance

While the considerations of the previous section yield an anomaly free continuum limit of the commutator of the Hamiltonian constraint with itself, it is also necessary that these considerations yield anomaly free continuum limits for the remaining constraint commutators. It turns out that (see Footnote i) B_{VSA}^{co} is closed under the action of diffeomorphisms and that the distributions $\mathcal{D}_{VSA}$ are deformations of diffeomorphism invariant elements in the algebraic dual space [13]. Therefore, similar to the case of the LM habitat, the algebra of diffeomorphism constraints is anomaly free and as an immediate consequence so is the VSA continuum limit of the commutator between a pair of diffeomorphism constraints. Related to this we note that (a) motivated by Thiemann's considerations of regulator covariance, Ref. [13] constructs the action of the finite triangulation operators $\hat{C}_\delta(N), \hat{D}_\delta(\vec{N}_i)$ so as to have the property that the finite triangulation operator actions at $\delta = \delta_1$ and $\delta = \delta_2$ are images of each other by a diffeomorphism and (b) that this property nicely dovetails with the 'deformation of diffeomorphism invariance' property of elements of $\mathcal{D}_{VSA}$.

The Poisson bracket between the Hamiltonian and diffeomorphism constraints is $\{C(N), D(\vec{M})\} = C(\mathcal{L}_{\vec{M}} N)$. Equivalently, under the action of any diffeomorphism ϕ generated by the diffeomorphism constraints, the Hamiltonian constraint $C(N)$ is mapped to $C(\phi_*(N))$ which translates to the condition $\hat{U}(\phi)\hat{C}(N)\hat{U}^\dagger(\phi) = \hat{C}(\phi_* N)$. Since $\hat{C}(N)$ is defined through the VSA continuum limit, we interpret this condition in terms of VSA continuum limits as:

$$\lim_{\delta \to 0} \hat{U}(\phi)\hat{C}_\delta(N)\hat{U}^\dagger(\phi) = \lim_{\delta \to 0} \hat{C}_\delta(\phi_* N). \qquad (49)$$

We refer to (49) as the condition of *diffeomorphism covariance*. Not surprisingly, the arbitrary choice of regulating coordinating patches for the evaluation of the quantum shift (see equation (39) and subsequent discussion), leads to the violation of this condition, the main culprit being the necessarily coordinate dependent evaluation of the density weighted lapse together with the fact that choice of the coordinate patch associated with a vertex v of $|c\rangle$ and the one associated with $\phi(v)$ of $|c'\rangle := \hat{U}(\phi)|c\rangle$ have, in general nothing to do with each other. This suggests that if we choose the coordinate patch $\{x\}_{c,v}$ associated with the former, we should 'move' this coordinate patch by ϕ so as to choose $\phi_*\{x\}_{c,v}$ for the latter. The problem with this is that there are many diffeomorphisms which map $|c\rangle$ to $|c'\rangle$ due to the 1 dimensional nature of the charge net graph. Remarkably, as shown in Ref. [17], by mildly restricting the vertex structure of charge nets in B_{VSA}^{co} to satisfy a certain 'non-degeneracy' property [17, 18], this problem can be alleviated and equation (49) then holds. [17]. However, this choice of 'diffeomorphism covariant' deformations generated by the action of a single Hamiltonian constraint leads to a divergent continuum limit for the commutator between a pair of Hamiltonian constraints. To see this recall that deformations generated by the Hamiltonian constraint at different values of δ are related by diffeomorphisms. Diffeomorphism covariance implicates the use of coordinate patches related by such diffeomorphisms. In the commutator the second constraint acts on deformations created by the first. As $\delta \to 0$, coordinate patches associated with these deformations become sick because the diffeomorphism scrunches edges into collinearity. This leads to a divergence of the commutator. The commutator continuum limit can then be rendered finite by enlarging the dependence of the vertex smooth functions to additional vertices of the charge net and specifying a certain benign 'short distance' behavior of the vertex smooth functions [13]. However, it then turns out that this continuum limit is not necessarily anomaly free. This happens because, roughly speaking, the requirement of diffeomorphism covariance acts as a 'magnifying glass' for diffeomorphically related deformations; as a result the continuum limit becomes extremely sensitive to the specification of the deformation at finite triangulation. It then turns out that by further specifying the deformations to be conical in a certain sense [17], the continuum limit of the commutator (46) is rendered anomaly free.

The final result, then is that the VSA continuum limit of the commutators between a pair of diffeomorphism constraints, between the diffeomorphism constraint and the Hamiltonian constraint, and between a pair of Hamiltonian constraints is well defined and anomaly free. Equation (48) is then replaced by the expression [17]:

$$\left(\frac{\hbar}{2\mathrm{i}}\frac{3}{4\pi}\right)^2 \lambda_v f_2(v,..,v) \sum_{I_v} \{N(x(v))\hat{e}^a_{I_v}\partial_a M(x(v)) - (N \leftrightarrow M)\}$$
$$\left(\frac{\sum_{J_v \neq I_v}\sum_{K_v \neq I_v} g_{ab}(\hat{e}^a_{K_v} - \hat{e}^a_{J_v})(\hat{e}^b_{K_v} - \hat{e}^b_{J_v})}{4(M-1)g_{ab}(\hat{e}^a_{I_v}\hat{e}^b_{I_v})}\right)^{\frac{1}{3}} 2\lambda_{v_{I_v}} \cos^2\frac{\theta_{I_{v_0}}}{2}\sum_i (q^i_{I_v})^2 \hat{e}^a_{I_v}\partial_a g(v).$$
(50)

We have reproduced the expression above from Ref. [17] to give a flavor of the complexity of the final result. An explanation of the various symbols appearing in this expression is out of the scope of the paper and we invite the interested reader to consult [17] for details.

5. Hamiltonian Constraint in Euclidean Gravity

In the previous section we showed how the weak coupling limit of Euclidean gravity could be quantized within the framework of LQG and led to a generally covariant quantum field theory in a precise sense. The key question is if the lessons learnt from analysis of $U(1)^3$ theory and other toy models described above could be applied to the $SU(2)$ theory itself and if one could obtain an anomaly free definition of the Hamiltonian constraint.

As we saw one of the key lessons that we learnt so far was the importance of a geometric interpretation of the Hamiltonian constraint. This interpretation was the key input in the choice of finite triangulation constraint. For Euclidean gravity in connection variables, a rather straightforward computation shows that

$$\{H[N], A_a^i(x)\tau_i\} = 2[\tau_i, L_{N\tilde{E}_i}A_a(x)] + 2[\tau_i, \mathcal{D}_a\Lambda_{(i)}(x)] \qquad (51)$$

where $H[N] = \int N \text{Tr}(F \wedge E \wedge E)$.

τ_i are the Pauli matrices and $\Lambda_{(i)}(x)$ are a triplet of gauge parameters defined as $\Lambda_{(i)}(x) = -N\tilde{E}_i^a A_a(x)$. That is, the action of the Hamiltonian constraint on the Ashtekar-Barbero connection can be understood in terms of a concomitant of phase space dependent diffeomorphism, a phase space dependent gauge transformation and an adjoint action by a fixed element of $su(2)$. The robustness of this interpretation came from a non-trivial observation by Ashtekar [22] who showed that precisely the same interpretation also held for action of $H[N]$ on densitized triads.

$$\{H[N], \tilde{E}_i^a(x)\tau^i\} = [\tau_i, L_{N\tilde{E}_i}\tilde{E}^a(x)] + [\tau_i, \mathcal{R}_{\lambda_{(i)}}\tilde{E}] \qquad (52)$$

$\mathcal{R}$ is the (infinitesimal) $SU(2)$ rotation of the triad.

The choice of finite triangulation constraint should be such that its action of spin-network states of the theory precisely captures the geometric action of classical constraint on the classical fields. A proposal for the Hamiltonian constraint based on these ideas is given in [23], where some preliminary evidence was presented in the favor of its off-shell anomaly freedom. We thus believe that the goal of defining quantum dynamics for Euclidean Loop Quantum Gravity which passes one of the key consistency checks of anomaly freedom is on the horizon.

6. Conclusions

In our view, the outstanding problem in LQG is a satisfactory definition of its quantum dynamics generated by the Hamiltonian constraint $\hat{C}(N)$. The tension between the local nature of the fields comprising $C(N)$ and the non-locality of some of the basic operators of the theory necessitates the approximation of the former in terms of the latter thus yielding an approximant to $\hat{C}(N)$. It is then necessary to define $\hat{C}(N)$ through a limiting procedure of better and better approximants, the operator $\hat{C}(N)$ arising as a continuum limit of its approximants in a suitable topology. Note that the approximant operators $\hat{C}_\delta(N)$ are defined on the *continuum* kinematic Hilbert space of the theory; while the 'regulating' structures (namely the triangulations T_δ of the manifold) become finer, this Hilbert space is unchanged. This is a crucial difference with Lattice Gauge Theory regularizations in which the Hilbert space itself lives on the lattice and the continuum limit involves an ever finer triangulation structure (namely the lattice) *and* its associated Hilbert space. It is this feature which makes the LQG treatment very dependent on the finite triangulation choices. In this Chapter we reviewed recent attempts to constrain these choices by imposing the condition that the quantum theory encode the most robust and beautiful property of general relativity, namely its *general covariance*, which via the HKT analysis [1] lies in the structure of its classical constraint algebra. These attempts are based on the key insight that the action of the Hamiltonian constraint, which generates *temporal* deformations of the Cauchy slice can be understood in terms of the action of triad dependent *spatial* diffeomorphisms [13, 14, 22, 23]. While the results are encouraging (see also [24]), there is still much to do. One lesson learnt [17] is that the more one (legitimately) demands from the theory, the more are the choices constrained. There is yet much to demand from the attempted constructions of $\hat{C}(N)$.

Chief amongst these demands is that the algebra generated by the continuum constraints be represented on a representation space i.e. a suitable *habitat*. This implies that the action of any finite string of constraint operators be well defined. In contrast the current state of the art for the Hamiltonian constraint[n] only constructs the continuum limit action of a single such constraint and the continuum limit action of its commutator; it does not yet construct the continuum limit action of the product of two constraints i.e. it constructs $\lim_{\delta \to 0} \lim_{\delta' \to 0} (\hat{C}_\delta(N)\hat{C}_{\delta'}(M) - M \leftrightarrow N)$ as opposed to $(\lim_{\delta \to 0} \lim_{\delta' \to 0} \hat{C}_\delta(N)\hat{C}_{\delta'}(M)) - (\lim_{\delta \to 0} \lim_{\delta' \to 0}(M \leftrightarrow N))$. The requirement of a genuine representation of the *entire* constraint algebra generated by the basic Poisson brackets between the constraints is far more stringent than that of an anomaly free continuum limit only of the basic commutators.

Yet another demand, based on physical intuition, is of a resolution between a conceptual tension between the extremely 'local' deformations generated by the Hamiltonian constraint and the propagating deformations of classical spacetime corresponding to gravitational waves [25]. Since the solutions of classical PFT describe the propagation of scalar waves, 1+1 PFT once again provides an invaluable testing ground to probe and resolve this tension and thereby further constrain/improve the available choices in defining a quantum dynamics for LQG [26]. Yet another demand is one of local Lorentz invariance in the quantum dynamics; unfortunately, we do not even have a clear articulation of this demand in the context of the underlying discreteness of LQG.

There exist other exciting lines of thought towards a definition of the quantum dynamics for LQG [27–33]. We mention two examples. In [30], the authors seek a quantization of the Hamiltonian constraint in 3-dimensional gravity which leads to the well known equations of Ponzano-Regge model. They show that such a quantum constraint indeed does exist, whose kernel matches the well studied solutions of Ponzano-Regge theory. This proposal has also been extended to four dimensions [31] and it would be of interest to see if there is any overlap with the material

---

[n]While the LM habitat supports a representation of the constraint algebra with density 1 Hamiltonian constraints, the representation *trivializes* in the sense described in Section 1.

presented in this Chapter. The recent work [32] also uses a 'spin foam-Regge calculus' inspired line of thought to directly construct an operator corresponding to the spatial curvature. Since the difference between the Euclidean and Lorentzian Hamiltonian constraints can also be encoded in the spatial curvature, this offers an alternative to the Thiemann trick described in Section 3.

In conclusion, we hope to have convinced the reader that the question of the existence of a satisfactory definition of the quantum dynamics of LQG can be confronted with a rich family of ideas and strategies and that there is room for optimism for a resolution of this question in the foreseeable future.

Acknowledgments

We thank Casey Tomlin for his generosity in allowing us to use the figures he created. We thank CT and Miguel Campiglia for going through a draft version of this work. We thank Abhay Ashtekar, Fernando Barbero, Hanno Sahlmann, Thomas Thiemann and Eduardo Villaseñor for their encouragement. MV thanks CT for a very productive collaboration [13] and MC for his invaluable help and insights in developing the material in [6], Section 2.1. AL thanks Adam Henderson and CT for collaborations [14], [15].

References

[1] S. Hojman, K. Kuchař and C. Teitelboim, *Annals Phys.* **96**, 88 (1976).
[2] D. Giulini and D. Marolf, *Class. Quant. Grav.* **16**, 2479 (1999).
[3] J. Lewandowski and D. Marolf, *Phys. Rev. D* **7**, 299 (1998).
[4] A. Laddha and M. Varadarajan, *Phys. Rev. D* **83**, 025019 (2011); T. Thiemann, *Lessons for Loop Quantum Gravity from Parameterized Field Theory*, e-Print: arXiv:1010.2426 [gr-qc].
[5] A. Ashtekar, T. Pawlowski and P. Singh, *Phys. Rev. D* **74**, 084003 (2006).
[6] M. Campiglia and M. Varadarajan, eprint
[7] A. Ashtekar, J. Lewandowski, D. Marolf, J. Mourao and T. Thiemann, *J. Math. Phys.* **36**, 6456 (1995).
[8] T. Thiemann, *Modern Canonical Quantum General Relativity* (Cambridge Monographs on Mathematical Physics).
[9] A. Ashtekar and J. Lewandowski, *Class. Quant. Grav.* **21**, R53 (2004).
[10] T. Thiemann, *Class. Quant. Grav.* **15**, 839 (1998).
[11] A. Perez, *Phys. Rev. D* **73**, 044007 (2006).
[12] R. Gambini, J. Lewandowski, D. Marolf, J. Pullin, *Int. J. Mod. Phys. D* **7**, 97 (1999).
[13] C. Tomlin and M. Varadarajan, *Phys. Rev. D* **87**, 044039 (2013).
[14] A. Henderson, A. Laddha and C. Tomlin, *Phys. Rev. D* **88**, 044028 (2013).
[15] A. Henderson, A. Laddha and C. Tomlin, *Phys. Rev. D* **88**, 044029 (2013).
[16] L. Smolin, *Class. Quant. Grav.* **9**, 883 (1992).
[17] M. Varadarajan, *Phys. Rev. D* **87**, 044040 (2013).
[18] N. Grot and C. Rovelli, *J. Math. Phys.* **37**, 3014 (1996).
[19] K. Kuchař, *Phys. Rev. D* **39**, 1579 (1989).
[20] V. Husain and K. Kuchař, *Phys. Rev. D* **42**, 4070 (1990).

[21] A. Laddha and M. Varadarajan, *Class. Quant. Grav.* **28**, 195010 (2011).
[22] A. Ashtekar, *Unpublished notes* (2012).
[23] A. Laddha, arxiv:14010931 (2014).
[24] J. Lewandowski and C.-Y. Lin, arxiv:1606.01830 (2016).
[25] L. Smolin, arxiv:gr-qc/9609034 (1996).
[26] M. Varadarajan, arxiv:1609.06034 (2016).
[27] E. Alesci, C. Rovelli, *Phys. Rev. D* **82**, 044007 (2010).
[28] A. Perez, D. Pranzetti, *Class. Quant. Grav.* **27**, 145009 (2010).
[29] B. Dittrich, T. Thiemann, *Class. Quant. Grav.* **23**, (2006).
[30] V. Bonzom, L. Freidel, *Class. Quant. Grav.* **28**, 195006 (2011).
[31] V. Bonzom, *Phys. Rev. D* **84**, 024009 (2011).
[32] E. Alesci, M. Assanioussi, J. Lewandowski, *Phys. Rev. D* **89**, 124017 (2014).
[33] J. Lewandowski and H. Sahlmann, *Phys. Rev. D* **91**, 044022 (2015).

Chapter 3

Spinfoam Gravity

Eugenio Bianchi

Institute for Gravitation and the Cosmos & Physics Department,
Penn State, University Park, PA 16802, USA

1. Introduction

Spins in quantum mechanics and the action of general relativity share a simple and surprising relation. This relation is at the roots of spinfoam gravity [1, 2]. The Wigner $6j$ symbol is an elementary object that appears in the theory of 'composition of angular momenta' in quantum mechanics. It is the simplest non-trivial invariant under rotations that can be built from Clebsch–Gordan coefficients only [3]. It turns out that this familiar quantity is related to the action of general relativity in 3 spacetime dimensions. In the limit of large spins $j_i \gg 1$, the following asymptotic formula holds [4]

$$\begin{Bmatrix} j_1 & j_2 & j_3 \\ j_4 & j_5 & j_6 \end{Bmatrix} \approx \frac{(8\pi G_N \hbar)^{3/2}}{\sqrt{-48\pi \, \mathrm{i}\, V}} \, e^{+\frac{\mathrm{i}}{\hbar} S} + \text{c.c..} \qquad (1)$$

Here S is a function obtained from the 3-dimensional Einstein-Hilbert action for a compact region $\mathcal{R}$

$$I_{\mathcal{R}}[g_{\mu\nu}] = \frac{1}{16\pi G_N} \int_{\mathcal{R}} d^3x \, \sqrt{g} R + \frac{1}{8\pi G_N} \int_{\partial \mathcal{R}} d^2x \, \sqrt{h} K \qquad (2)$$

as follows: the action is evaluated on the flat Euclidean metric, $S = I_{\mathcal{R}}[\eta_{\mu\nu}]$, and the region $\mathcal{R}$ is chosen so that its induced geometry is the one of a flat Euclidean tetrahedron. Under these conditions, the Einstein-Hilbert action determines the building-block of the so-called Regge action S [5, 6]. S depends only on a finite number of variables, specifically the lengths $\ell_1, \ldots, \ell_6$ of the six edges of the tetrahedron. The quantity $V = \int_{\mathcal{R}} d^3x \sqrt{g}$ in (1) is the volume of the tetrahedron expressed as a function of the edge-lengths. The relation between the *spins* j_i and the edge-lengths ℓ_i is

$$\ell_i = (j_i + 1/2)\, 8\pi G_N \hbar. \qquad (3)$$

The asymptotic formula (1) holds in the classically allowed region in which a tetrahedron with edges of lengths ℓ_i exists. Large spins $j_i \gg 1$ correspond to a classical

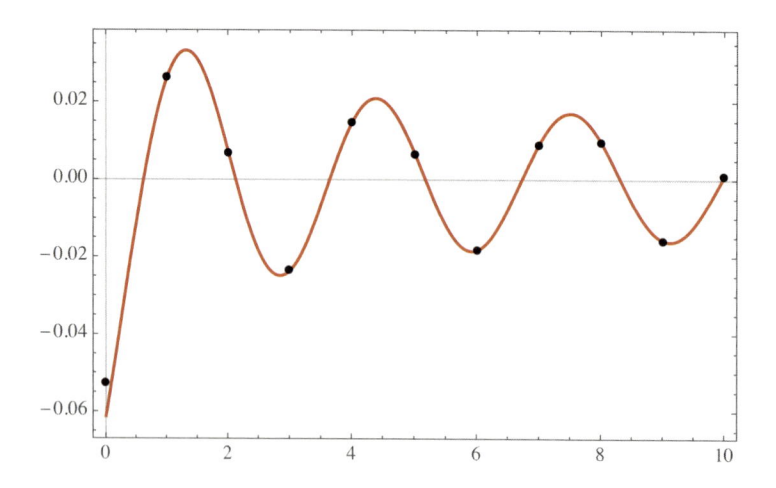

Fig. 1. The $6j$ symbol $\begin{Bmatrix} 9 & 9 & 9 \\ 9 & 9 & j \end{Bmatrix}$ as a function of j (dots) and the Ponzano-Regge approximation (continuous line).

limit $\hbar \to 0$ with the edge-lengths ℓ_i fixed. Figure 1 shows how accurate formula (1) is.

This surprising relation discovered by Ponzano and Regge in 1968 provides the simplest and oldest example of spinfoam model for quantum gravity, a realization of the path-integral over spacetime geometries [7, 8]

$$Z = \int \mathcal{D}g_{\mu\nu} \; e^{\frac{i}{\hbar} S[g_{\mu\nu}]} \qquad (4)$$

in terms of a sum over spins. The analogous quantity for Lorentzian General Relativity in 4 spacetime dimensions has long been searched [9–30] and found only in 2007 [31–34].

Formally, the path integral over spacetime geometries with an initial and a final boundary

$$W[q_{ab}^{(1)}, q_{ab}^{(2)}] = \int_{q_{ab}^{(1)}, q_{ab}^{(2)}} \mathcal{D}g_{\mu\nu} \; e^{\frac{i}{\hbar} S[g_{\mu\nu}]} \qquad (5)$$

is a solution of Hamiltonian constraint equation $\hat{\mathcal{C}} \, \Psi[q_{ab}] = 0$, i.e. a physical state in canonical quantum gravity. It turns out that the boundary space of states in spinfoam gravity coincides with the Hilbert space of loop quantum gravity discussed in *Chapters 1 and 2*. As a result the spinfoam path integral provides a covariant tool to solve the Hamiltonian. In fact spinfoam gravity can be understood as the covariant formulation of canonical loop quantum gravity as originally proposed by Reisenberger and Rovelli in 1996 [9].

In this chapter we introduce spinfoam gravity starting from its classical formulation as a topological field theory with defects (Sections 2-6), we describe its structure in terms of cellular quantum geometries (Sections 7-9) and we summarize recent developments (Sections 10-11). We restrict attention to the case of vanishing cosmological constant $\Lambda = 0$. The generalization to $\Lambda \neq 0$ has been derived recently [35–38].

Throughout this chapter we set $\hbar = 1$ and keep Newton's constant G_N explicit.

2. Topological Field Theory and Gravity

Topological field theories are field theories with no local degree of freedom: the only dynamical degrees of freedom have a global nature and capture the topological invariants of the manifold the theory is defined on [39–41]. These theories share with general relativity the invariance under diffeomorphisms and provide a classical starting point for the formulation of spinfoam gravity.

Consider a 4-dimensional manifold with topology $\mathcal{M} = M \times \mathbb{R}$ with M a compact 3-manifold. A topological field theory of the BF type with the Lorentz group $SO(1,3)$ as gauge group is defined by the action [15]

$$S_{top}[B,\omega] = \int_{\mathcal{M}} B_{IJ} \wedge F^{IJ}. \tag{6}$$

Here $\omega^{IJ} = \omega^{IJ}_\mu(x)dx^\mu$ is a Lorentz connection, $F^{IJ} = d\omega^{IJ} + \omega^I{}_K \wedge \omega^{KJ}$ its curvature and $B^{IJ} = B^{IJ}_{\mu\nu}(x)\,dx^\mu \wedge dx^\nu$ a two-form with values in the adjoint representation of Lorentz group. We also denote η_{IJ} the Minkowski metric with signature $(-+++)$ and ϵ^{IJKL} the Levi-Civita tensor, $\epsilon^{0123} = +1$. The theory described by the action (6) is manifestly invariant under diffeomorphisms $\text{Diff}(\mathcal{M})$ and under local Lorentz transformations. The action S_{top} has also another local symmetry, topological invariance: the action is invariant under shifts of the B field by the covariant derivative of a one-form Λ^{IJ},

$$B^{IJ} \to B^{IJ} + d\Lambda^{IJ} + \omega^I{}_K \wedge \Lambda^{KJ} + \omega^J{}_K \wedge \Lambda^{KI}, \tag{7}$$

as can be shown by integration by parts and using the Bianchi identity $dF^{IJ} + \omega^I{}_K \wedge F^{KJ} = 0$. Requiring the stationarity of the action with respect to variations δB and $\delta\omega$ we find that the classical solutions satisfy the equations of motion

$$F = 0 \quad \text{and} \quad dB^{IJ} + \omega^I{}_K \wedge B^{KJ} + \omega^J{}_K \wedge B^{KI} = 0. \tag{8}$$

The first equation tells us that the Lorentz connection ω is locally flat, and therefore locally can be written as a pure gauge configuration. The second equation (together with the shift symmetry equation (7) and the fact that locally all closed forms are exact) tells us that the B field can be written locally as $B^{IJ} = d\Lambda^{IJ} + \omega^I{}_K \wedge \Lambda^{KJ} + \omega^J{}_K \wedge \Lambda^{KI}$ for some 1-form Λ^{IJ}. Therefore locally, all solutions of the BF theory field equations are equal modulo gauge transformations and shifts: the theory has no local degrees of freedom.

General relativity can be formulated in the same language as the theory described above. The coupling of gravity to fermionic matter fields is best described via the introduction of the Lorentz group $SO(1,3)$ as internal gauge group. The fundamental variables of the theory are the Lorentz connection $\omega^{IJ} = \omega_\mu^{IJ}(x)dx^\mu$ and a coframe field $e^I = e_\mu^I(x)dx^\mu$. The spacetime metric is a derived quantity given by $g_{\mu\nu}(x) = \eta_{IJ}\, e_\mu^I(x) e_\nu^J(x)$. Einstein equations are the equations of motion obtained from the first order action for gravity [42, 43]

$$S_{grav}[e,\omega] = \frac{1}{16\pi G_N}\int_{\mathcal{M}} \frac{1}{2}\epsilon_{IJKL} e^I \wedge e^J \wedge F^{KL} - \frac{1}{\gamma} e_I \wedge e_J \wedge F^{KL}, \qquad (9)$$

plus the action of matter $S_{mat}[\psi, e, \omega]$. The theory is invariant under diffeomorphisms Diff($\mathcal{M}$) and under local Lorentz transformations as for the action in equation (6). The difference is that now there is no analogue of the shift symmetry equation (7): the theory has infinitely many dynamical degrees of freedom, two per point, and the equations of motion are non trivial,

$$\epsilon_{IJKL} e^J \wedge F^{KL} = 0 \quad \text{and} \quad e^I \wedge \left(de^J + \omega^J{}_K \wedge e^K \right) = 0. \qquad (10)$$

The first equation is the Einstein equation for pure gravity and the second the vanishing of the torsion $T^I = de^I + \omega^I{}_J \wedge e^J$. Note that the second term in the action equation (9) does not affect the classical equations of motion. The coupling constant γ is the Barbero-Immirzi parameter [44, 45] appearing in the canonical formulation of general relativity in real Ashtekar variables [46, 47] (*see Chapter 1*).

The topological theory defined by equation (6) can be understood as general relativity with all its degrees of freedom frozen so that the Lorentz connection is flat, $F = 0$. In general relativity this solution has the physical interpretation of a Minkowski spacetime. By identifying the B field with the exterior product of two cotetrads as suggested by the relation between the actions (6) and (9), i e by setting

$$B^{IJ} = \frac{1}{16\pi G_N}\left(\frac{1}{2}\epsilon^{IJ}{}_{KL} e^K \wedge e^L - \frac{1}{\gamma} e^I \wedge e^J \right), \qquad (11)$$

we generate also a solution of the topological theory. For instance by choosing the gauge such that $e^I = dx^I$ and $\omega^{IJ} = 0$, we have $B^{IJ} = \frac{1}{16\pi G_N}\Big(\frac{1}{2}\epsilon^{IJ}{}_{KL} dx^K \wedge dx^L - \frac{1}{\gamma} dx^I \wedge dx^J \Big)$ that solves equation (8). However, because of the symmetry (7), the physical interpretation of this solution is now rather different. This is most easily shown by considering the 4-volume of a spacetime region

$$\mathcal{V} = -\int \frac{1}{4!}\epsilon_{IJKL} e^I \wedge e^J \wedge e^K \wedge e^L. \qquad (12)$$

By inverting (11) we find

$$e^I \wedge e^J = -16\pi G_N \frac{\gamma^2}{1+\gamma^2}\left(\frac{1}{2}\epsilon^{IJ}{}_{KL} B^{KL} + \frac{1}{\gamma} B^{IJ} \right), \qquad (13)$$

and upon substitution in $\mathcal{V}$, equation (12) gives us the spacetime volume expressed in terms of the B field. However B fields that differ by a covariant derivative are physically equivalent because of the topological symmetry (7). Therefore in the topological theory different volumes, and more generally different spacetime geometries determined by B, are identified.

Conversely, general relativity can be formulated in terms of the topological theory (6) with the addition of a constraint on the B field that breaks the symmetry (7) and unfreezes the degrees of freedom of the Lorentz connection. The requirement is that there exist a one-form e^I such that B has the form (11). This condition is best expressed using equation (13): this equation tells us that the linear combination

$$\Sigma^{IJ} = \frac{1}{2}\epsilon^{IJ}{}_{KL} B^{KL} + \frac{1}{\gamma} B^{IJ} \tag{14}$$

is a simple two-form, i.e. it can be expressed as the exterior product of two one-forms. The requirement that Σ^{IJ} is simple is called *simplicity constraint* and it can equivalently be stated as the condition $\Sigma^{IJ} \wedge \Sigma^{KL} = \frac{1}{4!}(\epsilon_{MNPQ}\Sigma^{MN}\wedge\Sigma^{PQ})\epsilon^{IJKL}$ (quadratic simplicity constraint). The Plebanski action for general relativity for instance consists of the topological action $S_{top}[B,\omega]$ plus a term that imposes the vanishing of the quadratic simplicity constraint [48–50].

3. Classical Spinfoam Gravity: Degrees of Freedom and Foams

A classical spinfoam model is a topological field theory of the type (6) with a finite number of dynamical degrees freedom associated to a network of topological defects. The defects are introduced by equipping the 4-manifold $\mathcal{M}$ with a cellular decomposition.

A cellular decomposition is a way to present a manifold as composed of simple elementary pieces, cells with the topology of a ball. The simplest example is a triangulation, the decomposition of a manifold into 4-simplices, tetrahedra, triangles, segments and points. Here we consider decompositions that allow more general adjacency relations between cells [51].

Let $\mathcal{B}_n$ be the open ball of dimension n, with $n = 0, \ldots, 4$. We denote by $\bar{\mathcal{B}}_n$ its closure and consider a set of homeomorphisms $\phi_i : \bar{\mathcal{B}}_n \to \mathcal{M}$ that send the n-ball into a subset of $\mathcal{M}$. We assume that the subset $\phi_i(\partial \bar{\mathcal{B}}_n)$ is the disjoint union of a finite number of $\phi_i(\mathcal{B}_m)$. We denote by $\Delta_n \equiv \{\phi_i(\mathcal{B}_n) \,|\, i = 1, \ldots, I_n\}$ the set of cells of dimension n. If $\mathcal{M}$ is the disjoint union of cells

$$\mathcal{M} = \bigcup_{n=0}^{4} \bigcup_{i=1}^{I_n} \{\Delta_n\}_i, \tag{15}$$

these data provide a cellular decomposition of the manifold $\mathcal{M}$. We require the cellular decomposition to be simple: two n-cells share at most one $(n-1)$-cell on their boundary. In this case the two cells are said to be adjacent. A manifold equipped with a cellular decomposition is also called a cellular manifold. We call Δ_2 the 2-skeleton of the decomposition.

The spinfoam action consists of two terms, a 4-dimensional topological term of the form (6) and a term given by the integral of a 2-form over a branched surface in $\mathcal{M}$, the 2-skeleton Δ_2 of a cellular decomposition. The second term breaks the topological invariance of the first and unfreezes a finite number of degrees of the connection. In order to describe a truncation of general relativity, the second term is chosen so to impose the simplicity of Σ^{IJ}, equation (14) and the recovery of the area 2-form $e^I \wedge e^J$, equation (13). Different spinfoam models correspond to different proposals for the implementation of the simplicity of Σ^{IJ}. Here we describe the spinfoam action corresponding to the Engle-Pereira-Rovelli-Livine — Freidel-Krasnov model [31–34] generalized to arbitrary cellular decompositions [52–54].

In two dimensions every 2-form is simple, therefore simplicity of Σ^{IJ} is automatic. The only remaining requirement to be imposed on Σ^{IJ} regards the induced signature of the 2-dimensional surface. We require that Δ_2 is everywhere space-like by introducing a reference vector t^I that is timelike, $\eta_{IJ} t^I t^J = -1$. The spinfoam action for gravity is then given by

$$S[B,\omega,\lambda] = \int_{\mathcal{M}} B_{IJ} \wedge F^{IJ} + \int_{\Delta_2} \lambda_I t_J \Sigma^{IJ} \quad (16)$$

where λ^I is a 0-form playing the role of Lagrange multiplier imposing the linear simplicity constraint, i.e. the vanishing of the pull-back to Δ_2 of $t_J \Sigma^{IJ}$.

Using the reference vector t^I we can decompose the B^{IJ} field into its dual magnetic part B^I and its dual electric part E^I,

$$B^I = B^{IJ} t_J, \qquad E^I = \frac{1}{2} \epsilon^{IJKL} B_{JK} t_L. \quad (17)$$

The linear simplicity constraint $t_J \Sigma^{IJ} = 0$ on Δ_2 is then equivalent to the relation

$$B^I = \gamma E^I \quad \text{on} \quad \Delta_2, \quad (18)$$

between the electric and the magnetic parts, pulled back on the 2-skeleton Δ_2. This equation plays a central role in the definition of the quantum theory. Variation of the spinfoam action (16) with respect to the B^{IJ} field imposes that the Lorentz connection ω is locally flat, $F^{IJ} = 0$ in the bulk of $\mathcal{M} - \Delta_2$. The constraint (18) on the electric and magnetic parts of the B field however allow a curvature $F^{IJ} \neq 0$ supported on Δ_2.

The 4-manifold with boundary $\mathcal{M}' = \mathcal{M} - \Delta_2$ is path-connected but not simply-connected: there are closed paths in $\mathcal{M}'$ that encircle elements of the 2-skeleton and are non-contractible. As a result the first homotopy group $\pi_1(\mathcal{M} - \Delta_2)$ is non trivial. A presentation of this group can be obtained by introducing the notion of two-complex Δ^* dual to the cellular decomposition Δ.

A combinatorial two-complex $\mathcal{C} = (F, E, V, \partial)$ is defined by a finite set F of elements f called *faces*, a finite set E of elements e called *edges*, a finite set V of elements v called *vertices*, and a boundary relation ∂ that associates to each edge an ordered couple of vertices $\partial e = (s_e, t_e)$ and to each face f a cyclic sequence of edges, $\partial f = (e_1, \ldots, e_{n_f})$. The two-complex Δ^* dual to the cellular decomposition Δ has a vertex per 4-cell, an edge per 3-cell and a face per 3-cell of Δ. Two vertices are connected by an edge if they are dual to two adjacent 4-cells. Non-contractible loops in $\mathcal{M}' = \mathcal{M} - \Delta_2$ correspond to cyclic sequences of edges that bound a face f of Δ^* dual to a 2-cell in Δ_2.

The degrees of freedom of spinfoam gravity are best described in terms of the 2-complex $\mathcal{C} = \Delta^*$ dual to the cellular decomposition Δ. While the curvature locally vanishes in $\mathcal{M}' = \mathcal{M} - \Delta_2$, the parallel transport around a non-contractible loop can be non-trivial. Vertices and edges of the 2-complex Δ^* can be embedded in $\mathcal{M}$ as points p_v in the interior of 4-cells and paths ℓ_e connecting two 4-cells. The holonomy of the Lorentz connection along a path e is given by

$$G_e = P \exp i \int_{\ell_e} \omega \,. \tag{19}$$

Note that because of the local flatness of ω the holonomy G_e is invariant under diffeomorphisms of $\mathcal{M}$ that move the path ℓ_e while preserving the 1-skeleton of Δ, and in particular the simply-connected region defined by the two adjacent 4-cells. While the holonomy G_e can be set to the identity by a Lorentz gauge transformation, the product of holonomies around a face $G_{\partial f} = G_{e_1} \cdots G_{e_{n_f}}$ is non-trivial. Analogously we consider the flux of the B field through a 2-cell t_f dual to a face f of the 2-complex,

$$B_f^{IJ} = \int_{t_f} B^{IJ} \,. \tag{20}$$

The flux B_f^{IJ} is invariant under topological transformation of the type (7) that preserve the boundary of t_f, i.e. the 1-skeleton of Δ. As a result, it is also invariant under diffeomorphism of $\mathcal{M}$ that preserve the boundary of t_f.

A 2-complex $\mathcal{C}$ together with an assignment of holonomies h_e to its edges and fluxes B_f to its faces is called a *foam*. Together with the linear simplicity constraint (18) they provide the classical building blocks of spinfoam gravity.

4. Unitary Representations of the Rotation and the Lorentz Group

The groups $SU(2)$ and $SL(2,\mathbb{C})$ are respectively the double cover of the rotation group $SO(3)$ and of the part of the Lorentz group connected to the identity, $SO^\uparrow(3,1)$. These two groups play a key role in spinfoam gravity: unitary representations are associated to the faces of a foam and provide the *spins* of spinfoam gravity. In this section we summarize the relevant mathematical notions involved in this construction.

Unitary representations of the group $SU(2)$ on a Hilbert space V are generated by three Hermitian operators L^i, $i = 1, 2, 3$ obeying the commutation relations

$$[L^i, L^j] = i\epsilon^{ij}{}_k L^k \,.$$

In the following we also use the vector notation $\vec{x} = x^i$ and $\vec{x} \cdot \vec{y} = \delta_{ij} x^i y^j$. Irreducible representations $V^{(j)}$ are labeled by a half-integer $j = 0, \frac{1}{2}, 1, \ldots$, the spin and are finite dimensional $\dim V^{(j)} = 2j + 1$. We follow the standard notation and call

$$|j, m\rangle \in V^{(j)} \tag{21}$$

an orthonormal basis of simultaneous eigenstates of the Casimir operator $\vec{L}^2$ and of a component $L_z = \vec{L} \cdot \vec{z}$. The eigenvalues are $j(j+1)$ and $m = -j, \ldots, +j$ respectively.

Unitary representations of the group $SL(2, \mathbb{C})$ on a Hilbert space $\mathcal{V}$ are infinite dimensional and are generated by six Hermitian operators $J^{IJ} = -J^{JI}$, $I, J = 0, 1, 2, 3$. Let us introduce a unit time-like vector t^I, and define the generator of Lorentz transformations that leave t^I invariant (rotations) as

$$L^I = \frac{1}{2} \epsilon^I{}_{JKL} J^{JK} t^L , \qquad (22)$$

and the generator of boosts of t^I as

$$K^I = J^{IJ} t_J . \qquad (23)$$

Notice that $t_I L^I = t_I K^I = 0$ so that, in coordinates such that $t^I = (1, 0, 0, 0)$, we have $L^I = (0, L^i)$ and $K^I = (0, K^i)$. These generators of $SL(2, \mathbb{C})$ obey the following commutation relations:

$$[L^i, L^j] = +i \, \epsilon^{ij}{}_k \, L^k , \qquad (24)$$
$$[L^i, K^j] = +i \, \epsilon^{ij}{}_k \, K^k , \qquad (25)$$
$$[K^i, K^j] = -i \, \epsilon^{ij}{}_k \, L^k . \qquad (26)$$

Unitary irreducible representations $\mathcal{V}^{(p,j)}$ of $SL(2, \mathbb{C})$ (the principal series [55]) are labeled by a real number p and a half-integer j. As $SU(2)$ is a subgroup of $SL(2, \mathbb{C})$, they are also unitary representations of $SU(2)$, but reducible. In particular they decompose into irreducible representations as follows:

$$\mathcal{V}^{(p,j)} = V^{(j)} \oplus V^{(j+1)} \oplus V^{(j+2)} \oplus \cdots . \qquad (27)$$

In $\mathcal{V}^{(p,j)}$ the two invariant Casimir operators C_1 and C_2 have eigenvalues

$$C_1 = \frac{1}{2} J_{IJ} J^{IJ} = \vec{K}^2 - \vec{L}^2 = p^2 - j^2 + 1 , \qquad (28)$$
$$C_2 = \frac{1}{8} \epsilon_{IJKL} J^{IJ} J^{KL} = \vec{K} \cdot \vec{L} = p j . \qquad (29)$$

We denote by

$$|(p, j); j', m\rangle \in \mathcal{V}^{(p,j)} \qquad (30)$$

with $j' \geq j$ an orthonormal basis of simultaneous eigenstates of $\vec{L}^2$ and $L_z = L^I z_I$, with $t^I z_I = 0$.

The decomposition (27) allows us to identify the vector $|j'm\rangle$ that transforms under the representation j' of the rotation group with the vector $|(p, j); j', m\rangle$ that transforms in the representation j' of the little group of the Lorentz group that leaves the time-like vector t^I invariant. In spinfoam gravity we consider a map Y_γ that identifies the representation $V^{(j)}$ of $SU(2)$ with the lowest-spin block in the decomposition (27). More explicitly, calling γ the ratio between p and $j+1$, the map is defined by

$$Y_\gamma : V^{(j)} \to \mathcal{V}^{(\gamma(j+1),j)} \qquad (31)$$

$$|j,m\rangle \mapsto |(\gamma(j+1),j);j,m\rangle . \qquad (32)$$

This map has the following notable property: it determines a Hilbert space $\mathcal{V}_\gamma = \mathrm{Im} Y_\gamma$ where the matrix elements of the operator $K^i - \gamma L^i$ vanish,

$$\langle(\gamma(j+1),j);j,m'|K^i - \gamma L^i|(\gamma(j+1),j);j,m''\rangle = 0 . \qquad (33)$$

In spinfoam gravity the map Y_γ and the Hilbert space $\mathcal{V}_\gamma$ play a special role as they provide a solution of the linear simplicity constraint with Barbero-Immirzi parameter γ.

5. Boundary Variables and the Loop Quantum Gravity Hilbert Space

Consider a 4-manifold $\mathcal{M}$ with a 3-dimensional boundary M. The cellular decomposition Δ induces a decomposition Δ_3 on the boundary. The boundary variables of the classical theory (16) are the Lorentz connection ω together with its conjugate momentum, the B field. The imposition of the linear simplicity constraint (18) on Δ_2 reduces the symplectic form $\delta B^{IJ} \wedge \delta \omega_{IJ}$ to $\delta E^I \wedge \delta A_I$ where $A_I = \omega_{IJ} t^J + \gamma \frac{1}{2} \epsilon_{IJKL} \omega^{JK} t^L$ is the real Ashtekar connection and E^I its conjugate momentum, the dual electric field [44, 46, 47]. Accordingly, the Lorentz gauge group $SO(1,3)$ reduces to the rotation subgroup $SO(3)$ that leaves t^I invariant. We work with the covering groups $SL(2,\mathbb{C})$ and $SU(2)$.

The quantum theory is formulated in terms of gauge-invariant functionals of the real Ashtekar connection. Local flatness in $M' = M - \Delta_1$ imposes that the state $\Psi[A]$ depends only on coordinates on the moduli space $\mathrm{Hom}\big(\pi_1(M'), SU(2)\big)/SU(2)$. A convenient set of coordinates is obtained by noticing that the 2-complex $\mathcal{C} = \Delta^*$ induces on the boundary M a graph $\Gamma = (\Delta_3)^*$ consisting of N nodes n dual to the 3-cells Δ_3 and L links l dual to its 2-cells [56, 57]. We define the holonomy along links l of Γ as

$$h_l = P \exp \mathrm{i} \int_l \vec{A} \cdot \frac{\vec{\sigma}}{2} , \qquad (34)$$

and the electric flux through 2-cells t_l of Δ_2 as

$$\vec{L}_l = \int_{t_l} \vec{E} , \qquad (35)$$

where $A^I = (0, A^i)$ and $E^I = (0, E^i)$. As a result we recover the holonomy-flux algebra of loop quantum gravity [58] (*see Chapter 1*). The Hilbert space of states for given cellular decomposition Δ and boundary Δ_3 is

$$\mathcal{H}_\Gamma = L^2(SU(2)^L/SU(2)^N) , \qquad (36)$$

with states depending on the connection only via the holonomy, $\Psi[A] = \psi(h_l[A])$. An orthonormal basis $\mathcal{H}_\Gamma$ is provided by the spin-network basis. The basic ingredients are $SU(2)$ representation matrices

$$D^{(j)}(h)^m{}_{m'} = \langle j,m|U(h)|j,m'\rangle \tag{37}$$

and intertwiners $i^{m_1\cdots m_{N_n}}$, i.e. invariant tensors in the tensor product of $SU(2)$ representations. It is useful to have a physical picture of what intertwiners are as they play a central role in what follows. They can be understood as the collective state $|i\rangle$ of a system of N spins that is invariant under rotations. For instance, we can consider the ground state of the Hamiltonian $H = \vec{L}_{\text{tot}}^2$, where $\vec{L}_{\text{tot}} = \vec{L}_1 + \cdots + \vec{L}_N$ is the total spin of the system. The ground state of H is in general degenerate and, when $H = 0$, the associated eigenspace is the Hilbert space of intertwiners $\text{Inv}(V^{(j_1)} \otimes \cdots \otimes V^{(j_N)})$. An orthonormal basis $|i_k\rangle$ of this space defines a set of invariant tensors via the formula

$$|i_k\rangle = \sum_{m_1\cdots m_N} i_k^{m_1\cdots m_N} |j_1 m_1\rangle \cdots |j_N m_N\rangle. \tag{38}$$

Spin-network states are states in $\mathcal{H}_\Gamma$ determined by an assignment of a spin j_l to each link $l \in \Gamma$ and of intertwiner i_n to each node $n \in \Gamma$. In the holonomy representation they are given by

$$\psi_{j_l,i_n}(h_l) = \left(\bigotimes_{l\in\Gamma} D^{(j_l)}(h_l)\right) \cdot \left(\bigotimes_{n\in\Gamma} i_n\right) \tag{39}$$

where the dot $\cdot$ stands for contraction of tensors following the pattern of the graph Γ. More abstractly, we write $|\Gamma, j_l, i_n\rangle$ and $\psi_{j_l,i_n}(h_l) = \langle h_l|\Gamma, j_l, i_n\rangle$.

In loop quantum gravity the electric flux operator $\vec{L}_l$, equation (35) acts as a left-invariant vector field on functions of the holonomy. In spinfoam gravity, besides the electric flux $\vec{L}_l$, we have also a magnetic flux $\vec{K}_l$ defined by

$$\vec{K}_l = \int_{t_l} \vec{B} \tag{40}$$

where $B^I = (0, B^i)$, see equation (17). Together they provide a representation of the algebra of the Lorentz group, equation (24). The linear simplicity constraint (18) expressed in terms of electric and magnetic fluxes is

$$\vec{K}_l = \gamma \vec{L}_l. \tag{41}$$

Now we show how to embed the space of spin-networks in a larger space that carries a representation of $\vec{K}_l$, $\vec{L}_l$ and solves the linear simplicity constraint.

Unitary representation matrices of a $SL(2,\mathbb{C})$ group element G in representation $\mathcal{V}^{(p,j)}$ are denoted

$$D^{(p,j)}(G)^{j'm'}{}_{j''m''} = \langle (p,j), j', m'|U(G)|(p,j), j'', m''\rangle. \tag{42}$$

The map Y_γ, equation (31), allows us to associate to every function in $L^2(SU(2))$ a function on the Lorentz group via

$$D^{(j)}(h)^m{}_{m'} \;\to\; D^{(\gamma(j+1),j)}(G)^{jm}{}_{jm'}(G) = \langle j,m|Y_\gamma^\dagger U(G) Y_\gamma|j,m'\rangle. \tag{43}$$

In particular, we can map the spin-network basis $\psi_{j_l,i_n}(h_l)$ into functions on the Lorentz group,

$$\psi_{j_l,i_n}^{\gamma}(G_l) = \left(\bigotimes_{l\in\Gamma} D^{(\gamma(j+1),j)}(G_l)\right) \cdot \left(\bigotimes_{n\in\Gamma} Y_\gamma\, i_n\right) \qquad (44)$$

where we have defined the γ-simple intertwiners $Y_\gamma\, i_n$ via the components of $Y_\gamma|i_k\rangle$ of equation (38). Because of the result (33), matrix elements of $\vec{K}_l - \gamma \vec{L}_l$ always vanish on this basis of functions. Therefore they provide a solution of the linear simplicity constraint.

6. Spinfoam Partition Function and the Vertex Amplitude

The path integral over the B field for the topological field theory (6) simply imposes the flatness of the connection,

$$\int \mathcal{D}B^{IJ}\, e^{i\int B_{IJ}\wedge F^{IJ}} = \delta[F(\omega)]. \qquad (45)$$

In the presence of a cellular decomposition Δ we can focus on a region that contains a single 2-cell t_f. The delta function imposes that the holonomy of the connection along a path that encircles t_f vanishes. It is useful to use the language of the dual 2-complex $\mathcal{C} = \Delta^*$ introduced before. The closed path around a 2-cell is a sequence of edges e belonging to the boundary of a face f of the 2-complex. The delta function can then be written in terms of holonomies G_e along edges, equation (19) and can be expressed in terms of representations of the Lorentz group as

$$\mathcal{A}_f^{top}(G_e) = \delta\left(\prod_{e\in\partial f} G_e\right) = \sum_{k=0,\frac{1}{2},\dots}\int_0^\infty dp\, (p^2+k^2)\, \mathrm{Tr}\left(\prod_{e\in\partial f} D^{(p,k)}(G_e)\right). \qquad (46)$$

On the other hand for the spinfoam action (16), because of the simplicity constraint, it is not the case that the connection is imposed to be flat. Focusing again on a closed path that encircles a single *defect* $t_f \in \Delta_2$ we find a face amplitude

$$\mathcal{A}_f(G_e) = \sum_j (2j+1)\, \mathrm{Tr}\left(\prod_{e\in\partial f} Y_\gamma^\dagger\, D^{(\gamma(j+1),j)}(G_e)\, Y_\gamma\right). \qquad (47)$$

As discussed in (44) the map Y_γ provides a solution of the linear simplicity constraint. The path integral over the full manifold $\mathcal{M}$ is obtained as a product over the amplitudes $\mathcal{A}_f(G_e)$ associated to faces of the dual 2-complex, integrated over the edge holonomies. In the presence of a boundary we can hold fixed the holonomies G_l along links on the boundary and define the transition amplitude

$$W_\mathcal{C}(G_l) = \int dG_e \prod_f \mathcal{A}_f(G_e). \qquad (48)$$

The transition amplitude allows us to evolve a loop quantum gravity state once written in the form equation (44).

Using the spin-network basis we can express the transition amplitude in terms of a sum over spins associated to faces and intertwiners associated to edges of the 2-complex $\mathcal{C}$. The resulting expression is

$$W_{\mathcal{C}}(j_l, i_n) = \sum_{j_f} \sum_{i_e} \Big(\prod_f (2j_f + 1) \Big) \prod_v \big\{ Y_\gamma\, i_e \big\}_v \qquad (49)$$

where the only non-trivial term is the vertex amplitude [34]

$$\big\{ Y_\gamma\, i_e \big\}_v = \int \prod_{e=2}^{N_v} dG_e \bigotimes_{e=1}^{N_v} U(G_e) Y_\gamma i_e \qquad (50)$$

associated to vertices of the 2-complex. This completes the mathematical definition of the partition function and transition amplitude for spinfoam gravity.

7. Cellular Quantum Geometry: A Single Atom of Space

At the classical level the electric part of the field B^{IJ} endows the cellular decomposition with a geometric interpretation. In this section we describe the phase space of convex polyhedra, its quantization in terms of $SU(2)$ intertwiners and the relation to the constrained topological quantum field theory defined by the spinfoam action.

7.1. *Minkowski Theorem and the Phase Space of Polyhedra*

A tetrahedron is the convex hull of four points in 3-dimensional Euclidean space $\mathbb{R}^3$. Its geometry can be described using a *triad* of edge-vectors $\vec{e}_i$ ($i = 1, 2, 3$). For instance, the volume of the tetrahedron is given by $V = \frac{1}{3!} |\vec{e}_1 \cdot (\vec{e}_2 \times \vec{e}_3)|$. From the triad we can compute the normal to the plane supporting a face of the tetrahedron, for instance $\vec{L}_3 = \frac{1}{2} \vec{e}_1 \times \vec{e}_2$. The normals $\vec{L}_a$, with $a = 1, 2, 3, 4$, are normalized to the area A_a of the associated face and can be chosen to be outward-pointing. Notice that they sum up to zero, as it happens for any closed surface.

A remarkable property of the face-normals is that they can be used as fundamental variables: a set of four vectors $\vec{L}_a$ satisfying the closure condition

$$\vec{L}_1 + \vec{L}_2 + \vec{L}_3 + \vec{L}_4 = 0 \qquad (51)$$

completely describes the geometry of a tetrahedron.[a] The norm of the vector $\vec{L}_a$ is the area of the face a of the tetrahedron, so that we can write

$$\vec{L}_a = A_a \vec{n}_a \qquad (52)$$

where $\vec{n}_a$ is the unit outward-pointing normal to the face a. The scalar product

$$\vec{L}_a \cdot \vec{L}_b = A_a A_b \cos \theta_{ab} \qquad (53)$$

---

[a] A counting of the number of independent variables up to rotations is in order: we have 4×3 vector components, -3 components from the closure condition, -3 rotations, equals 6 independent variables. This number matches the number of edge-lengths that one can use to describe a tetrahedron.

measures the dihedral angle θ_{ab} between two faces of the tetrahedron. Similarly, the triple product of any three normals is related to the volume of the tetrahedron by the formula

$$V = \frac{\sqrt{2}}{3}\sqrt{|\vec{L}_1 \cdot (\vec{L}_2 \times \vec{L}_3)|}. \tag{54}$$

The space of tetrahedra with faces of fixed areas A_a has a remarkable property: it has the structure of a phase space with respect to a natural choice of rotationally-invariant Poisson brackets induced by the canonical brackets $\{n^i, n^j\} = \varepsilon^{ij}{}_k\, n^k$ for unit vectors on the sphere. Consider two functions $f(\vec{L}_a)$ and $g(\vec{L}_a)$ on the space of shapes of tetrahedra with faces of fixed areas A_a. The Poisson brackets

$$\left\{f(\vec{L}_a), g(\vec{L}_a)\right\} = \sum_{a=1}^{4} \vec{L}_a \cdot \left(\frac{\partial f}{\partial \vec{L}_a} \times \frac{\partial g}{\partial \vec{L}_a}\right) \tag{55}$$

make this space into a phase space. The phase space of a tetrahedron with fixed areas is 2-dimensional and a set of canonical variables $\{q, p\} = 1$ is given by

$$q = \text{angle between } \vec{L}_1 \times \vec{L}_2 \text{ and } \vec{L}_3 \times \vec{L}_4, \tag{56}$$
$$p = |\vec{L}_1 + \vec{L}_2|. \tag{57}$$

Every geometric property of the tetrahedron, such as the volume V, can be understood as a function of q and p.

There are two elegant mathematical results that allow us to extend the previous construction from tetrahedra to convex polyhedra in 3-dimensional Euclidean space. The first result is a theorem of Minkowski's [59] that states that the areas A_a and the unit-normals $\vec{n}_a$ to the faces of the polyhedron fully characterize its shape,[b] see Figure 2. We define the vectors $\vec{L}_a = A_a\, \vec{n}_a$ and call $\mathcal{P}_N$ the *space of shapes of polyhedra* with N faces of given areas A_a,

$$\mathcal{P}_N = \left\{\vec{L}_a, a = 1\,..\,N \,|\, \vec{L}_1 + \cdots + \vec{L}_N = 0,\ \|\vec{L}_a\| = A_a\right\}/SO(3). \tag{58}$$

The second is a result of Kapovich and Millson's that states that the set $\mathcal{P}_N$ has naturally the structure of a *phase space* [60]. The Poisson brackets between two functions $f(\vec{L}_a)$ and $g(\vec{L}_a)$ on $\mathcal{P}_N$ are

$$\{f, g\} = \sum_{a=1}^{N} \vec{L}_a \cdot \left(\frac{\partial f}{\partial \vec{L}_a} \times \frac{\partial g}{\partial \vec{L}_a}\right). \tag{59}$$

As in the case of the tetrahedron, these brackets arise (via symplectic reduction) from the rotationally-invariant Poisson brackets between functions $f(\vec{L}_a)$ on $(S^2)^N$. Thus we have that convex polyhedra with N faces of given areas form a $2(N-3)$ dimensional phase space [53].

---

[b]More precisely, given a set of N positive numbers A_a, and N unit-vectors $\vec{n}_a$ satisfying the condition $\sum_a A_a \vec{n}_a = 0$, there always exists a convex polyhedron having these data as areas and normals to its faces. Moreover, up to rotations $SO(3)$, the polyhedron is unique.

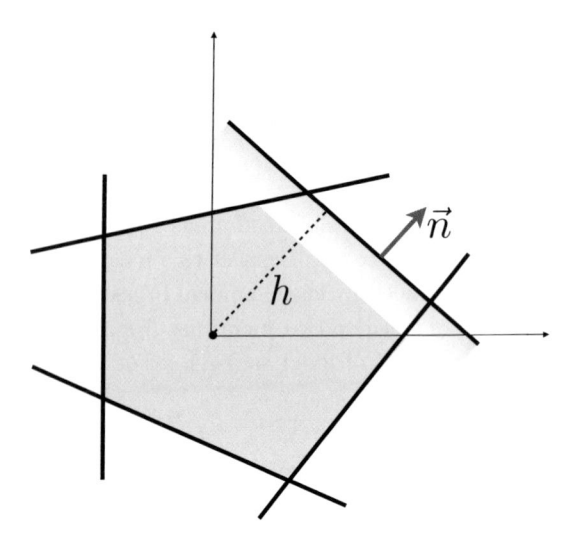

Fig. 2. A convex polyhedron can be obtained starting from N planes passing through the origin of 3d Euclidean space. Moving each plane in the direction of its normal $\vec{n}_a$ defines a convex hull, a polyhedron. Adjusting the distance h_a of a plane from the origin changes the areas A_a of the faces of the polyhedron. This procedure can be used to build a polyhedron with faces of given areas A_a satisfying the closure condition $\sum_a A_a \vec{n}_a = 0$. The Minkowski theorem states that such polyhedron exists and, up to rotations, is unique. A variational algorithm to reconstruct the polyhedron can be found in [53].

Canonical variables on this phase space can be chosen as follows: consider the set of vectors $\vec{p}_i = \sum_{a=1}^{i+1} \vec{L}_a$, where $i = 1, \cdots, N-3$; we define the coordinate q_i as the angle between the vectors $\vec{p}_i \times \vec{L}_{i+1}$ and $\vec{p}_i \times \vec{L}_{i+2}$, and the momentum variable $p_i = \|\vec{p}_i\|$ as the norm of the vector $\vec{p}_i$. From (59), it follows that these are canonically conjugate variables, $\{q_i, p_j\} = \delta_{ij}$.

Every geometric quantity, e.g. the length of an edge or the volume of the polyhedron [61], is a function of the canonical variables (q_i, p_i). The problem of determining this function is well-defined but not immediate to solve. The reason is that we have to reconstruct first the shape of the polyhedron from the normals to its faces, or equivalently from the point in phase space [53]. In general, the problem can be solved numerically using Lasserre's algorithm [62]. In the case of a pentahedron $N=5$ the problem has been solved analytically, and an expression of the volume $V(q_1, q_2, p_1, p_2)$ as a function in phase space is available [63]. It is interesting to notice that the classical dynamics of this system is strongly chaotic [63, 64].

7.2. *Spin-geometry and Quantum Polyhedra*

In quantum mechanics a spin system identifies a quantum direction in space. A remarkable idea proposed by Penrose in 1971 is that the angles between these quantum directions define a geometry, Penrose's 'spin-geometry', and can provide the elementary building-block of quantum space [65–68]. This model for an *atom of space* has later been shown to coincide with the notion of 'quantum polyhedron', the quantization of the classical system described in the previous section [53, 69–71].

Consider the simple quantum mechanical system described by an intertwiner $|i\rangle \in \text{Inv}(V^{j_1} \otimes \cdots \otimes V^{j_N})$, equation (38). The observables of the system are the rotational invariant operators that can be built from the angular momenta $\vec{L}_a$ only ($a = 1, \ldots, N$). As the state $|i\rangle$ of our system is rotationally invariant,[c] we have

$$(\vec{L}_1 + \cdots + \vec{L}_N)|i\rangle = 0, \qquad (60)$$

a quantum closure condition analogous to equation (51). The geometric interpretation of the observables $\vec{L}_a$ comes from identifying them with the normals to planes passing through a point in 3-dimensional Euclidean space. In particular, the Penrose 'metric' operator is given by the scalar product $\vec{L}_a \cdot \vec{L}_b$, up to a scale with dimensions of $length^4$. This scale can be fixed by making reference to equation (13): when the linear simplicity constraint is (18) is satisfied, we have that the 2-form built out of the tetrad can be written in terms of the electric part of the B^{IJ} field,

$$\frac{1}{2}\epsilon_{IJKL}e^J \wedge e^K t^L = 8\pi G_N \gamma E_I. \qquad (61)$$

When integrated over a 2-cell of Δ_2, equation (35), we find that the integrated area-element is $8\pi G_N \gamma \vec{L}_a$. We define the Penrose operator by

$$\hat{g}_{ab} = (8\pi G_N \gamma)^2 \vec{L}_a \cdot \vec{L}_b. \qquad (62)$$

It measures the angle [68] θ_{ab} between the a-planes and the b-plane via the formula $\theta_{ab} = \arccos(\hat{g}_{ab}/\sqrt{\hat{g}_{aa} \hat{g}_{bb}})$. A basis of intertwiner states $|i_{k_1 \cdots k_{N-3}}\rangle$ is a basis of eigenstates of a maximal commuting set of operators $\hat{g}_{ab}$. For instance, the spectrum of $\hat{g}_{12}$ is given by

$$\hat{g}_{12}|i_{k_1 \cdots k_{N-3}}\rangle = (8\pi G_N \gamma)^2 \frac{k_1(k_1+1) - j_1(j_1+1) - j_2(j_2+1)}{2} |i_{k_1 \cdots k_{N-3}}\rangle. \qquad (63)$$

As explained in the previous section, the point of intersection of the N planes can be inflated into a polyhedron by moving the planes away from the origin. This defines a polygon on each plane, i.e. a face of the polyhedron. The norm of the operator $\vec{L}_a$ measures the area A_a of this face

$$A_a = \sqrt{\hat{g}_{aa}} = 8\pi G_N \gamma \sqrt{\vec{L}_a \cdot \vec{L}_a}. \qquad (64)$$

Its eigenvalues are immediate to compute and every state $|i\rangle$ in our Hilbert space is an eigenstates of the area operator

$$A_a |i\rangle = 8\pi G_N \gamma \sqrt{j_a(j_a+1)} |i\rangle. \qquad (65)$$

---

[c]The proof is immediate: a finite rotation of the system is generated by the unitary operator $U(\vec{\alpha}) = \exp(i\vec{\alpha} \cdot \vec{L}_{\text{tot}})$; the invariance of the state under rotations is $U(\vec{\alpha})|i\rangle = |i\rangle$ for all rotation parameters $\vec{\alpha}$. Expanding at the linear order in small $\vec{\alpha}$ one recovers the closure condition.

The spectrum of the area is discrete and gapped, with a Planck scale gap $a_0 = 8\pi G_N \gamma \sqrt{3}/2$ corresponding to the minimum non-trivial spin $j_a = 1/2$. This formula matches the spectrum of the area operator found in canonical loop quantum gravity [67, 72–74].

The system that we have described plays the role of atom of space in spinfoam gravity. It can be understood as a quantum polyhedron as it can be obtained by quantizing a classical dynamical system: a convex polyhedron with canonical Poisson brackets. This is analogous to the case of the hydrogen atom, a purely quantum system that can be defined via the quantization of a classical particle in a Keplerian orbit. In the next section we discuss coherent states and the semiclassical behavior of the quantum system.

7.3. *Heisenberg Uncertainty Relations for Quantum Geometry*

Different components of the angular momentum do not commute $[L^i, L^j] = i\,\varepsilon^{ij}{}_k\, L^k$. As a result the dispersions ΔL^i on any spin state satisfy the uncertainty relations[d]

$$\Delta L^i \, \Delta L^j \geq \frac{1}{2}\left|\epsilon^{ij}{}_k \langle L^k \rangle\right|. \tag{66}$$

This is also the behavior of the quantum directions $\vec{L}_a$: the atom of space has a non-commutative quantum geometry. The phenomenon is most clearly illustrated in terms of the Penrose metric operator $\hat{g}_{ab}$. Consider three quantum planes a, b, c, and the angle between a, b and a, c. The associated Penrose operators do not commute

$$[\hat{g}_{ab}, \hat{g}_{ac}] = i\,(8\pi G_N \gamma)^4\, \vec{L}_a \cdot (\vec{L}_b \times \vec{L}_c) \tag{67}$$

and the commutator measures the linear independence of the three quantum planes. The Heisenberg uncertainty relation for the quantum geometry reads

$$\Delta \hat{g}_{ab} \, \Delta \hat{g}_{ac} \geq \frac{1}{2}\,(8\pi G_N \gamma)^4 \left|\langle \vec{L}_a \cdot (\vec{L}_b \times \vec{L}_c) \rangle\right|. \tag{68}$$

As a result the shape of a quantum polyhedron is fuzzy: if we try to determine with precision the angle between the planes a and b, then we lose control of the angle between the planes a and c unless the three are coplanar.

At the classical level we saw that q_i and p_i are canonical variables on the phase space of polyhedra. At the quantum level they correspond to operators with canonical commutation relations, $[\hat{q}_i, \hat{p}_j] = i\,\delta_{ij}$. Their dispersions satisfy uncertainty relations that are simpler than the ones we have seen above

$$\Delta \hat{q}_i \, \Delta \hat{p}_i \geq \frac{1}{2}. \tag{69}$$

The geometric interpretation is particularly clear in the case of the tetrahedron ($N = 4$): the states $|i_k\rangle$ that we use as a basis of the Hilbert space have definite

---

[d] As usual $\Delta A \equiv \sqrt{\langle A^2 \rangle - \langle A \rangle^2}$, where $\langle A \rangle \equiv \langle s|A|s \rangle$ is the expectation value of the operator A on the state $|s\rangle$.

angle between two faces of the tetrahedron, i.e. $\Delta \hat{p}_i = 0$; as a result the angle $\hat{q}_i$ between two opposite edges of the tetrahedron has maximal dispersion.

Coherent states provide an over-complete basis of the Hilbert space of the quantum polyhedron such that the uncertainty relations (69) are saturated [75, 76]. The simplest way to introduce them is to start from Bloch coherent states for a spin system, i.e. states that saturate the uncertainty relation (66) [77, 78]. The state $|j, j\rangle$ pointing in the z direction has $\langle L_z \rangle = j$, $\langle L_x \rangle = \langle L_y \rangle = 0$, and $\Delta L_z = 0$, $\Delta L_x = \Delta L_y = \sqrt{j/2}$. As a result it saturates the uncertainty relation $\Delta L_x \Delta L_y = \frac{1}{2}\langle L_z \rangle$. A coherent spin pointing in a different direction can be obtained by simply rotating the state $|j, j\rangle$ in the direction $\vec{n}$,

$$|j, \vec{n}\rangle \equiv U(R)|j, j\rangle = \sum_{m=-j}^{+j} \phi_m(\vec{n}) |j, m\rangle, \qquad (70)$$

where R is a rotation from the z direction to $\vec{n}$ and $\phi_m(\vec{n}) = \langle j, m | U(R) | j, j \rangle$. This state points in the direction $\vec{n}$,

$$\langle \vec{L} \rangle = j \vec{n} \qquad (71)$$

and has dispersion $\Delta(\vec{L} \cdot \vec{n}') = \sqrt{\frac{1-(\vec{n}\cdot\vec{n}')^2}{2}j}$. As a result, the relative dispersion vanishes in the limit of large spin,

$$\frac{\Delta(\vec{L} \cdot \vec{n}')}{|\langle \vec{L} \rangle|} \to 0 \quad \text{as} \quad j \to \infty. \qquad (72)$$

Moreover this class of states provides a resolution of the identity in $V^{(j)}$,

$$\mathbb{1}_j = \frac{2j+1}{4\pi} \int_{S^2} d\vec{n} \, |j, \vec{n}\rangle\langle j, \vec{n}|. \qquad (73)$$

Coherent states for quantum polyhedra $|i(\vec{n}_a)\rangle$, also called coherent intertwiners, are defined as the rotational invariant projection of N coherent spins $|j_a, \vec{n}_a\rangle$ satisfying the closure constraint $\sum_a j_a \vec{n}_a = 0$. Their explicit expression is [30]

$$|i(\vec{n}_a)\rangle = \sum_m \Phi_{m_1 \cdots m_N}(\vec{n}_a) |j_1, m_2\rangle \cdots |j_N, m_N\rangle, \qquad (74)$$

with

$$\Phi_{m_1 \cdots m_N}(\vec{n}_a) = \int_{SU(2)} dh \prod_{a=1}^{N} \langle j_a, m_a | U(h) | j_a, \vec{n}_a \rangle. \qquad (75)$$

These states are peaked on the polyhedron with normals $\vec{n}_a$, and the relative dispersion of geometric observables vanish in the limit $j_a \to \infty$: the classical limit arises at large quantum numbers, i.e. large spins. This regime corresponds to a size of the polyhedron that is large compared to the Planck scale, for instance the area of a face being much larger than the area gap $a_0 = 8\pi G_N \gamma \sqrt{3}/2$. Formally large j corresponds to the limit $8\pi G_N \gamma \to 0$ while keeping fixed the physical area of a face. In this limit the Heisenberg uncertainty relations (68) become trivial.

The shape of a classical polyhedron is completely coded in the canonical variables (q_i, p_i) on phase space, in particular the normals $\vec{n}_a$ can be computed from them.

It is useful to write the coherent states as functions on phase space $|i(q_i, p_i)\rangle \equiv |i(\vec{n}_a(q_i, p_i))\rangle$. The resolution of the identity on the Hilbert space of quantum polyhedra can then be written as an integral on phase space as [75]

$$\mathbb{1} = \int_{\mathcal{P}_N} d\mu(q_i, p_i) \, |i(q_i, p_i)\rangle\langle i(q_i, p_i)| \, . \tag{76}$$

This formula shows that we can write any quantum state of an atom of space as a superposition of coherent quantum polyhedra.

It is interesting to connect these recent developments with the original idea proposed by Penrose [65, 66]. The spin-geometry theorem states that there exist collective spin states such that, in the classical limit, the expectation value of the Penrose metric reproduces the scalar products of a set of N vectors $\vec{v}_a$ in 3d Euclidean space, $\langle \hat{g}_{ab} \rangle = \vec{v}_a \cdot \vec{v}_b$, and the relative dispersions vanish. The coherent states for an atom of space discussed above provide a concrete example of such states and play a central role in spinfoam gravity.

8. Cellular Quantum Geometry: Coherent Spin-networks

The atoms of space discussed in the previous section arise naturally as building blocks of boundary states in spinfoam gravity, see Section 5.

Consider a graph Γ consisting of N nodes connected by L links. To each link ll we associate a state $|j_l, m_l\rangle|j_l, m'_l\rangle$ of two spins with the same j_l. Each of the two spins lives at an endpoint of the link. The Hilbert space of the system is simply given by a tensor product over the links of factors $V^{(j_l)} \otimes V^{(j_l)}$. Equivalently we can organize the spins in groups sitting at nodes n of the network, $\bigotimes_{l|n \in \partial l} V^{(j_l)}$. We are interested in configurations of this system such that the spins sitting at each node are in a rotationally-invariant state. These states form a Hilbert space:

$$\mathcal{H}_{\Gamma, j_l} = \bigotimes_{n \in \Gamma} \mathcal{H}_n \tag{77}$$

with $\mathcal{H}_n = \mathrm{Inv}(\otimes_{l|n \in \partial l} V^{(j_l)})$: the system consists of an atom of space at each node of the graph. We can describe the state of the system in terms of the states i_n of each atom of space at a node n, with the graph Γ coding which nodes are connected by a link and thus share the same spin j_l. The state $|\Gamma, j_l, i_n\rangle$ that we have described is a spin-network state and it belongs to the loop quantum gravity Hilbert space $\mathcal{H}_\Gamma$ associated to the boundary of the cellular decomposition Δ, equation (36),

$$\mathcal{H}_\Gamma = \bigoplus_{\{j_l\}} \bigotimes_{n \in \Gamma} \mathcal{H}_n \, . \tag{78}$$

The geometric picture arising from this construction is that a spin-network state can be thought of as a collection of quantum polyhedra. Adjacent polyhedra share a face with matching areas, but not necessarily matching shapes. This structure is called a twisted geometry [79, 80].

Given a Riemannian metric q_{ab} on a 3-manifold M equipped with a cellular decomposition Δ_3, we can consider the discretization of the metric on Δ_3. This procedure generalizes the notion of triangulation to more general building-blocks. Conversely, we can consider the manifold with defects $M' = M - \Delta_1$ and the class of metrics q'_{ab} that have vanishing Riemann tensor everywhere in the bulk of M' and induce a linear structure on Δ_1 so that its elements are geodesics that support the curvature. The smooth metric q_{ab} on M can now be thought of as a large-scale approximation of one of these piecewise-flat metrics. A piecewise-flat metric on the cellular manifold (M, Δ_3) is fully determined by the description of the polyhedral geometry of each cell, the adjacency relation of the cells and matching conditions for the shape of interfaces between cells. This set of data corresponds to a dual graph $\Gamma = (\Delta_3)^*$, an assignment of areas j_l to the links of the graph and an assignment of unit-vectors $\vec{n}_{s_l}, \vec{n}_{t_l}$ corresponding to normals to faces that satisfy the closure and shape-matching conditions. The spin-network state

$$\psi_{j_l, i_n(\vec{n}_{s_l})}(h_l) = \left(\bigotimes_{l \in \Gamma} D^{(j_l)}(h_l) \right) \cdot \left(\bigotimes_{n \in \Gamma} i_n(\vec{n}_{s_l}) \right) \qquad (79)$$

$$= \int_{SU(2)} \left(\prod_{n \in \Gamma} dg_n \right) \prod_{l \in \Gamma} \langle j_l, \vec{n}_{t_l} | U(g_{s_l} h_l g_{t_l}) | j_l, \vec{n}_{s_l} \rangle \cdot \qquad (80)$$

with coherent intertwiners $i_n(\vec{n}_{s_l})$, equation (74), is peaked exactly on this piecewise-flat geometry.

In general relativity initial data on a spatial slice M correspond to the assignment of a Riemannian metric q_{ab} describing the intrinsic geometry of M and a symmetric tensor p^{ab}, the extrinsic curvature of M into the 4-manifold $\mathcal{M} = M \times \mathbb{R}$. In the canonical formulation these two quantities are canonical conjugate variables. In the presence of a cellular decomposition Δ, the extrinsic curvature can be discretized as done for the metric: the piecewise-flat version of the extrinsic curvature p^{ab} corresponds to an assignment of time-like four-vectors t_n^I, one per cell of the decomposition Δ_3. Given two nearby cells n and n' we can compute the Lorentzian dihedral angle

$$\Theta_{nn'} = \frac{-\eta_{IJ} t_n^I t_{n'}^J}{\sqrt{(-\eta_{KL} t_n^K t_n^L)(-\eta_{K'L'} t_{n'}^{K'} t_{n'}^{L'})}}. \qquad (81)$$

This boost angle ranges in $(-\infty, +\infty)$ and measures the extrinsic curve between the two neighboring cells. A coherent spin-network is peaked both on the intrinsic geometry of (M, Δ_3) and on its extrinsic geometry. A useful Gaussian ansatz for a coherent spin-network is

$$\psi_{j_l, \Theta_l, i_n(\vec{n}_{s_l})}(h_l) = \sum_{j_l} e^{-\frac{1}{2} \sum_{ll'} M_{ll'} (j_l - j_l^0)(j_{l'} - j_{l'}^0)} e^{i \sum_l \Theta_l j_l} \psi_{j_l, i_n(\vec{n}_{s_l})}(h_l) \qquad (82)$$

where $M_{ll'}$ is a positive-definite matrix that encodes correlations [81]. This ansatz is expected to provide an effective description of coherent spin-network states peaked on large spins $j_l^0 \gg 1$. Heat-kernel coherent states provide an elegant description of states peaked on the intrinsic and the extrinsic geometry of a cellular decomposition [82, 83]. They reduce to the Gaussian expression (82) in the appropriate limit. The matrix $M_{ll'}$ results to be diagonal, therefore they lack long-range correlations. Proposals of coherent spin-networks with long-range correlations are currently being studied.

9. Vertex-amplitude Asymptotics and Regge Gravity

The spinfoam dynamics is fully coded into the vertex amplitude $\left\{Y_\gamma i_e\right\}_v$, equation (50). It describes the dynamics of a single 4-cell of Δ. To understand its role it is useful to compute the amplitude it associates to a semiclassical boundary geometry.

The boundary of a 4-cell σ consists in a collection of 3-cells. We denote $v = \sigma^*$ the spinfoam vertex dual to the cell and Γ_v the graph that codes the adjacency relations between the 3-cells on the boundary of σ, i.e. $\Gamma_v = (\partial\sigma)^*$. We denote n the N nodes of Γ_v. To the graph Γ_v we associate a boundary state $\psi_{\Gamma_v, j_l, i_n}(h_l)$ and compute its amplitude.

We choose the boundary state to describe the classical geometry of the boundary of a polyhedron in 4-dimensional Minkowski space. The Lorentzian version of the Minkowski theorem of Section 7.1 [84] states that, given a set of time-like vectors t_n^I satisfying the closure condition $\sum_{n=1}^N t_n^I = 0$, there is a unique convex Lorentzian polyhedron σ with 3-dimensional faces of volume $V_n = \sqrt{-\eta_{IJ}t_n^I t_n^J}$ and unit normal $\hat{t}_n^I = V_n^{-1}t_n^I$. The boundary of the Lorentzian polyhedron consists of a collection of N Euclidean polyhedra. We call N_- and N_+ the number of past-pointing and future-pointing time-like vectors and say that the 4-dimensional polyhedron σ describes the transition from N_- to N_+ Euclidean polyhedra. The 4-dimensional Lorentzian geometry of the polyhedron σ induces an assignment of areas A_l and 3-normals $\vec{n}_{s_l}$ of the boundary polyhedra. We use these data as labels for the boundary state.

The vertex amplitude $\mathcal{A}_v(j_{ab}, i_a(\vec{n}_{ab}))$ for this boundary state is simply given by the evaluation of equation (50) on the set of intertwiners $i_n(\vec{n}_{s_l})$,

$$\mathcal{A}_v(j_l, i_n(\vec{n}_{s_l})) = \left\{Y_\gamma i_e\right\}_v = \int \prod_{n=2}^N dG_n \prod_{l\in\Gamma_v} \langle j_l, \vec{n}_{s_l}|Y_\gamma^\dagger U(G_{s_l}G_{t_l}^{-1})Y_\gamma|j_l, \vec{n}_{t_l}\rangle. \tag{83}$$

This expression has a remarkable asymptotic behavior in the limit of large spins j_l. Consider a uniform rescaling $j_l \to \lambda j_l$. In the limit $\lambda \gg 1$ the following asymptotic formula holds [85, 86]

$$\mathcal{A}_v(\lambda j_l, i_n(\vec{n}_{s_l})) = \frac{1}{\mathcal{N}\lambda^{12}} e^{+i\lambda S} + \text{c.c.} \tag{84}$$

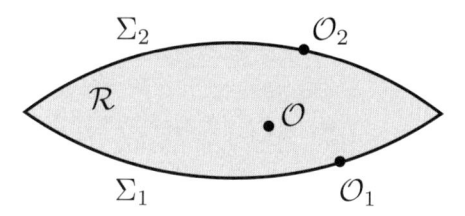

Fig. 3. A spacetime region $\mathcal{R}$ with space-like boundary $\Sigma_1 \cup \Sigma_2$. The operators $\mathcal{O}$, $\mathcal{O}_1$, $\mathcal{O}_2$ are bulk and boundary observables.

where

$$S = \sum_{l \in \Gamma_v} \gamma j_l \Theta_l = \frac{1}{8\pi G_N} \sum_{l \in \Gamma_v} A_l \Theta_l. \tag{85}$$

is the action Regge gravity [5] for a 4-dimensional polyhedron and we have used the expression $A_l = 8\pi G_N \gamma j_l$ for the asymptotic behavior of the area spectrum, equation (65). This formula provides a generalization of the Ponzano-Regge formula [4] (1) to 4-dimensional Lorentzian general relativity. Notice how the Barbero-Immirzi parameter γ has disappeared from the final asymptotic formula, consistently with the independence of classical Einstein equations from this parameter, equation (10).

10. Reconstructing a Semiclassical Spacetime

In Section 8 we have discussed coherent spin-network states. These states are peaked on the intrinsic and extrinsic geometry of a cellular decomposition of space. In this section we discuss how to reconstruct a semiclassical spacetime in spinfoam gravity.

Consider a 4-manifold $\mathcal{M}$ with a Lorentzian metric $g_{\mu\nu}$. We are interested in a finite spacetime region $\mathcal{R}$ with a purely spatial boundary $\Sigma = \partial \mathcal{R} = \Sigma_1 \cup \Sigma_2$. The metric $g_{\mu\nu}$ induces an intrinsic and an extrinsic geometry on the past and the future boundaries denoted Σ_1 and Σ_2, Figure 3. Here we are interested in the converse problem of reconstructing the geometry in the bulk of $\mathcal{R}$ from data on the boundary.

A cellular decomposition Δ of the region $\mathcal{R}$ induces a cellular decomposition Δ_3 of its boundary. The dual description consists in a 2-complex $\mathcal{C} = \Delta^*$ and a graph $\Gamma = (\Delta_3)^*$ associated to its boundary. In spinfoam gravity we associate a state $\psi_\Gamma^\gamma(G_l)$ to the boundary graph, equation (44) and an amplitude $W_\mathcal{C}(G_l)$ to the 2-complex, equation (48) [87, 88]. It is useful to think of the spinfoam amplitude as a linear functional on boundary states and adopt the bra-ket notation

$$\langle W_\mathcal{C} | \psi_\Gamma \rangle = \int dG_l \, W_\mathcal{C}(G_l) \, \psi_\Gamma^\gamma(G_l). \tag{86}$$

The expectation value of a boundary operator $\mathcal{O}_1$, such as the Penrose operator $\hat{g}_{ab}$ of equation (62), is given by

$$\langle \mathcal{O}_1 \rangle = \frac{\langle W_\mathcal{C} | \mathcal{O}_1 | \psi_\Gamma \rangle}{\langle W_\mathcal{C} | \psi_\Gamma \rangle}. \tag{87}$$

A semiclassical boundary state as the one discussed in equation (82) has the property of being peaked on a classical intrinsic and extrinsic geometry of the boundary as measured by boundary operators.

The spinfoam path-integral provides a sum over quantum spacetime geometries in $\mathcal{R}$. These geometries can be probed by computing the expectation values of bulk operators $\mathcal{O}$, Figure 3. We say that the boundary state determines a semiclassical spacetime when the expectation values of the bulk operators satisfy the equations of motion of the classical spinfoam gravity action (16), i.e. a discrete version of Einstein equations, and have small dispersion.

Current studies of the reconstruction of a classical spacetime geometry from spinfoam gravity have focused on saddle-point methods: a semiclassical boundary state determines the dominating contributions of the spinfoam action and the evaluation of expectation values of observables is given by the saddle point of the action [89–95]. The asymptotic result (84) falls in this class [85, 86]. Similarly, in spinfoam cosmology [96–98] one considers boundary states peaked on homogeneous 3-geometries and shows that a semiclassical cosmological spacetime solving Friedmann equations is reconstructed in the bulk.

A semiclassical spacetime consists of a classical configuration of the geometry together with quantum fluctuations. Correlations of fluctuations can be probed by computing for instance the connected 2-point function of Penrose metric operators

$$\mathcal{C}_{abcd} = \langle \hat{g}_{ab}\, \hat{g}_{cd} \rangle - \langle \hat{g}_{ab} \rangle \langle \hat{g}_{cd} \rangle \,. \tag{88}$$

These correlations have been studied using the saddle-point approximation and shown to match the graviton propagator of perturbative quantum gravity [99–107].

Going beyond the saddle-point approximation requires new methods. A natural tool is the numerical implementation of the spinfoam sum [108], equation (49), together with renormalization group techniques [109, 110] (*see Chapter 5.*) Note that, for fixed cellular decomposition Δ, the spinfoam path integral has only a finite (although large) number of degrees of freedom. Moreover the theory has no ultraviolet divergences because of the discrete nature of the spectrum of $SU(2)$ representations appearing in the spinfoam sum (49). The technical problem is similar to the one of lattice gauge theory, but the physical picture is rather different: the discreteness of the theory is not the artifact of a lattice regularization but a prediction of the cellular quantum geometry. The quantization of electric fluxes reflects itself into the discrete spectrum of the area operator with a Planck scale gap $a_0 = 8\pi G_N \gamma \sqrt{3}/2$. Note that, while there are no ultraviolet divergences, the theory defined in this chapter has infrared divergences corresponding to configurations with large spins j_l [111]. In the presence of a positive cosmological constant the theory is also infrared finite [35–38]. Moreover, the theory described in this chapter concerns only the spinfoam dynamics at fixed cellular decomposition Δ, or equivalently at fixed 2-complex $\mathcal{C} = \Delta^*$. The inclusion of a sum over 2-complexes is a natural extension and a needed one in order to match the full Hilbert space of loop quantum gravity. Group field theory [112, 113] provides a natural assignment of weights to different 2-complexes together with a Feynman diagrammatic scheme (*see Chapter 4*). Numerical techniques developed in the framework of causal dynamical triangulations [114] are also expected to be relevant, although this direction is yet to be explored.

11. Conclusions

Spinfoam gravity has its roots in the formulation of general relativity as a topological field theory with constraints. It is a theory with a finite but large number of degrees of freedom associated with Wilson loops of the Lorentz connections around 2-dimensional defects in a cellular decomposition of a 4-manifold. Remarkably, the theory provides a covariant formulation of canonical loop quantum gravity and equips the quantum geometry of loop quantum gravity with a spacetime picture. Moreover, spinfoams bring into loop quantum gravity techniques previously developed in the context of Regge's discrete approach to gravity, at the classical and the quantum levels. These techniques have been used to show how a semiclassical spacetime can be reconstructed from a sum over quantum geometries and provide an avenue for studying the dynamics of loop quantum gravity beyond the semiclassical regime.

References

[1] A. Perez, "The spin foam approach to quantum gravity," *Living Rev. Rel.* **16** (2013) 3, 1205.2019.
[2] C. Rovelli and F. Vidotto, *Covariant Loop Quantum Gravity*. Cambridge Monographs on Mathematical Physics (Cambridge University Press, (2014)).
[3] L. Landau and E. Lifshits, *Quantum Mechanics Non-Relativistic Theory* (vol. 3, first ed., Pergamon, UK, (1958)).
[4] G. Ponzano and T. Regge, "Semiclassical limit of Racah coefficients," *Spectroscopic and Group Theoretical Methods in Physics*, ed. F. Block (North Holland, Amsterdam, (1968)).
[5] T. Regge, "General relativity without coordinates," *Nuovo Cim.* **19** (1961) 558–571.
[6] T. Regge and R. M. Williams, "Discrete structures in gravity," *J. Math. Phys.* **41** (2000) 3964–3984, gr-qc/0012035.
[7] C. W. Misner, "Feynman quantization of general relativity," *Rev. Mod. Phys.* **29** (1957) 497–509.
[8] S. W. Hawking, "Quantum Gravity and Path Integrals," *Phys. Rev. D* **18** (1978) 1747–1753.
[9] M. P. Reisenberger and C. Rovelli, "'Sum over surfaces' form of loop quantum gravity," *Phys. Rev. D* **56** (1997) 3490–3508, gr-qc/9612035.
[10] J. W. Barrett and L. Crane, "Relativistic spin networks and quantum gravity," *J. Math. Phys.* **39** (1998) 3296–3302, gr-qc/9709028.
[11] F. Markopoulou and L. Smolin, "Causal evolution of spin networks," *Nucl. Phys. B* **508** (1997) 409–430, gr-qc/9702025.
[12] J. C. Baez, "Spin foam models," *Class. Quant. Grav.* **15** (1998) 1827–1858, gr-qc/9709052.

[13] L. Freidel and K. Krasnov, "Spin foam models and the classical action principle," *Adv. Theor. Math. Phys.* **2** (1999) 1183–1247, hep-th/9807092.

[14] J. W. Barrett and L. Crane, "A Lorentzian signature model for quantum general relativity," *Class. Quant. Grav.* **17** (2000) 3101–3118, gr-qc/9904025.

[15] J. C. Baez, "An Introduction to spin foam models of quantum gravity and BF theory," *Lect. Notes Phys.* **543** (2000) 25–94, gr-qc/9905087.

[16] R. De Pietri, L. Freidel, K. Krasnov, and C. Rovelli, "Barrett-Crane model from a Boulatov-Ooguri field theory over a homogeneous space," *Nucl. Phys. B* **574** (2000) 785–806, hep-th/9907154.

[17] A. Perez and C. Rovelli, "A Spin foam model without bubble divergences," *Nucl. Phys. B* **599** (2001) 255–282, gr-qc/0006107.

[18] A. Perez and C. Rovelli, "Spin foam model for Lorentzian general relativity," *Phys. Rev. D* **63** (2001) 041501, gr-qc/0009021.

[19] A. Perez and C. Rovelli, "3+1 spinfoam model of quantum gravity with space - like and time - like components," *Phys. Rev. D* **64** (2001) 064002, gr-qc/0011037.

[20] R. E. Livine and D. Oriti, "Barrett-Crane spin foam model from generalized BF type action for gravity," *Phys. Rev. D* **65** (2002) 044025, gr-qc/0104043.

[21] L. Crane, A. Perez, and C. Rovelli, "Perturbative finiteness in spin-foam quantum gravity," *Phys. Rev. Lett.* **87** (2001) 181301.

[22] S. Alexandrov and E. R. Livine, "SU(2) loop quantum gravity seen from covariant theory," *Phys. Rev. D* **67** (2003) 044009, gr-qc/0209105.

[23] E. R. Livine and D. Oriti, "Implementing causality in the spin foam quantum geometry," *Nucl. Phys. B* **663** (2003) 231–279, gr-qc/0210064.

[24] L. Freidel and D. Louapre, "Diffeomorphisms and spin foam models," *Nucl. Phys. B* **662** (2003) 279–298, gr-qc/0212001.

[25] A. Perez, "Spin foam models for quantum gravity," *Class. Quant. Grav.* **20** (2003) R43, gr-qc/0301113.

[26] M. Bojowald and A. Perez, "Spin foam quantization and anomalies," *Gen. Rel. Grav.* **42** (2010) 877–907, gr-qc/0303026.

[27] J. W. Cherrington and J. D. Christensen, "Positivity in Lorentzian Barrett-Crane models of quantum gravity," *Class. Quant. Grav.* **23** (2006) 721–736, gr-qc/0509080.

[28] S. Alexandrov, "Spin foam model from canonical quantization," *Phys. Rev. D* **77** (2008) 024009, 0705.3892.

[29] E. Bianchi and L. Modesto, "The Perturbative Regge-calculus regime of loop quantum gravity," *Nucl. Phys. B* **796** (2008) 581–621, 0709.2051.

[30] E. R. Livine and S. Speziale, "A New spinfoam vertex for quantum gravity," *Phys. Rev. D* **76** (2007) 084028, 0705.0674.

[31] J. Engle, R. Pereira, and C. Rovelli, "Flipped spinfoam vertex and loop gravity," *Nucl. Phys. B* **798** (2008) 251–290, 0708.1236.

[32] L. Freidel and K. Krasnov, "A new spin foam model for 4d gravity," *Class. Quant. Grav.* **25** (2008) 125018, 0708.1595.

[33] R. Pereira, "Lorentzian LQG vertex amplitude," *Class. Quant. Grav.* **25** (2008) 085013, 0710.5043.

[34] J. Engle, E. Livine, R. Pereira, and C. Rovelli, "LQG vertex with finite Immirzi parameter," *Nucl. Phys. B* **799** (2008) 136–149, 0711.0146.

[35] M. Han, "4-dimensional spin-foam model with quantum Lorentz group," *J. Math. Phys.* **52** (2011) 072501, 1012.4216.

[36] W. J. Fairbairn and C. Meusburger, "Quantum deformation of two four-dimensional spin foam models," *J. Math. Phys.* **53** (2012) 022501, 1012.4784.

[37] M. Han, "Cosmological constant in LQG vertex amplitude," *Phys. Rev. D* **84** (2011) 064010, 1105.2212.

[38] H. M. Haggard, M. Han, W. Kamiński, and A. Riello, "SL(2,C) Chern-Simons theory, a non-planar graph operator, and 4D loop quantum gravity with a cosmological constant: Semiclassical geometry," *Nucl. Phys. B* **900** (2015) 1–79, 1412.7546.

[39] G. T. Horowitz, "Exactly soluble diffeomorphism invariant theories," *Commun. Math. Phys.* **125** (1989) 417.

[40] E. Witten, "Topological quantum field theory," *Commun. Math. Phys.* **117** (1988) 353.

[41] M. Atiyah, "Topological quantum field theories," *Inst. Hautes Etudes Sci. Publ. Math.* **68** (1989) 175–186.

[42] R. Hojman, C. Mukku, and W. A. Sayed, "Parity violation in metric torsion theories of gravitation," *Phys. Rev. D* **22** (1980) 1915–1921.

[43] S. Holst, "Barbero's Hamiltonian derived from a generalized Hilbert-Palatini action," *Phys. Rev. D* **53** (1996) 5966–5969, gr-qc/9511026.

[44] J. F. Barbero G., "Real Ashtekar variables for Lorentzian signature space times," *Phys. Rev. D* **51** (1995) 5507–5510, gr-qc/9410014.

[45] G. Immirzi, "Quantum gravity and Regge calculus," *Nucl. Phys. Proc. Suppl.* **57** (1997) 65–72, gr-qc/9701052.

[46] A. Ashtekar, "New variables for classical and quantum gravity," *Phys. Rev. Lett.* **57** (1986) 2244–2247.

[47] A. Ashtekar, "New Hamiltonian formulation of general relativity," *Phys. Rev. D* **36** (1987) 1587–1602.

[48] J. F. Plebanski, "On the separation of Einsteinian substructures," *J. Math. Phys.* **18** (1977) 2511–2520.

[49] R. Capovilla, T. Jacobson, and J. Dell, "General relativity without the metric," *Phys. Rev. Lett.* **63** (1989) 2325.

[50] M. P. Reisenberger, "A left-handed simplicial action for Euclidean general relativity," *Class. Quant. Grav.* **14** (1997) 1753–1770, gr-qc/9609002.

[51] R. Oeckl, *Discrete Gauge Theory: From Lattices to TQFT*. 2005.

[52] W. Kaminski, M. Kisielowski, and J. Lewandowski, "Spin-foams for all loop quantum gravity," *Class. Quant. Grav.* **27** (2010) 095006, 0909.0939. [Erratum: *Class. Quant. Grav.* **29**, 049502 (2012)].

[53] E. Bianchi, P. Dona, and S. Speziale, "Polyhedra in loop quantum gravity," *Phys. Rev. D* **83** (2011) 044035, 1009.3402.

[54] Y. Ding, M. Han, and C. Rovelli, "Generalized spinfoams," *Phys. Rev. D* **83** (2011) 124020, 1011.2149.

[55] Ruhl, *The Lorentz Group and Harmonic Analysis* (Benjamin (1970)).

[56] E. Bianchi, "Loop quantum gravity a la Aharonov-Bohm," *Gen. Rel. Grav.* **46** (2014) 1668, 0907.4388.

[57] L. Freidel, M. Geiller, and J. Ziprick, "Continuous formulation of the loop quantum gravity phase space," *Class. Quant. Grav.* **30** (2013) 085013, 1110.4833.

[58] A. Ashtekar and J. Lewandowski, "Background independent quantum gravity: A status report," *Class. Quant. Grav.* **21** (2004) R53, gr-qc/0404018.

[59] H. Minkowski, "H. Allgemeine Lehrsätze über die konvexe Polyeder," *Nachr. Ges. Wiss., Göttingen*, (1897) 198–219.

[60] M. Kapovich and J. J. Millson, "The symplectic geometry of polygons in Euclidean space," *J. Diff. Geom.* **44**, 3 (1996) 479–513.

[61] E. Bianchi, "The length operator in loop quantum gravity," *Nucl. Phys. B* **807** (2009) 591–624, 0806.4710.

[62] J. B. Lasserre, "An analytical expression and an algorithm for the volume of a convex polyhedron in Rn," *J. Optim. Theor. Appl.* **39**, pp. 363–377.

[63] H. M. Haggard, "Pentahedral volume, chaos, and quantum gravity," *Phys. Rev. D* **87** (2013), no. 4, 044020, 1211.7311.

[64] C. E. Coleman-Smith and B. Mller, "A Helium atom of space: Dynamical instability of the isochoric pentahedron," *Phys. Rev. D* **87** (2013), no. 4, 044047, 1212.1930.

[65] R. Penrose, "Angular momentum: an approach to combinatorial space-time," *Quantum Theory and Beyond* ed. T. Bastin (Cambridge University Press, Cambridge, 1971) 151–180.

[66] R. Penrose, "On the nature of quantum geometry," *Magic Without Magic: JA Wheeler Festschrift* (1972).

[67] C. Rovelli and L. Smolin, "Discreteness of area and volume in quantum gravity," *Nucl. Phys. B* **442** (1995) 593–622, gr-qc/9411005. [Erratum: *Nucl. Phys. B* **456** 753 (1995)].

[68] S. A. Major, "Operators for quantized directions," *Class. Quant. Grav.* **16** (1999) 3859–3877, gr-qc/9905019.

[69] A. Barbieri, "Quantum tetrahedra and simplicial spin networks," *Nucl. Phys. B* **518** (1998) 714–728, gr-qc/9707010.

[70] E. Bianchi and H. M. Haggard, "Discreteness of the volume of space from Bohr-Sommerfeld quantization," *Phys. Rev. Lett.* **107** (2011) 011301, 1102.5439.

[71] E. Bianchi and H. M. Haggard, "Bohr-Sommerfeld quantization of space," *Phys. Rev. D* **86** (2012) 124010, 1208.2228.

[72] A. Ashtekar and J. Lewandowski, "Quantum theory of geometry. 1: Area operators," *Class. Quant. Grav.* **14** (1997) A55–A82, gr-qc/9602046.

[73] A. Ashtekar and J. Lewandowski, "Quantum theory of geometry. 2. Volume operators," *Adv. Theor. Math. Phys.* **1** (1998) 388–429, gr-qc/9711031.

[74] A. Ashtekar, A. Corichi, and J. A. Zapata, "Quantum theory of geometry III: Non-commutativity of Riemannian structures," *Class. Quant. Grav.* **15** (1998) 2955–2972, gr-qc/9806041.

[75] F. Conrady and L. Freidel, "Quantum geometry from phase space reduction," *J. Math. Phys.* **50** (2009) 123510, 0902.0351.

[76] L. Freidel and E. R. Livine, "The fine structure of SU(2) Intertwiners from U(N) Representations," *J. Math. Phys.* **51** (2010) 082502, 0911.3553.

[77] B. S. S. John R. Klauder, *Coherent States: Applications in Physics and Mathematical Physics* (World Scientific, (1985)).

[78] A. M. Perelomov, *Generalized Coherent States and Their Applications* (Springer-Verlag, (1986)).

[79] L. Freidel and S. Speziale, "Twisted geometries: A geometric parametrisation of SU(2) phase space," *Phys. Rev. D* **82** (2010) 084040, 1001.2748.

[80] B. Dittrich and J. P. Ryan, "Phase space descriptions for simplicial 4d geometries," *Class. Quant. Grav.* **28** (2011) 065006, 0807.2806.

[81] E. Bianchi, E. Magliaro, and C. Perini, "Coherent spin-networks," *Phys. Rev. D* **82** (2010) 024012, 0912.4054.

[82] T. Thiemann, "Gauge field theory coherent states (GCS): 1. General properties," *Class. Quant. Grav.* **18** (2001) 2025–2064, hep-th/0005233.

[83] B. Bahr and T. Thiemann, "Gauge-invariant coherent states for loop quantum gravity. II. Non-Abelian gauge groups," *Class. Quant. Grav.* **26** (2009) 045012, 0709.4636.

[84] W. M. Wieland, "A new action for simplicial gravity in four dimensions," *Class. Quant. Grav.* **32** (2015), no. 1, 015016, 1407.0025.

[85] J. W. Barrett, R. J. Dowdall, W. J. Fairbairn, H. Gomes, and F. Hellmann, "Asymptotic analysis of the EPRL four-simplex amplitude," *J. Math. Phys.* **50** (2009) 112504, 0902.1170.

[86] J. W. Barrett, R. J. Dowdall, W. J. Fairbairn, F. Hellmann, and R. Pereira, "Lorentzian spin foam amplitudes: Graphical calculus and asymptotics," *Class. Quant. Grav.* **27** (2010) 165009, 0907.2440.

[87] C. Rovelli, *Quantum gravity*. 2004.

[88] R. Oeckl, "General boundary quantum field theory: Foundations and probability interpretation," *Adv. Theor. Math. Phys.* **12** (2008) 319–352, hep-th/0509122.

[89] F. Conrady and L. Freidel, "On the semiclassical limit of 4d spin foam models," *Phys. Rev. D* **78** (2008) 104023, 0809.2280.

[90] V. Bonzom, "Spin foam models for quantum gravity from lattice path integrals," *Phys. Rev. D* **80** (2009) 064028, 0905.1501.

[91] E. Magliaro and C. Perini, "Emergence of gravity from spinfoams," *Europhys. Lett.* **95** (2011) 30007, 1108.2258.

[92] F. Hellmann and W. Kaminski, "Geometric asymptotics for spin foam lattice gauge gravity on arbitrary triangulations," 1210.5276.

[93] M. Han and T. Krajewski, "Path integral representation of Lorentzian spinfoam model, asymptotics, and simplicial geometries," *Class. Quant. Grav.* **31** (2014) 015009, 1304.5626.

[94] M. Han, "Semiclassical analysis of spinfoam model with a small Barbero-Immirzi parameter," *Phys. Rev. D* **88** (2013) 044051, 1304.5628.

[95] M. Han, "Covariant loop quantum gravity, low energy perturbation theory, and Einstein gravity with high curvature UV corrections," *Phys. Rev. D* **89** (2014), no. 12, 124001, 1308.4063.

[96] E. Bianchi, C. Rovelli, and F. Vidotto, "Towards spinfoam cosmology," *Phys. Rev. D* **82** (2010) 084035, 1003.3483.

[97] E. Bianchi, T. Krajewski, C. Rovelli, and F. Vidotto, "Cosmological constant in spinfoam cosmology," *Phys. Rev. D* **83** (2011) 104015, 1101.4049.

[98] E. Magliaro, A. Marciano, and C. Perini, "Coherent states for FLRW space-times in loop quantum gravity," *Phys. Rev. D* **83** (2011) 044029, 1011.5676.

[99] C. Rovelli, "Graviton propagator from background-independent quantum gravity," *Phys. Rev. Lett.* **97** (2006) 151301, gr-qc/0508124.

[100] E. Bianchi, L. Modesto, C. Rovelli, and S. Speziale, "Graviton propagator in loop quantum gravity," *Class. Quant. Grav.* **23** (2006) 6989–7028, gr-qc/0604044.

[101] E. R. Livine and S. Speziale, "Group Integral Techniques for the Spinfoam Graviton Propagator," *JHEP* **11** (2006) 092, gr-qc/0608131.

[102] E. Alesci and C. Rovelli, "The Complete LQG propagator. I. Difficulties with the Barrett-Crane vertex," *Phys. Rev. D* **76** (2007) 104012, 0708.0883.

[103] E. Alesci and C. Rovelli, "The Complete LQG propagator. II. Asymptotic behavior of the vertex," *Phys. Rev. D* **77** (2008) 044024, 0711.1284.

[104] E. Alesci, E. Bianchi, and C. Rovelli, "LQG propagator: III. The New vertex," *Class. Quant. Grav.* **26** (2009) 215001, 0812.5018.

[105] E. Bianchi, E. Magliaro, and C. Perini, "LQG propagator from the new spin foams," *Nucl. Phys. B* **822** (2009) 245–269, 0905.4082.

[106] C. Rovelli and M. Zhang, "Euclidean three-point function in loop and perturbative gravity," *Class. Quant. Grav.* **28** (2011) 175010, 1105.0566.

[107] E. Bianchi and Y. Ding, "Lorentzian spinfoam propagator," *Phys. Rev. D* **86** (2012) 104040, 1109.6538.

[108] B. Bahr and S. Steinhaus, "Quantum cuboids and the EPRL-FK path integral for quantum gravity," 1508.07961.

[109] B. Bahr, B. Dittrich, F. Hellmann, and W. Kaminski, "Holonomy spin foam models: Definition and coarse graining," *Phys. Rev.* D **87** (2013), no. 4, 044048, 1208.3388.

[110] B. Dittrich, M. Martin-Benito, and S. Steinhaus, "Quantum group spin nets: refinement limit and relation to spin foams," *Phys. Rev.* D **90** (2014) 024058, 1312.0905.

[111] A. Riello, "Radiative corrections to the Lorentzian Engle-Pereira-Rovelli-Livine and Freidel-Krasnov spinfoam graviton," *Phys. Rev.* D **89** (2014), no. 6, 064021, 1310.2174.

[112] L. Freidel, "Group field theory: An overview," *Int. J. Theor. Phys.* **44** (2005) 1769–1783, hep-th/0505016.

[113] A. Baratin and D. Oriti, "Group field theory and simplicial gravity path integrals: A model for Holst-Plebanski gravity," *Phys. Rev.* D **85** (2012) 044003, 1111.5842.

[114] J. Ambjorn, A. Goerlich, J. Jurkiewicz, and R. Loll, "Nonperturbative quantum gravity," *Phys. Rept.* **519** (2012) 127–210, 1203.3591.

Chapter 4

Group Field Theory and Loop Quantum Gravity

Daniele Oriti

Max Planck Institute for Gravitational Physics (Albert Einstein Institute),
Am Mühlenberg 1, 14476 Golm, Germany

1. GFT from LQG Perspective: The Underlying Ideas

1.1. *A Historical Prelude*

Group field theories can be approached from different angles, coming from different lines of research in quantum gravity. Historically, their first appearance [2, 3] came as a development of tensor models [4] (themselves a generalization of matrix models [5], which provided a successful quantization of (pure) 2d gravity), allowing to make contact with state sum formulations of 3d quantum gravity (Ponzano-Regge and Turaev-Viro model), whose relation with simplicial quantum gravity [6], was already known, and more generally topological BF theory in any dimension. These first models were obtained by taking the simplest tensor model for 3d simplicial gravity and: (1) replacing the domain set for the tensor indices with a group manifold ($SU(2)$); (2) adding a gauge invariance property to the field (tensor), with the effect of introducing a gauge connection on the lattices generated by the perturbative expansion of the model. The triviality of the kinetic and interaction kernels (simple delta functions on the group) in the GFT action resulted in the amplitudes matching those of BF theory discretized on the same lattices (imposing flatness of the connection). Written in terms of group representations, the same amplitudes took the form of the state sums mentioned above. This is the first way to understand group field theories: GFTs can be seen as tensor models enriched by algebraic data with a quantum geometric interpretation (allowing a nice encoding of discrete gravity degrees of freedom), or, equivalently, as more general class of combinatorially non-local field theories of tensorial type. The relation between state sum models of topological field theory, and their GFT formulation, and loop quantum gravity was soon pointed out in [7] (where the link to the dynamical triangulations approach [8] was also mentioned): the boundary states of such models matched the newly developed loop representation for quantum gravity [9]. Indeed, spin networks (introduced in LQG immediately afterwards) that are discussed in *Chapter 1* describe also the

Hilbert space of GFT models. The latter then acquired a nice interpretation from the LQG perspective: GFTs are quantum field theories for spin networks, providing them with a covariant dynamics. This covariant definition started being developed a few years later [10–12]. Indeed, interest in GFTs as quantum gravity models was enhanced with the realization that they provide a complete definition of spin foam models for 4d gravity [13] discussed in *Chapter 3:* they capture the same quantum amplitudes as Feynman amplitudes, organizing them coherently in a sum over spin foam complexes, arising from the perturbative expansion of the field theory. It is in this context that most developments have taken place in subsequent years, bringing in particular an improved understanding of the quantum geometry encoded in several interesting spin foam models for 4d gravity. Finally, in recent years we have witnessed a renaissance of tensor models [14], with important mathematical results concerning their combinatorial structures. Since they share the same combinatorial structures as GFTs, this has triggered further developments in GFTs. Indeed, it is by combining the tools coming from tensor models and the quantum geometric understanding provided by loop quantum gravity and spin foam models that, we believe, we have the best chance to make progress on the remaining open issues of quantum gravity, within the GFT framework.

1.2. *A Brief Definition*

A (single-field) GFT is a theory of a field $\varphi : G^{\times d} \to \mathbb{C}$ defined on d copies of a group manifold G, with action

$$S(\varphi,\varphi^*) = \int [dg_I][dg'_J]\varphi^*(g_I)\mathcal{K}(g_I,g'_J)\,\varphi(g'_J) +$$
$$+ \sum_i \frac{\lambda_i}{D_i!} \int [dg_{I1}]...[dg_{JD_i}]\varphi^*(g_{I1})\ldots\mathcal{V}_i(g_{I1},...,g_{JD_i})\cdots\varphi(g_{JD_i}) \quad (1)$$

where i labels the possible interaction terms (weighted by coupling constants), each involving D_i fields (or their complex conjugates), in turn depending on d group elements each. A specific GFT model is defined by a choice of group G, dimension d, kinetic and interaction kernels $\mathcal{K}$ and $\mathcal{V}_i$. The crucial feature of GFT models, as opposed to ordinary (local) QFTs on space-time, beside the physical interpretation of all the ingredients, is that the interaction kernels, field arguments are related in a *combinatorially non-local* way, i.e. each field is correlated to the others only through *some* of its arguments. The specific combinatorial pattern of such correlations is another defining property of specific models, as we will discuss (see [15] for an extensive treatment of the combinatorial aspects of GFTs).

This combinatorial non-locality becomes manifest in the quantum theory, defined in perturbation theory by the partition function:

$$Z = \int \mathcal{D}\varphi\mathcal{D}\varphi^*\, e^{-S(\varphi,\varphi^*)} = \sum_\Gamma \frac{\prod_i \lambda_i^{n_i(\Gamma)}}{sym(\Gamma)} \mathcal{A}_\Gamma \quad (2)$$

where Γ denotes GFT Feynman diagrams, $sym(\Gamma)$ the order of their automorphism group, $n_i(\Gamma)$ the number of interaction vertices of type i, and $\mathcal{A}_\Gamma$ is the corresponding Feynman amplitude, obtained by convolution of interaction kernels with propagators. Again, the explicit form of the amplitudes $\mathcal{A}_\Gamma$ and the combinatorial structure of the diagrams Γ depend on the model considered. The combinatorial non-locality of the interaction terms, however, generically implies that GFT Feynman diagrams are not graphs but cellular complexes of arbitrary topology.

The combinatorial structure of GFT fields, action, Feynman diagrams, quantum states and amplitudes, are the same as those of tensor models [4, 14] *enriched* by the additional group-theoretic data. Because of them, GFTs are quantum field theories whose basic quanta are spin network vertices, i.e. nodes with d open links, labelled by the same algebraic data of LQG states, and their Feynman amplitudes $\mathcal{A}_\Gamma$ are generically spin foam amplitudes. Spin foam amplitudes are, in turn, dual to lattice gravity path integrals, so GFTs combine the main ideas of dynamical triangulations (quantum gravity as a sum over random lattices) [8] and of quantum Regge calculus [6] (quantum gravity as a sum over geometric data assigned to a given lattice).

In the following, we will highlight structures shared with other ways of doing loop quantum gravity, as well as points of departure and new concepts brought in by the GFT reformulation. We will also discuss how GFTs cast the problem of defining a background independent theory of quantum gravity based on LQG ideas in a QFT language. This allows the use of several powerful tools, to realize concretely the notion of 'quantum atoms of space' and to treat spacetime, indeed, like a condensed matter quantum system, suggesting new lines of developments.

2. GFT Kinematics: Hilbert Space and Observables

2.1. *Fock Space of Quantum States*

The Hilbert space of states for single-field GFTs is a Fock space built out of a fundamental 'single-atom' Hilbert space $\mathcal{H}_v = L^2(G^{\times d})$: $\mathcal{F}(\mathcal{H}_v) = \bigoplus_{V=0}^\infty sym\left\{\left(\mathcal{H}_v^{(1)} \otimes \mathcal{H}_v^{(2)} \otimes \cdots \otimes \mathcal{H}_v^{(V)}\right)\right\}$, where sym indicates symmetrization with respect to the permutation group S_V [16]. This encodes bosonic statistics for field operators (other possibilities can be considered [17, 18])

$$[\hat{\varphi}(\vec{g}), \hat{\varphi}^\dagger(\vec{g}')] = \mathbb{I}_G(\vec{g}, \vec{g}') \quad [\hat{\varphi}(\vec{g}), \hat{\varphi}(\vec{g}')] = [\hat{\varphi}^\dagger(\vec{g}), \hat{\varphi}^\dagger(\vec{g}')] = 0 \quad (3)$$

where $\mathbb{I}_G(\vec{g}, \vec{g}') \equiv \prod_{i=1}^d \delta(g_i(g_i')^{-1})$, and we used the notation $\vec{g} = (g_1, .., g_d)$.

In quantum gravity models G is chosen to be the local gauge group of gravity in the appropriate dimension and signature, i.e. $G = SU(2), SL(2,\mathbb{R})$ in 3d and $G = Spin(4), SL(2,\mathbb{C})$ in 4d (or their $SU(2)$ subgroup, for connecting with LQG).

Each Hilbert space $\mathcal{H}_v$ provides the space of states of a single "quantum" of the GFT field, a quantum gravity 'atom'. It can be understood as a fundamental spin network vertex, represented by a node with d outgoing links (ending up in 1-valent nodes), labelled by group elements, or as a 3-cell (polyhedron) with d boundary faces. This is just a pictorial representation. Whether the states represent quantum gravity spin network vertices or geometric polyhedra depends on the type of data they carry and the dynamics they satisfy. For $G = SU(2)$, and with the closure condition $\varphi(g_I) = \varphi(hg_I)$ $\forall h \in G$ imposed on the fields, however, the polyhedral interpretation is justified and the same is true for $G = SL(2,\mathbb{C})$ and $G = Spin(4)$ with simplicity constraints and closure conditions correctly imposed. In particular, for $d = 4$, the GFT quanta represent quantum tetrahedra and their Hilbert space is $\mathcal{H}_v = \bigoplus_{J_i \in \mathbb{N}/2} Inv\left(\mathcal{H}^{J_1} \otimes ... \otimes \mathcal{H}^{J_4}\right)$, where each $\mathcal{H}^{J_i}$ is the Hilbert space of an irreducible unitary representation of $SU(2)$ labeled by the half-integer J_i.

2.2. Quantum Observables

Kinematical observables are functionals of the field operators $\mathcal{O}\left(\hat{\varphi}, \hat{\varphi}^\dagger\right)$. Of special importance are polynomial observables, whose evaluation in the vacuum state defines GFT n-point functions [19]. Any convolution of a finite number of GFT field operators with appropriate kernels would define one such observable, as in any quantum field theory. The peculiarity of GFTs, with respect to ordinary QFTs, is the possibility for these kernels to have a richer combinatorial structure, involving a non-local pairing of field arguments, i.e. relating only a subset of the d arguments of a given GFT field with a subset of the arguments of a different one. Of particular interest for LQG are 'spin network observables':

$$O_{\Psi=(\gamma, J^{(ab)}_{(ij)}, \iota_i)}(\hat{\varphi}^\dagger) = \left(\prod_{(i)} \int [dg_{ia}]\right) \Psi_{(\gamma, J^{(ab)}_{(ij)}, \iota_i)}(g_{ia} g_{jb}^{-1}) \prod_i \hat{\varphi}^\dagger(g_{ia}), \quad (4)$$

where g_{ia} (resp. g_{jb}) (with $a, b = 1, ..., d$) are group elements related to the arguments of the field associated to the vertex i (resp. j), so that a pair of indices (a, b) denotes each of the edges connecting two vertices i and j, $\Psi_{(\gamma, J^{(ab)}_{(ij)}, \iota_i)}(G^{ab}_{ij})$ is a spin network functional labelled by a closed graph γ with representations $J^{(ab)}_{(ij)}$ associated to its edges, and intertwiners ι_i to its vertices. The bosonic statistics implies that Ψ is symmetric under permutations of vertex labels. This observable creates a spin network state associated to a graph γ out of the Fock vacuum.

2.3. GFT as Second Quantized Reformulation of the LQG Kinematics

We now discuss in what sense GFT provides a second quantized formalism for spin networks and how can one directly link (a certain version of) canonical LQG and GFT [16].

By 'LQG kinematical Hilbert space' we intend, here, a Hilbert space constructed out of states associated to closed graphs and such that, for each graph γ, we have $\mathcal{H}_\gamma = L^2\left(G^E/G^V, d\mu = \prod_{e=1}^{E} d\mu_e^{Haar}\right)$ (here $G = SU(2)$), where e are the links of the graph (E is their total number), with a graph-based scalar product defining the

Haar measure on each link μ_e^{Haar}. The same Hilbert space can be represented also in the flux basis, via the non-commutative Fourier transform [20]. The union for all graphs of such Hilbert spaces is, of course, not a Hilbert space. In the LQG and spin foam literature, one finds different ways in which these graph-based Hilbert spaces can be organized to define the Hilbert space of the theory. The GFT Hilbert space achieves this result by 'decomposing them into elementary building blocks'.

The basic idea is to consider any wave function in $\mathcal{H}_\gamma$, where γ is a graph with V nodes, as an element of $\mathcal{H}_V = L^2\left((G^{\times d}/G)^{\times V}, d\mu = \prod_{v=1}^{V} \prod_{i=1}^{d} d\mu_{Haar,i}^v\right)$, satisfying special restrictions. The latter is the space of V spin network vertices, each possessing d outgoing open links, and the extra restrictions enforce the gluing of suitable pairs of such open links to form the links of the graph γ. A function Ψ_γ can be obtained from a wavefunction $\phi_V \in \mathcal{H}_V$ as

$$\Psi_\Gamma(G_{ij}^{ab}) = \prod_{[(ia),(jb)]} \int_G d\alpha_{ij}^{ab} \, \phi_V(\ldots, g_{ia}\,\alpha_{ij}^{ab}, \ldots, g_{jb}\alpha_{ij}^{ab}, \ldots) = \Psi_\Gamma(g_{ia}(g_{jb})^{-1}), \tag{5}$$

with the same notation as in 4. This defines an embedding of elements of $\mathcal{H}_\gamma$ into $\mathcal{H}_V$. The same construction can be phrased in the flux and spin representations. Moreover, the scalar product of two quantum states in $\mathcal{H}_V$ associated in this way to the same graph agrees with the one computed in $\mathcal{H}_\gamma$. This means that $\mathcal{H}_\gamma$ is embedded faithfully in $\mathcal{H}_V$. Obviously $\mathcal{H}_V$ also contains states associated to open graphs, with links of open spin network vertices not glued to any other.

The physical picture behind $\mathcal{H}_V$ is that of a 'many-atom' Hilbert space, with each 'quantum gravity atom' corresponding to a Hilbert space $\mathcal{H}_v = L^2\left(G^{\times d}/G\right)$. An orthonormal basis $\psi_{\vec{\chi}}(\vec{g})$ in each $\mathcal{H}_v$ is given by the spin network wave functions for individual spin network vertices (labelled by spins and angular momentum projections associated to their d open edges, and intertwiner quantum numbers):

$$\vec{\chi} = \left(\vec{J}, \vec{m}, \mathcal{I}\right) \quad \to \quad \psi_{\vec{\chi}}(\vec{g}) = \langle \vec{g} | \vec{\chi} \rangle = \left[\prod_{a=1}^{d} D_{m_a n_a}^{J_a}(g_a)\right] C_{n_1..n_d}^{J_1...J_d, \mathcal{I}} \; . \tag{6}$$

The Hilbert space is then extended to include arbitrary numbers of QG atoms $\mathcal{H}_{GFT} = \bigoplus_{V=0}^{\infty} \mathcal{H}_V$ and can be turned into the Fock space already introduced by standard methods [16], introducing the fundamental GFT field operators

$$\hat{\varphi}(g_1,..,g_d) \equiv \hat{\varphi}(\vec{g}) = \sum_{\vec{\chi}} \hat{\varphi}_{\vec{\chi}} \, \psi_{\vec{\chi}}(\vec{g}) \qquad \hat{\varphi}^\dagger(g_1,..,g_d) \equiv \hat{\varphi}^\dagger(\vec{g}) = \sum_{\vec{\chi}} \hat{\varphi}_{\vec{\chi}}^\dagger \, \psi_{\vec{\chi}}^*(\vec{g}) \; .$$

Similarly, quantum observables can be turned from first quantized operators acting on the many-atom Hilbert spaces $\mathcal{H}_V$ to second quantized operators on the Fock space, following again standard procedures. Given the matrix elements $\mathcal{O}_{n,m}(\vec{\chi}_1, ..., \vec{\chi}_m, \vec{\chi}'_1, ..., \vec{\chi}'_n)$ (or the correspondent functions in the group or flux basis) of the relevant operator $\widehat{\mathcal{O}_{n,m}}$ in a basis of open spin network vertices, take the appropriate convolutions of such functions with creation and annihilation operators, according to which spin network vertices are acted upon by the operator and which spin network vertices result from the same action, to obtain its second quantized counterpart. The result will thus be a linear combination of polynomials of creation and annihilation operators, i.e. of GFT field operators, thus a GFT observable:

$$\widehat{\mathcal{O}_{n,m}} \;\to\; \langle \vec{\chi}_1,....,\vec{\chi}_m | \widehat{\mathcal{O}_{n,m}} | \vec{\chi}'_1,...,\vec{\chi}'_n \rangle = \mathcal{O}_{n,m}\left(\vec{\chi}_1,...,\vec{\chi}_m,\vec{\chi}'_1,...,\vec{\chi}'_n\right) \;\to\;$$

$$\widehat{\mathcal{O}_{n,m}}\left(\hat{\varphi},\hat{\varphi}^\dagger\right) = \int [d\vec{g}_i][d\vec{g}'_j]\, \hat{\varphi}^\dagger(\vec{g}_1)..\hat{\varphi}^\dagger(\vec{g}_m)\mathcal{O}_{n,m}\left(\vec{g}_1,..,\vec{g}_m,\vec{g}'_1,..,\vec{g}'_n\right) \hat{\varphi}(\vec{g}'_1)..\hat{\varphi}(\vec{g}'_n)\,.$$

2.4. Similarities and Differences with the LQG Hilbert Space

The kinematical Hilbert space of GFT is analogous to the one in LQG in the sense that its quantum states are the same type of functions on group manifolds, associated to graphs, and characterized by the same representation labels, group or Lie algebra elements. Thus they also encode quantum gravity degrees of freedom in purely combinatorial and algebraic structures. However, there are also key differences. First of all, there is no embedding of GFT states into a continuous manifold, but they are associated to abstract graphs. This means that there is a priori no action of diffeomorphisms, nor knotting degrees of freedom. Thus they also differ from the s-knot states of the diffeo-invariant Hilbert space of canonical LQG. The only symmetry follows from choice of quantum statistics, i.e. symmetry under permutations of vertex labelings. From this point of view, the GFT state space takes the combinatorial and algebraic nature of the degrees of freedom of quantum space to be fundamental, and no continuum intuition is assumed. In fact, there is no attempt to define a continuum limit at this kinematical level, if not in the sense of a limit of infinite number of QG atoms (akin to a thermodynamic limit in condensed matter). In particular, in contrast to *Chapter 5*, no cylindrical equivalence among GFT states is imposed, and, in contrast to *Chapter 1*, graph links labeled with trivial connection or zero representation label are not neglected (as atoms with zero momentum in condensed matter). Moreover, while GFT states associated to graphs with different numbers of nodes are by definition orthogonal, GFT states associated to different graphs with the same number of nodes are not, contrary to LQG states. One could say that graph structures are given less relevance than in standard LQG, because they are reduced to specific correlations among the fundamental GFT quanta and quantum states for the same number of quanta but with different correlations overlapping. At the same time, the physical relevance of graph structures is somewhat enhanced, because no link with spin zero is removed and because the number of graph nodes is turned into a new physical observable. Thus we have similarities, but also differences. One motivation to accept these differences, drop some features of the LQG Hilbert space (e.g. those coming from a continuum embedding) and embrace the GFT one with its fundamental discreteness is that the latter has a clear Fock structure, giving direct meaning to the notion of 'QG atom of quantum space', and making powerful analytical tools available.

3. The Quantum Dynamics

Given the general set-up, one needs to specify a GFT model, that is: a group manifold, the valence d of the fields, and a set of kinetic and interaction terms, including their encoded combinatorial patterns. In quantum field theory, given a topological space-time manifold, the ingredients selecting a specific model are: a locality principle, symmetries (which determine the field content and the general theory space), simplicity, all well grounded in phenomenology. The renormalization group is then used as a check of consistency, and as a way to generate effective dynamics at different scales. In GFT, at present, the only type of 'indirect phenomenology' we can rely on is the current theories of spacetime and fundamental interactions, and the basic insights of other attempts at constructing a quantum theory of spacetime.

Indeed, one can identify three main strategies, in the GFT literature, stemming from three main ways to approach the formalism, and three different directions in quantum gravity research, all converging somehow to the GFT formalism.

3.1. *GFT Dynamics from Canonical LQG*

The first is suggested by the picture of GFTs as second quantized theories of spin networks and of canonical LQG [16]. GFTs take seriously the basic insights of traditional LQG, based on canonical quantization of GR in the continuum, in particular the same type of quantum states (even if not quite the same Hilbert space). Should we choose also the same dynamics, encoded in some Hamiltonian constraint operator, as in *Chapter 2?* One may say that it is not so reasonable to expect that the fundamental dynamics of space-time at the Planck scale (and maybe beyond) is obtained simply by the operator version of the GR dynamics. GR may be, after all, just an effective theory, and indeed it is only tested at scales many orders of magnitude away from the Planckian. In fact, in GFTs many ingredients of the LQG state space, inspired directly from the continuum setting, are dropped. However, to derive a canonical quantum dynamics from the continuum classical dynamics is still a valid possibility, and conventional wisdom about effective theories and running of scales may not apply in a background independent context. Indeed, even continuum canonical LQG ends up with discrete, combinatorial, algebraic structures. One may then take these discrete structures as fundamental, and look for simplest definition of their dynamics, not necessarily taken from any continuum spacetime dynamics. Whatever route one decides to follow, if one has such a canonical operator dynamics, the GFT reformulation of the same (defining a specific GFT model) is straightforward, at least at a heuristic level.

First of all, one has to find the second quantized counterpart of the canonical dynamical operator. This could be directly an Hamiltonian constraint or a 'projection' operator $\widehat{P}$ onto solutions of the Hamiltonian constraint equation, such that: $\widehat{F}|\Psi\rangle \equiv \left(\widehat{P} - \widehat{I}\right)|\Psi\rangle = 0$. When written as an operator on the Fock space, such operator will decompose into operators whose action involves $2, 3, ..., (n+m)$ spin network vertices, weighted by coupling constants. This decomposition may well involve an infinite number of components, to be restricted by symmetry conditions or physical considerations (e.g. renormalization group flow). For each (n,m)-body component of $\widehat{P}$, we consider the matrix elements in a basis of single-vertex states and construct the second quantized projector operator (choosing normal ordering) as:

$$\widehat{F}|\Psi\rangle \equiv \sum_{n,m}^{\infty} \lambda_{n,m} \left[\sum_{\{\vec{\chi},\vec{\chi}'\}} \hat{\varphi}^{\dagger}_{\vec{\chi}_1}...\hat{\varphi}^{\dagger}_{\vec{\chi}_m} \, P_{n,m}(\vec{\chi}_1,...,\vec{\chi}_m,\vec{\chi}'_1,...,\vec{\chi}'_n) \, \hat{\varphi}_{\vec{\chi}'_1}...\hat{\varphi}_{\vec{\chi}'_n} \right. $$

$$\left. - \sum_{\vec{\chi}} \hat{\varphi}^{\dagger}_{\vec{\chi}} \hat{\varphi}_{\vec{\chi}} \right] |\Psi\rangle = 0$$

Even given the above GFT operator, the identification of the corresponding GFT action and partition function has to proceed in a rather heuristic manner. One would like to define a partition function Z for the canonical quantum LQG theory, that is for arbitrary states in the Fock space, thus arbitrary collections of spin network vertices (including those associated to closed graphs). The simplest choice would be an analog of the *microcanonical ensemble*, in which only states solving the canonical dynamical equation contribute: $Z_m = \sum_s \langle s| \, \delta(\widehat{F}) |s\rangle$, where s denotes an arbitrary complete basis of states in the Hilbert (Fock) space of the quantum theory. The GFT dynamics (of existing GFT models), however, corresponds to a quantum LQG dynamics of a more general, *grandcanonical* type

$$Z_g = \sum_s \langle s|e^{-(\widehat{F}-\mu\widehat{N})}|s\rangle \quad ,$$

where the sign of the chemical potential μ determines whether states with many or few spin network vertices are favored. To rewrite the above partition function as a GFT path integral, one introduces a basis of eigenstates of the GFT field operator:

$$Z_g = \sum_s \langle s|e^{-(\widehat{F}-\mu\widehat{N})}|s\rangle = \int \mathcal{D}\varphi \mathcal{D}\overline{\varphi}\, e^{-|\varphi|^2} \langle \varphi| e^{-(\widehat{F}-\mu\widehat{N})} |\varphi\rangle \quad .$$

This is a GFT path integral with quantum amplitude $e^{-|\varphi|^2} \langle \varphi| e^{-(\widehat{F}-\mu\widehat{N})} |\varphi\rangle \equiv e^{-S_{eff}}$ where the effective action S_{eff} is obtained from a *classical action* S_0 as:

$$S_{eff}(\varphi,\overline{\varphi}) = S(\varphi,\overline{\varphi}) + \mathcal{O}(\hbar) = \frac{\langle\varphi|\widehat{F}|\varphi\rangle}{\langle\varphi|\varphi\rangle} + \mathcal{O}(\hbar) \quad .$$

Quantum corrections may amount to new interaction kernels or to a redefinition of the coupling constants for the ones in S. For a given operator equation, then, the corresponding classical (and bare) GFT action is of the form:

$$S(\varphi,\varphi^{\dagger}) = m^2 \int d\vec{g}\, \varphi^*(\vec{g})\varphi(\vec{g}) - \sum_{n,m} \lambda_{n+m} \left[\int [d\vec{g}_i][d\vec{g}'_j]\, \varphi^*(\vec{g}_1)...\varphi^*(\vec{g}_m) \right.$$

$$V_{n+m}(\vec{g}_1,...,\vec{g}_m,\vec{g}'_1,...,\vec{g}'_n)\, \varphi(\vec{g}'_1)...\varphi(\vec{g}'_n)\Big] \qquad V_{n+m}(\vec{g}_1,...,\vec{g}_m,\vec{g}'_1,...,\vec{g}'_n)$$

$$= P_{n+m}(\vec{g}_1,...,\vec{g}_m,\vec{g}'_1,...,\vec{g}'_n)$$

so that the GFT interaction kernels (which, as we will see, are the spin foam vertex amplitudes) are nothing else than the matrix elements of the canonical projector operator, and a mass term incorporates the chemical potential.

As we stressed, the above is quite heuristic, as both the GFT path integral and the quantum statistical partition function of spin networks have to be properly defined. The advantage of the GFT reformulation of the canonical LQG dynamics is exactly that, *starting* from the GFT partition function in terms of the classical GFT action, it allows to use QFT tools to define properly the quantum dynamics, as we will discuss. A few general facts are however clear from the above correspondence. The GFT formulation of a LQG dynamics stands with respect to it as the field theory formulation of the dynamics of a many-body system with respect to its first quantized formulation in terms of a Schrödinger equation for many-body wave functions. The former, one expects, is the best set-up to tackle issues involving large numbers of degrees of freedom of the system. Next, the quantum dynamics can obviously be studied perturbatively around the Fock vacuum (thus in terms of the spin foam expansion, as we will see), but this, as in standard QFT, is expected to be a good approximation of the full dynamics only for processes involving very few quantum geometric degrees of freedom, i.e. only for physical situations in which the physical vacuum of interest is well approximated by the perturbative Fock vacuum. Finally, the sector of the theory corresponding to solutions of the canonical Hamiltonian constraint is the sub-sector of the full GFT quantum dynamics corresponding to a restriction to the micro canonical ensemble.

3.2. *GFT Dynamics from Spin Foams/Lattice Gravity Path Integrals*

The above strategy for the definition of GFT models, starting directly from a canonical quantum dynamics, has not been followed until now. The main strategy that has been followed starts from the definition of spin foam amplitudes [13] discussed in *Chapter 3*, but now encoding them in a GFT model. This has been mainly done in a simplicial context [1] (but see [15]).

Working in the simplicial complex means that one chooses d equal to the would-be space-time dimension, interprets the GFT fields (i.e. the quanta they create/annihilate) as $(d-1)$-simplices (quantum tetrahedra in $d=4$), with the arguments of the GFT fields attached to their $(d-2)$-faces. Next, one usually restricts possible interactions to a single one, describing $d+1$ such simplices glued pairwise across their faces to form (the boundary of) a d-simplex. The kinetic term describes the gluing of two such d-simplices across a shared $(d-1)$-simplex (the GFT quantum being propagated from one interaction vertex to the next). With this combinatorics, one has a general action

$$S_{GFT} = \int [dg_i][dg'_i]\, \varphi^*(g_i)\, \mathcal{K}(g_i, g'_i)\, \varphi(g'_i)$$
$$+ \frac{\lambda}{(d+1)!} \int [dg_{ij}]\, \varphi(g_{1j})....\varphi(g_{(d+1)j})\, \mathcal{V}(g_{ij}) \; + \; \text{c c.}$$

The GFT Feynman diagrams are then 2-complexes dual to (the 2-skeleta of) simplicial complexes. The GFT perturbative expansion will then give a prescription for summing over such complexes, weighted by quantum amplitudes. The 2-complex corresponding to each Feynman diagram is a collection of vertices connected by links, in turn bounding 2-cells. Each GFT interaction kernel assigns an amplitude to each vertex of such 2-complex, while the GFT propagator gives an amplitude to each link of the same, and encodes a prescription for connecting the variables entering the vertex amplitude to the ones appearing in the amplitude of a neighboring vertex. The complete Feynman amplitude for the 2-complex is obtained by summing over common variables in vertex amplitudes and propagators. Now, the vertex amplitude $\mathcal{V}$ corresponding to a given GFT model is nothing else than the spin foam vertex amplitude characterizing a given spin foam model, and the spin foam model is itself fully specified by the whole GFT Feynman amplitude. This correspondence between spin foam models and GFTs is one to one and fully generic [1, 12]: for any given spin foam model, there is a GFT model such that the spin foam amplitudes are reproduced as Feynman amplitudes, and any GFT model defines a spin foam model in its perturbative expansion. Thus any procedure for defining a spin foam model gives automatically a definition of a GFT model. Notice that, just like the spin foam amplitudes can be recast in different forms, there is a certain freedom in a GFT model to redefine kinetic and interaction terms, while maintaining the same expression for the whole Feynman amplitudes. These amplitudes can be written, just as the GFT field itself and the GFT action, in different variables: group elements, Lie algebra elements or group representations. In terms of group elements, the (spin foam) amplitudes take the form of lattice gauge theories [11, 13, 24, 25], the lattice being the 2-complex. In Lie algebra variables, the amplitudes take the form of discrete gravity path integrals on the same lattice (or on the dual triangulation) [20, 26]. In group representations, the amplitudes take the standard spin foam form [13].

In the simplicial context, spin foam constructions in 4d have followed one main route [1, 13]. This amounts to imposing constraints on classical variables, quantum states and quantum amplitudes (written in discrete path integral or spin foam form) for 4d BF theory with group $SL(2,\mathbb{C})$ or $Spin(4)$ discretized on a given simplicial complex. The constraints are the discretized version of the simplicity constraints that turn the BF dynamics into the gravitational one in the continuum [13, 27].

The constrained spin foam amplitudes become equivalently discrete path integrals for Plebanski gravity [26]. Moreover, the simplicity constraints amount also to a restriction in the decomposition of the representations of $SL(2,\mathbb{C})$ or $Spin(4)$ into representations of the diagonal $SU(2)$ subgroup, allowing a link with canonical LQG. The crucial issue becomes of course the correct discretization and quantum implementation of the simplicity constraints, which is where specific models differ.

This route can be followed directly at the GFT level, by imposing constraints on the GFT fields, representing, we recall, quantum tetrahedra. To illustrate the resulting GFT models, we give two examples in the Riemannian setting including the Barbero-Immirzi parameter γ (in the Lorentzian setting less is known and at present we only have one model, with minor variations: the Engle-Pereira-Rovelli-Livine (EPRL) model [28] or the BC model [29], depending on whether the Barbero-Immirzi parameter is included or not). The same models can be straightforwardly extended to combinatorial settings more general than the simplicial one [15, 30], but the geometric features of such generalized constructions are not yet fully understood. The first example is the version of the EPRL model [13, 28], resulting from a specific choice of imposing the constraints in the representation variables, and only at the level of the GFT kinetic term:

$$S_{\text{GFT}}^{\text{EPRL}} = \int [dg_i][dg_i'] \, \varphi^*(g_i) \, C^{-1}(g_i, g_i') \, \varphi(g_i')$$

$$+ \frac{\lambda}{5!} \int [dg_{ij}] \, \varphi(g_{1j}) .... \varphi(g_{5j}) \prod_{i \neq j, i,j=1}^{5} \delta(g_{ij}, g_{ji}) \; + \; \text{cc}.$$

where the interaction term is the same as in BF models and simply identifies the $Spin(4)$ group arguments of the GFT fields according to the combinatorics of faces in a 4-simplex, while the kinetic term is determined by:

$$C_{\text{EPRL}}(g_i, g_i') = \sum_{j_i^+, j_i^-, J_i \in \mathbb{N}/2} \left(\prod_{i=1}^{4} d_{j_i^+} d_{j_i^-} d_{J_i} \, \delta_{|1-\gamma|j_i^+, (1+\gamma)j_i^-} \, \delta_{J_i, j_i^+ + j_i^-} \right) \int dh_\pm dh'_\pm$$

$$\int \prod_i du_i \left[\prod_{i=1}^{4} \chi^{j_i^+}\left(g_i^+ h_+ u_i (h'_+)^{-1} (g_i'^+)^{-1}\right) \chi^{j_i^-}\left(g_i^- h_- u_i (h'_-)^{-1} (g_i'^-)^{-1}\right) \chi^{J_i}(u_i) \right],$$

where all the integrals are over $SU(2)$, the $Spin(4)$ group elements are decomposed into their selfdual/anti-selfdual components, (j_i^+, j_i^-) label irreducible unitary representations of $Spin(4)$, while J_i label irreps of the diagonal $SU(2)$ subgroup, and χ are the representation characters; here γ has to be a rational number. A second example is the version of the Baratin-Oriti (BO) model [26] resulting from a specific choice of imposing the constraints in Lie algebra (flux) variables, again only in the kinetic term:

$$S_{\text{GFT}}^{\text{BO}} = \int [dg_i][dg_i']dkdk' \, \varphi^*(g_i; k) \, C^{-1}(g_i, k; g_i', k') \, \varphi(g_i'; k')$$

$$+ \frac{\lambda}{5!} \int [dg_{ij}][dk_j]\varphi(g_{1j}; k_1) .... \varphi(g_{5j}; k_5) \prod_{i \neq j, i,j=1}^{5} \delta(g_{ij}, g_{ji}) + \text{c.c.}$$

with the same BF interaction term, and an additional set of variables $k_i \in S^3 \simeq SU(2)$, interpreted as unit normals to the tetrahedra, introduced to ensure the covariant imposition of the constraints, while the kinetic term is determined by:

$$C_{\text{BO}}(g_i, k; g_i', k') = \int [dx_i][dy_i] \int dh^\pm dh'^\pm \, \delta\left(k'k^{-1}\right) \prod_i \left[E_{g_i^+ h^+}(x_i^+) E_{g_i^- h^-}(x_i^-) \star \right.$$

$$\left. \delta_{-kx_i^- k^{-1}}(\beta x_i^+) \star \delta_{-x_i^+}(y_i^+) \delta_{-x_i^-}(y_i^-) \star E_{g_i'^+ h'^+}(y_i^+) E_{g_i'^- h'^-}(y_i^-) \right]$$

where $\star$ is the $\star$-product entering the definition of the Lie algebra representation in terms of variables $X_i = (x_i^+, x_i^-) \in \mathfrak{so}(4)$ and encoding the chosen quantization map for flux operators, $E_g(x)$ are the corresponding non-commutative plane waves, satisfying $E_g(x) \star E_h(x) = E_{gh}(x)$, and $\delta_{-y}(x) = \int du E_u(y) E_u(x)$ is the corresponding non-commutative delta function [20, 26]. This second model can easily be written down in group variables only (carrying out the Lie algebra integrations).

These choices of GFT action give rise to two different spin foam models, as the corresponding Feynman amplitudes are different. The general properties of the GFT definition of the spin foam models are the same, though: the spin foam amplitudes associated to a given spin foam 2-complex (or dual triangulation) are completed by a sum over complexes of any topology, generated perturbatively with the canonical combinatorial weights of a field theory, depending on the coupling constants and on the symmetry group of each complex, as in 2. They can be equivalently expressed in group variables (where they take the form of lattice gauge theories), Lie algebra variables (where they take the form of simplicial gravity path integrals) or group representations. This formulation of the quantum gravity dynamics is covariant, and it incorporates topology change in a very natural way. It is a form of discrete 3rd quantization of gravity, and a peculiar type of discrete realization of gravitational path integral, combining both sum over lattices as in dynamical triangulations [8], and a sum over discrete geometric data for each given triangulation, like in quantum Regge calculus [6]. Combined with the direct (heuristic) link between a GFT model and a canonical operator dynamics, the GFT encoding of spin foam models allows to link them directly with the canonical theory.

The 'spin foam' strategy is quite satisfactory from the point of view of encoding discrete geometry and making contact with discretized gravity. However, it leaves the definition of the *theory space*, e.g. the set of possible interactions that can/should be included in the theory, rather ambiguous. One could consider adding more interaction terms with different combinatorics, for example, not to mention that the same strategy of constraining BF theory could be extended to formulations of gravity more general than the Holst-Palatini one. One possible attitude toward these issues is to rely on the fundamental nature of simplicial geometry and on the renormalization group. One could argue that simplicial structures can be considered the most basic type of lattices on which to discretize geometry, and that one has to simply start from the simplest GFT action ensuring geometricity of the simplicial structures it generates, and then run the quantum dynamics and the renormalization group to generate all possible interactions compatible with that at different scales, the only constraint being renormalizability. Still, one may want to have a more principled definition of the GFT theory space [23], resting on basic assumptions, some GFT counterpart of QFT axiomatics.

3.3. GFT Dynamics from Tensorial Axiomatics

This leads to a third strategy for the construction of GFT models. It partly stems from the search for a good notion of locality for GFTs by extending the basic features of matrix models for 2d quantum gravity [5] to higher dimensions, stressing the interpretation of GFTs as (richer) tensor models [14]. First, to be able to interpret the GFT field as a proper tensor, one has to define a transformation under a unitary group $U^{\times d}$, like: $\varphi(g_1,...,g_d) \to \int [dg_i] U(g_1', g_1) \cdots U(g_d', g_d) \varphi(g_1,...,g_d)$, for $\int dg\, U(g_i', g_i)^* U(g_i, \tilde{g}_i) = \delta(g_i', \tilde{g}_i)$, which in turn requires the d arguments of the GFT field to be *labelled*. Given the tensorial transformation property, one can define *tensor invariant* interactions corresponding to invariant convolutions I_b of polynomials of GFT fields. Such invariants are indexed by *colored d-graphs* $\mathcal{B}$ constructed as follows: for each GFT field (resp. its complex conjugate) draw a white (resp. black) node with d outgoing links each labelled by d different colors, then connect all links with one another following the two conditions that a white (resp. black) node can only be connected to a black (resp. white) node and that only links with the same color can be connected. For real fields, one has a similar definition involving orthogonal transformations. This is the tensor invariance property generalizing the invariance of matrix models ($d = 2$), whose interactions can only be traces of matrix polynomials, replacing the notion of locality of usual QFT. This suggests a GFT theory space with arbitrary tensor invariant interactions and a kinetic term which may instead break the invariance (again, as in usual QFT):

$$S_{GFT} = \int [dg_i][dg_i']\, \varphi^*(g_i)\, \mathcal{K}(g_i, g_i')\, \varphi(g_i') + \sum_{b \in \mathcal{B}} t_b I_b(\varphi, \varphi^*) \qquad (7)$$

where t_b are the coupling constants corresponding to the various tensor invariant interactions. The use of tensor invariant interactions has a very nice property at the topological level. The Feynman diagrams of such GFTs can be represented as $(d+1)$-colored graphs, in which the d-colored graphs representing interactions are of course associated to vertices of the diagrams and an extra color is associated to the lines of propagation of the GFT field. $(d+1)$-colored graphs are in one to one correspondence with simplicial d-manifolds with at most conical singularities [14]. These topological properties give an additional motivation to follow this third strategy, while one has also to notice that, up to now, this strategy has offered less indications in the choice of field content and of kinetic/interaction kernels.

The same topological considerations show also an interesting link between models based on simplicial interactions and tensor invariant models. Starting from simplicial interactions, one notices that the data in a GFT Feynman diagram are sufficient to determine a 2-complex dual to a simplicial complex, but not to specify the full homology of the same complex, in particular the k-cells for $k < d - 2$. This is a problem, in particular if one wants to understand the symmetry properties of the amplitudes associated to the same Feynman diagrams [17, 31]. A solution to this problem [14, 18] is to add *colors* to simplicial GFTs. Instead of working with a single d-valent GFT field, one uses $d+1$ such fields φ_a, and their complex conjugates, labelled by a color index $a = 1, .., d + 1$, with a kinetic term for each of them, and a single simplicial interaction involving a convolution of all of them, one for each $(d-1)$-simplex in a d-simplex, plus the complex conjugate of the same interaction. Feynman diagrams will again be $(d+1)$-colored graphs, representing the 1-skeletal of simplicial pseudo-manifolds. Moreover [32], starting from such colored simplicial models and integrating out all but one of the GFT fields, one is left with a theory of a single GFT field with interactions indexed by colored d-graphs $\mathcal{B}$. This intriguing connection deserves to be further explored.

3.4. GFT Symmetries

Still, much remains to be understood about basic principles determining the GFT theory space. In particular, one would like to understand more about GFT symmetries. For GFTs aiming at quantum gravity, one would like to understand better especially the role of diffeomorphisms. Here as well there are three main lines of attack, obviously not exclusive, corresponding to the three GFT construction strategies above. Coming from the canonical theory, one would make sure that some Hamiltonian (and spatial diffeo) constraint operator is encoded in the quantum GFT dynamics by construction, and look for the sector in which only quantum states satisfying the corresponding operator equations are relevant. This route has not been followed up to now, if not indirectly via the spin foam route. Following the spin foam or discrete gravity route, one would first look for some discrete counterpart of diffeo symmetry satisfied by the GFT Feynman amplitudes, and then look for the field-theoretic definition of the same at the GFT level. Simplicial diffeos have been studied in both discrete gravity [33] and topological spin foam models [34], and their GFT counterpart has been identified [17]. This turns out to be a *global* quantum group symmetry, at the GFT level, and leads naturally to the same discrete WdW-like equation and recursion relations seen also at the spin foam level [35]. However, these simplicial diffeos are broken by the simplicity constraints, leaving open the issue of recovering them in a suitable approximation of gravitational 4d GFT models. A third complementary strategy follows again the insights of matrix models. There [5], the Virasoro algebra is found from a set of constraints satisfied by the partition function, in turn following from the Schwinger-Dyson equations of the n-point functions associated to loop observables. The latter can also be directly related to the Wheeler DeWitt equation of 2d Liouville gravity corresponding to the continuum limit of the same matrix models. An analogous result is found in simple tensor models [14, 36], for a generalization of the Witt algebra. However, we still miss the link between this symmetry algebra and higher-dimensional diffeomorphisms, and we need to generalize these results to GFTs with interesting quantum geometric data. In general, the treatment of GFT symmetries as in standard QFTs (Noether theorem, Ward identities, symmetry breaking) needs to be better developed [37].

4. The Continuum Limit of Quantum Geometry in GFT

GFTs give a picture of space-time as fundamentally discrete: the fundamental degrees of freedom that (should) make up space-time are GFT quanta (simplices or spin nets vertices), as brought to the forefront by the second quantized formulation, and their dynamical processes are the GFT Feynman diagrams, themselves cellular complexes replacing a smooth space-time manifold. Further, if the group manifold is chosen to be $SU(2)$ or another compact group, one obtains an additional discreteness of quantum geometric spectra, such as areas and volumes, as in standard LQG.

Still, our conventional description of gravity and space-time is in terms of a continuum field theory, general relativity, based on a smooth metric field on a smooth manifold (it is in this sense that the Regge geometries found in GFT Feynman amplitudes, which are continuous but piecewise flat, thus singular, are not enough). To reconcile this description with the one provided by GFTs is the problem of extracting an effective continuum description for the GFT degrees of freedom.

4.1. *Clarifying the Nature of the Problem*

Before we move on to discuss this area of GFT research, we would like to emphasize the difference with the problem of the classical limit. By *continuum approximation* in this context we mean a regime in which a large number of (interacting) fundamental degrees of freedom, i.e. GFT quanta, are considered, possibly in terms of some collective variables (and with additional physical constraints). By *classical approximation* we mean the regime in which the quantum nature of the same degrees of freedom (few or many) can be neglected. The two limits are different, do not need to commute at all, and have to be explored independently. In particular, it could well be that the quantum nature of the fundamental degrees of freedom is actually *needed* to achieve the correct continuum effective physics, i.e. the classical approximation may have to be taken *after* the continuum approximation. From this perspective, the semiclassical approximations in spin foam models [13] (e.g. the asymptotic analysis, showing dominance of Regge geometries for each spin foam 2-complex) and canonical LQG (e.g. coherent states peaking on discrete classical geometric data for any given spin network graph [38]), however interesting, do not ensure the existence of the right continuum limit, and may even not select the right configurations to consider to reach it. The continuum limit in GFTs is rather to be understood in analogy with the thermodynamical limit in usual QFTs (e.g. in condensed matter) as the regime of (formally) infinite fundamental atoms and of various cut-offs being (formally) removed.

The problem of the continuum in GFT can then be articulated in four main related aspects. One is the perturbative renormalizability of GFT models. The second is the study of their phase structure. The third is their non-perturbative, constructive definition. The fourth is the extraction of effective continuum physics from them, in the appropriate regime. We will discuss the last, more physical issue in the next section. Here we will deal with the first three, more formal issues.

In order to appreciate further their relevance, let us first discuss what role these aspects of GFTs play from the point of view of canonical LQG and of spin foam models. Both GFT renormalizability and constructibility are simply equivalent to the requirement of GFT models be (non-perturbatively) well-defined. As such, they are a re-phrasing of the requirement of having a well-defined dynamical constraint operator. Another issue concerning the definition of the canonical constraint operator is that of quantization ambiguities discussed in *Chapter 2*. These have been related to the issue of renormalizability in perturbative QG [39], and its GFT counterpart is again GFT renormalizability. Even if the theory was well-defined mathematically, important physical questions would remain: what is the flow of effective quantum dynamics across scales? what is the continuum phase structure? These issues are crucial both in the canonical and in the covariant setting. They can be tackled very effectively in the GFT formalism, thanks to the QFT setting, while it is not so clear (given the experience with the same issues in many-body quantum theory) that they can be tackled as effectively remaining in the canonical operator formalism. As for the spin foam context, if the spin foam expansion is understood as arising from expansion of canonical evolution operator, the problem is to make sense of this expansion, i.e. to define the canonical evolution operator rigorously, and becomes again the issue of finding a constructive definition of the GFT partition function. If spin foams are understood as lattice gravity path integrals, the problem of the continuum translates into the issue of refining/coarse graining them, and identify the flow of effective dynamics and its fixed points [25]. That is, the same problems that GFT recasts in QFT language by turning the same lattice gravity path integrals into Feynman amplitudes. Regardless of the precise point of view one takes about spin foam models, one thing is clear: there is simply no spin foam theory without a prescription of how to deal with spin foam amplitudes on different complexes. In other words, a complete definition of a spin foam theory is not simply the set of all possible spin foam amplitudes for all possible spin foam complexes, and relational physics is not to be extracted simply by choosing some given spin foam complex, adapted to the situation at hand. Rather, a complete definition is given by: this set of complexes *plus* a precise organization principle, a procedure to relate the amplitudes for different complexes. Because the set of all complexes has to include the ones made of infinite numbers of cells/links/faces (to account for an infinite number of degrees of freedom), this becomes the problem of defining the continuum limit. A formal definition of such organization principle specifies a formal definition of the theory, to be made rigorous. This principle could be a refinement limit [25] discussed in *Chapter 6*, or a prescription for summing over complexes. The latter could be chosen arbitrarily, but this introduces a further huge ambiguity. Rather, it should be selected by an appropriate set of principles (maybe motivated by the canonical perspective [21]). The GFT formalism provide exactly such prescription, a natural and clear organization principle for spin foam amplitudes, and for defining their continuum limit.

4.2. *A Survey of Current Research*

In pursuing the issue of renormalizability and constructibility of GFTs, the input from tensor models has proven crucial. In fact one main obstacle towards making sense of the continuum limit of GFTs is to control the sum over triangulations/spin foam 2-complexes they generate, which are much more intricate than standard QFT Feynman diagrams. Progress has been made, thanks to developments in tensor models. The introduction of colors [14, 18] gave control over the combinatorics and topology of GFT Feynman diagrams, and over their sum. One important result has been the understanding of the large-N limit of tensor models (where N is the size of the tensors) and topological GFTs (where N is the cut-off in representations) [40], which are the starting point for the construction of most 4d gravity models. It has been shown that the leading order in N corresponds to *melonic* diagrams [14, 32, 41], which are dual to triangulations of spheres with a peculiar combinatorial structure [14].

GFT renormalization started being tackled in [42] and is, since then, a very active research area [43]. Next to the mentioned combinatorial aspects, this involves a deeper understanding of the scaling of amplitudes in the large-N limit. Indeed, the large-N regime is the regime of many GFT degrees of freedom (for given combinatorics). From this point of view it represent the analog of the UV regime in usual QFTs. Conversely, the low-N regime would be the analog of the IR regime. Consistently with this picture, it is in this large-N regime that divergences in spin foam amplitudes are usually found. However, the geometric interpretation associated to it is not totally clear. Large spins in $SU(2)$ spin foam models correspond to large areas and volumes for the discrete structures they are based on, consistently with the quantum geometry of canonical LQG. From the point of view of simplicial geometry, then, the geometric notion of UV/IR is rather the opposite of the formal QFT one. Still, much caution should be exercised in interpreting in geometric terms the algebraic data and discrete structures of GFTs (and spin foams), before a proper continuum limit is established and we control how such data map to continuum geometric ones. Still, one can proceed guided only by the mathematical behavior of GFT amplitudes. Their scaling with N is the first thing that has been studied, with focus on (colored) topological models, in particular showing the suppression of singular topologies [44]. Still in the context of topological GFTs, remarkable calculations of radiative corrections were performed [45], and one interesting implication was that, in order to achieve renormalizability, these models need to be augmented by a kinetic term given by the Laplace-Beltrami operator on the group manifold

$$\int [dg_i]\, \varphi^*(g_i) \sum_{i=1}^{d} \Delta_{G_i}\, \varphi(g_i) \qquad .$$

While its quantum geometric meaning is still unclear, this is indeed a natural choice of kinetic term, and it has been later shown to make topological models superrenormalizable (at least in the Abelian case) [43]. Similar, and much more challenging computations have to be performed in a more systematic way in the context of 4d gravity models. While the BC model is super-renormalizable as well, at least for special choices of edge amplitudes [11, 29], little is known about models incorporating the Barbero-Immirzi parameter, like the EPRL and BO models. The simplest radiative corrections of the EPRL model have been investigated in [28] and, much more thoroughly, in [46], and are similarly dependent on the details of edge weights.

While these calculations are technically impressive and important, one would like a more systematic analysis of perturbative renormalizability, which in turn requires more control on the theory space. Indeed, up to now this type of systematic analysis has been carried out only in simplified models characterized by tensor invariant interactions and the mentioned Laplacian-type kinetic term (this class of models has been dubbed *tensorial GFTs*). While the use of tensor invariant interactions provides a notion of locality, the Laplacian term gives a clear notion of *scale* (indexed by its eigenvalues, i.e. by group representations, as expected from LQG and spin foam models). This allows to apply to the more involved GFT context the rigorous multi scale renormalization analysis of conventional QFTs. Results have been piling up rapidly. After the identification of several renormalizable models in various dimensions [43], the investigation of models incorporating the gauge invariance conditions characterizing topological GFT models (and crucial also for gravitational models in 4d) has started, and renormalizable models of the Abelian type were identified first [43, 47]. Lately, a non-Abelian GFT model with gauge invariance and interactions up to order six, extending the Boulatov model to include a Laplacian term, was also shown to be renormalizable [48]. This type of analysis is highly non-trivial, due again to the intricacies of cellular complexes, requiring an extension or adaptation of several notions and tools from standard renormalization theory: the notion of Wick ordering, the notion of connectedness and of 1-particle irreducibility, that of contraction of high subgraphs, using tools from crystallization theory such as *dipole moves*. The stage is now set for a similar systematic analysis of gravitational GFT models in 4d, incorporating also simplicity constraints. A renormalizability result in this context, for models of the EPRL or BO type, would be of paramount importance for the whole field of quantum gravity. Renormalizability (or perturbative finiteness, which is nicer in some respect but more problematic in others, as it makes it harder to identify the relevant channels of interactions at different scales) would imply that the given GFT model is perturbatively well-defined, as a QFT. It would represent a *truly background independent and renormalizable quantum (field) theory of spacetime*. Given that how spin foam models arise in GFT, perturbative GFT renormalizability would also give meaning to the spin foam sum over complexes.

The renormalization group also determines the flow of effective GFT dynamics across scales, and helps us to map out the phase structure of the theory. Detailed calculations of the beta functions for various tensorial GFTs have been performed [49], including non-Abelian ones [50]. While establishing general conclusions is very tricky [50], the argument can be made for asymptotic freedom (or safety) being rather generic in GFTs. Asymptotic freedom would of course be welcome news for GFT models of quantum gravity because it would confirm their being well-defined as QFTs. It would also suggest the existence of phase transitions at some point during the flow towards small values of N, i.e. small values of group representations, due to a corresponding growth in the GFT coupling constants. One powerful tool to study such flow, mapping out the GFT phase diagram, is the so-called Functional Renormalization Group. An extension of the FRG to matrix models has been developed [51], and the first definition and application of the FRG to tensorial GFTs has been provided recently [52]. Beside making available this powerful tool, it allowed to obtain interesting information about fixed points and the phase diagram of simple GFT models at both large and small N, and the first indications of a phase transition between a phase characterized (in mean field) by a vanishing GFT field, and a 'condensed' phase with non-vanishing expectation value of the same. The possible physical relevance of this scenario will be discussed in the next section.

The analysis of GFT phase transitions is in its infancy. The only other result is the proof of existence of a phase transition for any topological (BF) model in any dimension, in the melonic sector [53]. More is known in tensor models. Here the problem is tackled by explicit resummation of the partition function, in the appropriate regime, i.e. from a statistical rather than field-theoretic point of view. For both i.i.d. and dually weighted models, the critical behavior of the melonic sector as well as of subdominant orders (including a rigorous double scaling limit) has been studied, with explicit calculation of the critical behavior [32, 41] [54]. Once more, work on simpler models may pave the way for the analysis of GFTs.

Results on GFTs should be compared with analogous results in canonical LQG and in the lattice gravity approach to spin foam models. In the canonical setting, these issues have not been much explored, though, and what we know at present is only that inequivalent kinematical representations of the holonomy-flux algebra exist, beyond the Ashtekar Lewandowski one [55, 56]. Also, work on GFT renormalization should be carefully compared with renormalization of spin foam models treated as lattice gauge theories [25]. In fact, the same procedures of subtraction of subgraphs, adopted to establish perturbative GFT renormalizability, can be seen as lattice coarse graining procedure, from a different perspective.

To conclude, we mention results on constructive aspects of GFTs, aiming at a non-perturbative definition of GFTs, i.e. the 'summability' of the perturbative series (it is not going to be convergent, in general). The summability in the melonic (large-N) sector and of subdominant orders, all involving complexes of trivial topology, has been already discussed. Interesting results have been obtained also on the Borel summability of the whole GFT partition function for topological models [57], and these results, although their physical meaning is yet to be explored, are remarkable because they amount to being able to sum over all cellular topologies generated by the GFT perturbative expansion. More remains to be done, and once more results on constructive tensor models [58] will be an important reference.

5. Extracting Effective Continuum Physics from GFTs

5.1. *A Physical Perspective on the Continuum and on Phase Transitions in GFT*

As mentioned, the crucial problem for any background independent quantum theory of gravity is to recover classical GR on a smooth manifold in the continuum and classical limit. Many results have been obtained in the canonical setting as well as in spin foams (and in Regge calculus), e.g. on semiclassical states and asymptotic analyses of spin foam amplitudes. Most of them are immediate to import in GFT, since GFT is a reformulation of canonical LQG and a complete encoding of spin foams. These are significant assets. However, the important points we have already mentioned in the previous section should not be forgotten: first, the classical approximation is independent and possibly secondary to the continuum one; second, simple graphs/lattices (few degrees of freedom) could be useful to capture some limited information of continuum geometry and physics, in very special regimes, but effective continuum physics requires the limit of formally infinite number of fundamental degrees of freedom. This is also the reason why recent attempts to compute effective continuum physics using only spin foam amplitudes [59] are problematic from the GFT point of view: they are confined to a regime, associated to *simple* spin foam 2-complexes (the perturbative GFT regime at the lowest order), and to a class of states, *simple* spin network graphs (very few GFT quanta), very far from the non-perturbative, approximately continuum sector of the theory, and instead too close to the regime of fully degenerate geometry (GFT Fock vacuum). They give important insights, but the GFT perspective suggests that something different is needed.

Let us clarify. The GFT formulation of LQG and spin foam models provides tools to deal with many QG degrees of freedom, i.e. to control the superposition of states and amplitudes associated to different, refined lattices, just like in condensed matter systems the QFT language is useful to control many-particle physics and to extract effective dynamics from it. Like QFTs in condensed matter, one can deal with the GFT dynamics perturbatively, i.e. in its spin foam expansion. However, this should be expected to be the right language only as long as few degrees of freedom are involved, i.e. close to perturbative GFT vacuum. In turn this perturbative vacuum is physically a 'no-space state', degenerate in geometry and topology. One would expect non-degenerate geometries and effective continuum physics to arise far from such vacuum state and close to a different non-perturbative vacuum (a different phase) of the theory. This expectation is consistent with the canonical LQG picture, where the kinematical Ashtekar-Lewandowski vacuum is geometrically degenerate, and simple spin network excitations around it are not enough to generate a smooth geometry. There are two interpretations one could give to this 'need for a new, geometric phase'. The first treats this as a purely formal requirement, as simply implying that the theory is only physical in one appropriate phase (i.e. for the appropriate range of parameters), but with no meaning attached to other phases of the theory or to the phase transitions between them. This is of course a conceptually consistent picture and, for what we know, could be the right one (it is, for example, the picture behind most work on the dynamical triangulations approach to quantum gravity [8]). A second interpretation, however, could also be put forward. It supposes that the other phases of the system may have a physical existence as well, even though they happen not to describe our current geometric regime of physical universe. In this view, the transition between phases may also have physical meaning, that is it could correspond itself to a physical process of the quantum gravity building blocks of space-time. More specifically, one could suggest that the phase transition to the geometric phase is what replaces the big bang singularity and describes the 'origin' of our physical universe. Thus, one has a cosmological interpretation for the GFT phase transition to a non-degenerate, geometric non-perturbative GFT vacuum. This hypothesis is dubbed 'geometrogenesis' [60]. While the search for such phase transition at the more rigorous level continues as summarized in the previous section, it is worth exploring possible concrete scenarios for it in GFT, aiming for a more direct extraction of possible physical consequences, and better physical insights into the formalism. This also means exploring candidates for the new vacua. A further hypothesis can then be put forward: our geometric universe could be born from a *condensation* of quantum space atoms [22]. This would realize explicitly in the GFT setting the idea of space-time as a condensate and of an emergent universe, often discussed in the context of analog gravity models [61, 62].

5.2. *Some Recent Results*

Thus one turns the attention to a special class of states within the GFT Hilbert space: GFT condensates [63], i.e. quantum states characterized by a macroscopic occupation number for some given quantum observable [64], and involving a superposition of an arbitrary number of GFT quanta. In the simplest case, these are such that all the GFT quanta are in the same quantum state. One can show [63], using results in LQG and discrete gravity, that such condensate states, provided geometricity conditions (e.g. simplicity constraints) are imposed in the quantum dynamics, admit an interpretation as continuum homogeneous spaces, of the type used in cosmology. They are quantum states of the full theory, but they are characterized by a collective wave function depending only on the quantum geometric data associated to homogeneous anisotropic geometries. This characterization still leaves room for a variety of constructions [63, 65, 66], but the simplest condensate state one (the GFT analog of the Gross-Pitaevskii ansatz for Bose-Einstein condensates):

$$|\sigma\rangle := \mathcal{N}(\sigma)\exp(\hat{\sigma})|0\rangle \quad \text{with} \quad \hat{\sigma} := \int (dg)^4 \, \sigma(g_1,\ldots,g_4) \, \hat{\varphi}^\dagger(g_1,\ldots,g_4) \quad (8)$$

where $\sigma(kg_1,\ldots,kg_4) = \sigma(g_1,\ldots,g_4) \, \forall k \in Spin(4)$ or $SL(2,\mathbb{C})$, and $\mathcal{N}(\sigma)$ is a normalization factor, is already very interesting. This state is quite special, as it is a coherent state for the GFT field operator: $\hat{\varphi}(g_I)|\sigma\rangle = \sigma(g_I)|\sigma\rangle$, and using it as a (coarse grained) proxy of the true vacuum state is a sort of mean field approximation. Inserting this ansatz for the vacuum state of the theory in the quantum dynamics, that is in the Schwinger-Dyson equations, and making some further approximations, one gets an effective equation for the collective wave function σ [63]:

$$\int [dg'_i] \, \tilde{\mathcal{K}}(g_1,\ldots,g_4,g'_1,\ldots,g'_4)\sigma(g'_1,\ldots,g'_4) + \lambda \left. \frac{\delta \tilde{\mathcal{V}}[\varphi,\varphi^*]}{\delta \bar{\varphi}(g_1,\ldots,g_4)} \right|_{\varphi \to \sigma, \varphi^* \to \sigma^*} = 0.$$

which is nothing else than the classical equation of motion of the initial GFT model, up to any additional approximation needed to ensure consistency of the interpretation (symbolized by the ~ on the dynamical kernels). Given the interpretation of the collective wave function σ as a distribution over the space of continuum homogeneous geometries, this equation represents a non-linear extension of the WdW-like equation of loop quantum cosmology [67]. Extended quantum cosmology equations of this type have been previously suggested in [68, 69]. The crucial point is that such quantum cosmology equation is here *derived from the fundamental theory*, for a suitable class of states, and no minisuperspace reduction is carried out (it should be compared with similar approaches, like [70]); rather, (generalized) quantum cosmology emerges from the fundamental dynamics as a kind of hydrodynamics approximation. This derivation matches the general expectations we discussed about the continuum approximation of the theory and phase transitions. Moreover, the derivation is completely general (for 4d gravity models with geometricity conditions), and it applies for any choice of fundamental GFT (thus LQG/spin foam) dynamics, encoded in the GFT kernels $\mathcal{K}$ and $\mathcal{V}$. A simple example corresponding to using the Laplacian operator as $\mathcal{K}$ and neglecting the GFT interaction, for both Lorentzian and Riemannian gravity, also incorporating a massless scalar field, already shows how a semi-classical Friedmann equation could be obtained by such methods [63, 71]. Similar results hold [63] for different choices of GFT condensate states.

This equation can also be written down in terms of expectation values of the collective operators corresponding to cosmological variables [72]. A natural canonical pair, in terms of which one should try to write any effective semiclassical cosmological dynamics in a GFT setting, is given by (the expectation value of) the total flux $\hat{b}_I^i = i\kappa \int (dg)^4 \, \hat{\varphi}^\dagger(g_J) \frac{d}{dt}\hat{\varphi} \left(\exp\left(\tau_I^i t\right) g_J\right)\Big|_{t=0}$ (κ is a combination of Planck's and Newton's constants), from which geometric quantities like macroscopic areas and volumes are extracted, and the 'average holonomy' $\hat{\Pi}[g_I]^{\text{av.}} = \langle\hat{\Pi}[g_I]\rangle/\langle\hat{N}\rangle$, which contains extrinsic curvature information, computed from the extensive 'total holonomy' $\hat{\Pi}[g_I] = \int (dg)^4 \, \vec{\pi}[g_I] \, \hat{\varphi}^\dagger(g_J)\hat{\varphi}(g_J)$, for a choice of coordinates on the group $g = \sqrt{1-\vec{\pi}[g]^2}\,\mathbf{1} - i\vec{\sigma}\cdot\vec{\pi}[g]$, and the number operator $\hat{N}$. If these are the (expectation values of the) operators coming directly from the full theory, one sees immediately that the theory incorporates crucial quantum corrections to the usual classical variables. In fact, the total flux is, like its microscopic counterpart, non-commutative, while the standard cosmological variables correspond to its commutative limit $\hat{f}_I^i = i\kappa \int (dg)^4 \, \hat{\varphi}^\dagger(\pi[g_J])\frac{\partial}{\partial \pi_i^I}\hat{\varphi}(\pi[g_J])$. Likewise, from the average holonomy one can define a macroscopic connection variable: $\vec{\omega} := -\frac{\langle N\rangle^{1/3}\langle\vec{\Pi}\rangle}{|\langle\vec{\Pi}\rangle|} \arcsin \frac{|\langle\vec{\Pi}\rangle|}{\langle N\rangle}$. This second fact is important in several ways. To start with, one sees that the connection enters the effective cosmological dynamics necessarily with quantum holonomy corrections encoded in the sine function, as in loop quantum cosmology. Next, and most important, the effective holonomy carries a dependence on the number of fundamental cells forming the universe. Again, this is like the lattice-refinement scheme in loop quantum cosmology, but here not only the dependence on N is derived and not assumed, but this N is a new second quantized quantum observable of the theory, which enters necessarily both the kinematics and the effective cosmological dynamics [72]. Finally, the same effective dynamics will relate the expectation value of N and that of a, the scale factor, in such a way that cosmological holonomies end up depending on the scale factor as well, like in the so-called $\bar{\mu}$-scheme of LQC. For more details, we refer to the literature [63, 65, 71, 72].[a] It is clear, however, that one has now many promising paths to explore, concerning, for example: the detailed analysis of effective cosmology coming out of various fundamental GFT models (e.g. EPRL or BO), the study of improved ansatz for condensate states (with a more detailed encoding of topology, involving better correlations between GFT quanta, etc); most important, a new way to study cosmological perturbations, understood as fluctuations above the GFT condensate, for which the derivation of an effective field theory picture is the most pressing issue. Beside the many intriguing conceptual aspects of this new picture of cosmology emerging from the full QG theory, this seems also a promising avenue towards extracting testable predictions directly from the fundamental theory, putting it in contact with observations.

We close by mentioning other results aiming at the extraction of effective continuum physics from GFTs, and working also in the spirit of mean field theory, all obtained for GFT perturbations around (approximate) solutions of the classical GFT equations. They range from the derivation of an effective non-graph changing Hamiltonian constraint for spin networks starting from a GFT of topological type [73], to the extraction of classical equations for geometric phase space variables from the same GFT equations, using LQG coherent states as ansatz for the approx-

---

[a]There have been further significant advances in this area since this article was completed. See the 2016 papers in arXiv by De Cesare, Gielen, D. Oriti, A. Pithis, M. Sakellariadou, L. Sindoni, P. Tomov and E. Wilson-Ewing.

imate solutions [74]. This last work is very close to the extraction of cosmology from GFT condensates discussed above. Also, focusing on the effective dynamics of perturbations around classical GFT solutions, one could recast it in the form of non-commutative scalar field theories on a non-commutative flat space [75], at least for specific models and specific choices of perturbations. This is an intriguing result, which should now be reanalyzed in the context of GFT condensate cosmology.

6. Conclusions

We have provided a quick survey of the GFT formulation of quantum geometry and of its dynamics, and of some of the results in this area. We have tried to clarify the links (and differences) with loop quantum gravity and spin foam models, emphasizing that GFTs offer a new elegant description of a quantum gravity theory based on spin networks, and at the same time a completion of the spin foam dynamics. This QFT framework, in particular, offers new tools that may prove crucial for addressing the issue of the continuum approximation and for the extraction of effective continuum dynamics from the full theory. In the long run, the goals of this approach are clear. We aim for: (1) a reliable (class of) model(s) for 4d Lorentzian quantum gravity with matter, with a nice and well-understood encoding of quantum geometry at the microscopic scales; (2) a proof that the same (class of) model(s) is perturbative renormalizable and possibly constructively well-defined; (3) a detailed map of the continuum phase structure of the same model(s); (4) a detailed understanding of the effective cosmological equations for the very early Universe emerging from the fundamental quantum gravity dynamics, and a quantum gravity solution to cosmological puzzles (flatness and horizon problems, the cosmological constant, the cosmological singularity) within the full theory, including a theory of cosmological perturbations from first (quantum gravity) principles, to be tested against observations. This is an ambitious program, as any quantum gravity program has to be. The main asset of the GFT formalism, we believe, is the novel outlook it provides on canonical LQG as well as on spin foams, and the new tools we mentioned, with the simultaneous possibility to incorporate and take advantage of all the results obtained in such contexts, alongside those obtained in related formalisms like tensor models. It is this fruitful blend of solid, established results and techniques coming from different corners with a novel, promising new perspective that we hope will guarantee many more results and scientific surprises in the future, and a decisive progress for the whole field of quantum gravity.

References

[1] D. Oriti, in *Foundations of Space and Time*, G. Ellis, J. Murugan, A. Weltman (eds.) (Cambridge University Press, 2012), arXiv: 1110.5606 [hep-th].
[2] D. V. Boulatov, *Mod. Phys. Lett. A* **7** (1992) 1629-1646, [arXiv:hep-th/9202074].
[3] H. Ooguri, *Mod. Phys. Lett. A* **7** (1992) 2799, hep-th/9205090.

[4] J. Ambjorn, B. Durhuus, T. Jonsson, *Mod. Phys. Lett. A* **6** (1991) 1133-1146.
[5] P. Di Francesco, P. Ginsparg, J. Zinn-Justin, *Phys. Rept.* **254** (1995) 1-133, hep-th/9306153; G. Moore, N. Seiberg, M. Staudacher, *Nucl. Phys. B* **362** (1991) 665-709.
[6] T. Regge, R. M. Williams, *J. Math. Phys.* **41** (2000) 3964-3984, arXiv: gr-qc/0012035.
[7] C. Rovelli, *Phys. Rev. D* **48** (1993) 2702-2707, arXiv: hep-th/9304164.
[8] J. Ambjorn, J. Jurkiewicz, R. Loll, *Phys. Rept.* **519** (2012) 127, arXiv:1203.3591.
[9] A. Ashtekar, J. Lewandowski, *Class. Quant. Grav.* **21** (2004) R53-R152.
[10] R. De Pietri et al., *Nucl. Phys. B* **574** (2000) 785, arXiv: hep-th/9907154.
[11] A. Perez and C. Rovelli, *Phys. Rev. D* **63** (2001) 041501, arXiv: gr-qc/0009021.
[12] M. Reisenberger, C. Rovelli, *Class. Quant. Grav.* **18** (2001) 121, arXiv: gr-qc/0002095.
[13] A. Perez, *Liv. Rev. Rel.* **16** (2013) 3, arXiv:1205.2019 [gr-qc].
[14] R. Gurau and J. Ryan, *SIGMA* **8** (2012) 020, arXiv:1109.4812 [hep-th].
[15] D. Oriti, J. Ryan, J. Thurigen, to appear.
[16] D. Oriti, arXiv:1310.7786 [gr-qc].
[17] A. Baratin, F. Girelli, D. Oriti, *Phys. Rev. D* **83** (2011) 104051, arXiv:1101.0590.
[18] R. Gurau, *Comm. Math. Phys.* **304** (2011) 69, arXiv:0907.2582 [hep-th]; J. Ben Geloun, J. Magnen, V. Rivasseau, *Eur. Phys. J. C* **70** (2010) 1119, arXiv:0911.1719 [hep-th].
[19] A. Perez, C. Rovelli, in *Quanta of Maths*, eds. E. Blanchard, et al., 501-518 (American Mathematical Society (2011)) gr-qc/0104034.
[20] A.Baratin, D. Oriti, *Phys. Rev. Lett.* **105** (2010) 221302, arXiv:1002.4723 [hep-th]; A. Baratin et al., *Class. Quant. Grav.* **28** (2011) 175011, arXiv:1004.3450 [hep-th]; C. Guedes, D. Oriti, M. Raasakka, *J. Math. Phys.* **54** (2013) 083508, arXiv:1301.7750.
[21] T. Thiemann, A. Zipfel, *Class. Quant. Grav.* **31** (2014) 125008, arXiv:1307.5885 [gr-qc].
[22] D. Oriti, *PoS QG-PH* (2007) 030, arXiv: 0710.3276 [gr-qc].
[23] V. Rivasseau, *Fortsch. Phys.* **62** (2014) 81-107, arXiv:1311.1461 [hep-th].
[24] D. Oriti, R. M. Williams, *Phys. Rev. D* **63** (2001) 024022, arXiv: gr-qc/0010031.
[25] B. Dittrich, M. Martin-Benito, E. Schnetter, *New J. Phys.* **15** (2013) 103004, arXiv:1306.2987 [gr-qc].
[26] A. Baratin and D. Oriti, *Phys. Rev. D* **85** 044003 (2012), arXiv:1111.5842 [hep-th].
[27] R. De Pietri, L. Freidel, *Class. Quant. Grav.* **16** (1999) 2187, arXiv: gr-qc/9804071; S. Gielen, D. Oriti, *Class. Quant. Grav.* **27** (2010) 185017, arXiv:1004.5371 [gr-qc].
[28] J. Engle, E. Livine, R. Pereira, C. Rovelli, *Nucl. Phys. B* **799**, 136 (2008), arXiv:0711.0146; J. Ben Geloun, R. Gurau, V. Rivasseau, *Europhys. Lett.* **92** (2010) 60008, arXiv:1008.0354 [hep-th].
[29] A. Baratin, D. Oriti, *New J. Phys.* **13** (2011) 125011, arXiv:1108.1178 [gr-qc].
[30] W. Kaminksi, M. Kisielowski, J. Lewandowski, *Class. Quant. Grav.* **27** (2010) 095006, Erratum-ibid. 29 (2012) 049502, arXiv:0909.0939 [gr-qc].
[31] V. Bonzom, M. Smerlak, *Phys. Rev. Lett.* **108** (2012) 241303, arXiv:1201.4996 [gr-qc].
[32] V. Bonzom, R. Gurau, V. Rivasseau, *Phys. Rev. D* **85**, 084037 (2012), arXiv:1202.3637.
[33] B. Dittrich, PoS QGQGS2011 (2011) 012, arXiv:1201.3840 [gr-qc].
[34] L. Freidel, D. Louapre, *Nucl. Phys. B* **662** (2003) 279-298, arXiv: gr-qc/0212001.
[35] V. Bonzom, E. Livine, S. Speziale, *Class. Quant. Grav.* **27** (2010) 125002.
[36] R. Gurau, *Nucl. Phys. B* **852** (2011) 592, arXiv:1105.6072 [hep-th].
[37] J. Ben Geloun, *J. Math. Phys.* **53** (2012) 022901, arXiv:1107.3122 [hep-th]; J. Ben Geloun, *J. Phys. A* **44** (2011) 415402, arXiv:1106.1847 [hep-th]; A. Kegeles, D. Oriti, *J. Phys. A* **49** (2016) 135401, arXiv:1506.03320 [hep-th]; A. Kegeles, D. Oriti, arXiv:1608.00296 [gr-qc].

[38] T. Thiemann, *Class. Quant. Grav.* **23** (2006) 2063,s arXiv: gr-qc/0206037; D. Oriti, R. Pereira, L. Sindoni, *Class. Quant. Grav.* **29** (2012), 135002, arXiv: 1202.0526 [gr-qc].
[39] A. Perez, *Phys. Rev. D* **73** (2006) 044007, arXiv: gr-qc/0509118.
[40] R. Gurau, *Annales Henri Poincare* **13** (2012) 399-423, arXiv:1102.5759 [gr-qc].
[41] V. Bonzom et al., *Nucl. Phys. B* **853** (2011) 174-195, arXiv:1105.3122 [hep-th].
[42] L. Freidel, R. Gurau, D. Oriti, *Phys. Rev. D* **80** (2009) 044007, arXiv:0905.3772 [hep-th].
[43] J. Ben Geloun, V. Rivasseau, *Commun. Math. Phys.* (2012), arXiv:1111.4997; J. Ben Geloun, D. O. Samary, arXiv:1201.0176; J. Ben Geloun, arXiv:1306.1201.
[44] S. Carrozza, D. Oriti, *JHEP* **1206** (2012) 092, arXiv:1203.5082 [hep-th]; V. Bonzom, M. Smerlak, *Annales Henri Poincare* **13** (2012) 185-208, arXiv:1103.3961 [gr-qc].
[45] J. Ben Geloun, V. Bonzom, *Int. J. Theor. Phys.* **50** (2011) 2819, arXiv: 1101.4294.
[46] A. Riello, *Phys. Rev. D* **88** (2013) 024011, arXiv:1302.1781 [gr-qc].
[47] S. Carrozza, D. Oriti, V. Rivasseau, *Commun. Math. Phys.* **327** (2014) 603-641, arXiv: 1207.6734 [hep-th].
[48] S. Carrozza, D. Oriti, V. Rivasseau, *Commun. Math. Phys.* **330** (2014) 581-637, arXiv:1303.6772 [hep-th]; S. Carrozza, arXiv:1310.3736 [hep-th].
[49] J. Ben Geloun, arXiv:1205.5513 [hep-th]; J. Ben Geloun, arXiv:1210.5490 [hep-th].
[50] S. Carrozza, arXiv:1407.4615 [hep-th].
[51] A. Eichhorn, T. Koslowski, arXiv:1408.4127 [gr-qc].
[52] D. Benedetti, J. Ben Geloun, D. Oriti, to appear.
[53] A. Baratin et al., *Lett. Math. Phys.* **104** (2014) 1003-1017, arXiv:1307.5026 [hep-th].
[54] S. Dartois, R. Gurau, V. Rivasseau, *JHEP* **1309** (2013) 088 (2013), arXiv:1307.5281 [hep-th]; V. Bonzom, R. Gurau, J. Ryan, A. Tanasa, arXiv:1404.7517 [hep-th].
[55] T. Koslowski, H. Sahlmann, *SIGMA* **8** (2012) 026, arXiv:1109.4688 [gr-qc].
[56] B. Dittrich, M. Geiller, arXiv:1401.6441 [gr-qc].
[57] J. Magnen et al., *Class. Quant. Grav.* **26** (2009) 185012, arXiv:0906.5477 [hep-th].
[58] T. Delpouve, R. Gurau, V. Rivasseau, arXiv:1403.0170 [hep-th].
[59] E. Bianchi, C. Rovelli, S. Speziale, *Class. Quant. Grav.* **23** (2006) 6989-7028, arXiv: gr-qc/0604044; E. Bianchi, C. Rovelli, F. Vidotto, *Phys. Rev. D* **82** (2010) 084035, arXiv: 1003.3483 [gr-qc].
[60] D. Oriti, *Stud. Hist. Philos. Mod. Phys.* **46** (2014) 186, arXiv:1302.2849 [physics.hist-ph].
[61] C. Barcelo, S. Liberati, M. Visser, *Living Rev. Rel.* **14** (2011) 3, arXiv: gr-qc/0505065.
[62] B. L. Hu, *Int. J. Theor. Phys.* **44** (2005) 1785–1806, arXiv: gr-qc/0503067.
[63] S. Gielen, D. Oriti, L. Sindoni, *Phys. Rev. Lett.* **111** (2013) 031301, arXiv: 1303.3576 [gr-qc]; S. Gielen, D. Oriti, L. Sindoni, *JHEP* **1406** (2014) 013, arXiv:1311.1238 [gr-qc].
[64] L. Pitaevskii, S. Stringari, *Bose-Einstein Condensation* (OUP, 2003).
[65] L. Sindoni, arXiv:1408.3095 [gr-qc].
[66] D. Oriti, D. Pranzetti, J. Ryan, L. Sindoni, to appear.
[67] A. Ashtekar, P. Singh, *Class. Quant. Grav.* **28** (2011) 213001, arXiv: 1108.0893 [gr-qc].
[68] M. Bojowald et al., *Phys. Rev. D* **86** (2012) 124027, arXiv: 1210.8138 [gr-qc].
[69] G. Calcagni, S. Gielen, D. Oriti, *Class. Quant. Grav.* **29** (2012) 105005, arXiv: 1201.4151 [gr-qc].
[70] E. Alesci, F. Cianfrani, arXiv: 1210.4504 [gr-qc], arXiv:1402.3155 [gr-qc].
[71] S. Gielen, arXiv:1404.2944 [gr-qc]; G. Calcagni, arXiv:1407.8166 [gr-qc].
[72] S. Gielen, D. Oriti, arXiv:1407.8167 [gr-qc].

[73] E. Livine, D. Oriti, J. Ryan, *CQG* **28** (2011) 245010, arXiv:1104.5509 [gr-qc].
[74] D. Oriti, L. Sindoni, *New J. Phys.* **13** (2011) 025006, arXiv:1010.5149 [gr-qc].
[75] W. Fairbairn, E. Livine, *Class. Quant. Grav.* **24** 5277 (2007), arXiv:gr-qc/0702125; F. Girelli, E. Livine, D. Oriti, *Phys. Rev. D* **81** 024015 (2010), arXiv: 0903.3475 [gr-qc].

Chapter 5

The Continuum Limit of Loop Quantum Gravity: A Framework for Solving the Theory

Bianca Dittrich

Perimeter Institute for Theoretical Physics,
31 Caroline Street North Waterloo, Ontario Canada N2L 2Y5

1. Solving the Dynamics of Loop Quantum Gravity

As explained in *Chapter 1*, loop quantum gravity has provided a rigorous non-perturbative framework, in which to formulate the dynamics of quantum gravity. It allowed fascinating insights into quantum geometry and a possible structure of quantum space time. As discussed in *Chapter 2*, in order to specify the complete dynamics of the theory, one has to construct the so-called physical Hilbert space of wave functions satisfying the quantum constraints. In this process, we need to introduce the continuum limit (see *Chapter 4*). In the framework presented here, physical states — i.e., solutions of the quantum constraints, which give the equations of motions of the theory — are constructed by taking the refinement limit via a coarse graining procedure.

The conceptual underpinnings of this framework rely on the inductive limit Hilbert space construction used in loop quantum gravity to define the continuum Hilbert space [1], so far in the kinematical setting. We point out that this construction becomes more powerful if one allows for a generalization of the refinement maps that define the inductive limit Hilbert spaces. It leads to a framework in which physical states are computed in a truncation scheme, where the type of truncation is determined by the dynamics itself. This procedure allows for an understanding of the dynamics of quantum gravity on all scales — where a notion of scale is given by the coarseness or fineness of configurations. The different scales of the theory are connected via the cylindrical consistency condition inherent in the inductive limit construction. This replaces the notion of renormalization flow in theories with a background scale.

We start our considerations with a short explanation of the inductive limit construction in Section 2 and discuss the difference between kinematical and dynamical understanding of the continuum limit. In Section 3 we start with the task to construct the physical Hilbert space of the theory and explain that it necessitates the construction of the refinement limit for the dynamics of the theory. This results in an iterative coarse graining scheme, in which physical states — or amplitude maps — are constructed in a certain truncation, labelled by the coarseness or fineness of the discrete structures involved. The relation of this scheme with a renormalization flow is clarified in Section 4. Concrete realizations of this scheme in the form of (decorated) tensor network methods are briefly explained in Section 5. We then point out the powerful notion of diffeomorphism symmetry for discrete systems in Section 6. The realization of this diffeomorphism symmetry is necessary for the definition of physical states and also indicates that a continuum limit is reached. In this sense physical states can only be defined in the continuum limit. We end with a discussion and outlook of future developments in Section 7.

2. Continuum Limit in Canonical Loop Quantum Gravity

Let us begin by clarifying the need for a continuum limit in canonical loop quantum gravity. To this end we will briefly discuss how this continuum formulation is achieved (a complete discussion can be found in [2]). The key point is to use a inductive limit construction for the kinematical Hilbert space of loop quantum gravity. This construction needs the following ingredients

(1) A directed partially ordered set of labels, in the case of the Ashtekar Lewandowski (AL) representation [1, 3] given by a suitable set of graphs α embedded into the spatial manifold M. The partial ordering is induced by a set of refining operations (adding an edge, subdividing an edge, inverting an edge).
(2) Hilbert spaces $\mathcal{H}_\alpha$ associated to these labels.
(3) Embedding maps $\iota_{\alpha\alpha'} : \mathcal{H}_\alpha \to \mathcal{H}_{\alpha'}$ for each pair of labels with $\alpha \prec \alpha'$, i.e. α' is finer than α. These embedding maps have to satisfy the consistency condition $\iota_{\alpha'\alpha''} \circ \iota_{\alpha\alpha'} = \iota_{\alpha\alpha''}$ for any triple $\alpha \prec \alpha' \prec \alpha''$.

The inductive limit of Hilbert spaces is given by the

$$\mathcal{H} := \overline{\cup_\alpha \mathcal{H}_\alpha \big/ \sim} \tag{1}$$

where the equivalence relation is defined as follows: two elements $\psi_\alpha \in \mathcal{H}_\alpha$ and $\psi'_{\alpha'} \in \mathcal{H}_{\alpha'}$ are equivalent $\psi_\alpha \sim \psi'_{\alpha'}$ iff there exist a refinement α'' of α and α' such that $\iota_{\alpha\alpha''}(\psi_\alpha) = \iota_{\alpha'\alpha''}(\psi'_{\alpha'})$. In words two elements are equivalent if they become equal under refinement eventually.

The inner product on the Hilbert spaces $\mathcal{H}_\alpha$ has to be compatible with this equivalence relation, that is, it must be *cylindrically consistent*

$$\langle \psi_\alpha | \psi'_\alpha \rangle_\alpha = \langle \iota_{\alpha\alpha'}(\psi_\alpha) | \iota_{\alpha\alpha'}(\psi'_\alpha) \rangle_{\alpha'} . \tag{2}$$

Also observables, which are a priori given as family of observables $\mathcal{O} = \{\mathcal{O}_\alpha\}_\alpha$ defined on the Hilbert space $\mathcal{H}_\alpha$ have to be cylindrically consistent, that is

$$\iota_{\alpha\alpha'}(\mathcal{O}_\alpha \psi_\alpha) = \mathcal{O}_{\alpha'} \iota_{\alpha\alpha'}(\psi_\alpha) . \tag{3}$$

The conditions (2,3) make the inner product and the observables well defined on the continuum Hilbert space, given by the inductive limit of the Hilbert space $\mathcal{H}_\alpha$. On a practical level they ensure that any calculation done on a given graph α (or any other discrete structure) gives the same result as on any refined graph.

Thus the construction of the inductive limit enables one to test the theory 'along' discrete structures, such as the graphs α. It is however *not* the case that the states are unknown away from the discrete structure in question. In fact the embedding maps allow to reconstruct the states on an arbitrary refined graph α', starting from states on a coarse graph α. That is all additional degrees of freedom, associable to α' but not to α are being put into a specific state encoded in the embedding maps. It is natural to interpret this specific state as vacuum, in fact in the AL representation [3] discussed in *Chapter 1*, this state is given by the Ashtekar Lewandowski vacuum. It is given as (equivalence class represented by) the state associated to the empty graph $\alpha = \emptyset$ which carries a one–dimensional Hilbert space $\mathcal{H}_\emptyset = \mathbb{C}$. The equivalence class of this vacuum state is characterized by the chosen embeddings — turning this around, the nature of the vacuum state characterizes the embeddings.

As explained in *Chapter 1*, the basic field variables of loop quantum gravity are given by the Ashtekar–Barbero connection A_a^i and the triad densities E_i^a [4, 5]. The connection is integrated and exponentiated to holonomies, along the edges given by the graph α the triads give rise to flux operators. The Ashtekar–Lewandowski vacuum is a totally squeezed state that gives maximal uncertainty to the connection and is maximally peaked at vanishing triad variables, that is formally $\psi_{vac}(A) \equiv 1$. Thus the states in any $\mathcal{H}_\alpha$ are highly distributional — (spatial) geometry encoded in the triads is only excited along the graph α. Away from this graph, all expectation values and fluctuations of the (smeared) triads are vanishing.

2.1. *Kinematical Understanding of the Continuum Limit*

We can now discuss the continuum limit in canonical loop quantum gravity at the kinematical level: construction of states in the AL representation, or even alternative representations, that can be interpreted as describing continuum geometries. In the AL representation, the construction of coherent states has been explored [6] and coarse graining in the kinematical Hilbert space has also been considered [7]. However, here one works with a fixed graph and therefore keeps the distributional nature of the states with respect to the excitations of spatial geometry — that is, the states again describe a spatial metric that is almost everywhere totally degenerate.

It is also possible to construct alternative representations of the observable (holonomy–flux) algebra of loop quantum gravity. The first such alternative representation [8] changes the vacuum from being peaked on a totally degenerate spatial geometry to one that is peaked on a non-degenerate (background) geometry. In this Koslowski-Sahlmann representation, vacuum fluctuations of the triad are still vanishing. Note that the embedding maps for this representation are different from the one for the AL representations. In particular the Koslowski–Sahlmann vacuum is not invariant under (spatial) diffeomorphisms in contrast to the AL representation.

Another alternative representation, that is based on a (space-time) diffeomorphism invariant vacuum has been recently proposed [9] and can be understood as a dual of the AL representation. The vacuum is now again a totally squeezed state but is peaked on flat connections, and maximally uncertain in the triad variables. This vacuum is actually a physical state for BF theory, whose equation of motion demand vanishing curvature. We will therefore label this representation as the BF. This construction is based, as the AL representation, on an inductive limit. However the label set is not given anymore by graphs but by triangulations. The vertices (in $(2+1)$ dimensions) or edges (in $(3+1)$ dimensions) of this triangulation can support curvature excitations. Thus the states can be interpreted as piecewise flat geometries. (Note that in $(3+1)$ dimensions this flatness is with respect to the AL representation, whereas in the $(2+1)$ dimensions[a] the flatness is with respect to the 3D spin connection.) In this sense this BF representation avoids a key problem of the AL representation, which is that AL states describe geometries which are almost everywhere totally degenerate.

Thus, whereas the AL embeddings impose the vanishing of ('finer') triad operators, the BF embedding maps impose the vanishing of ('finer') curvature operators (built from holonomies). These embedding maps coincide with 'naive time evolution maps' that arise in BF theory. In $(2+1)$ dimensions BF theory describes the dynamics of general relativity, and therefore the BF vacuum defines a physical state, giving rise to a physical Hilbert space. This illustrates an important point — namely, that eventually the embedding maps should be chosen by the dynamics of the system.

The BF representation has been also generalized — via a quantum group deformation of the underlying gauge group — to a vacuum peaked on a homogeneously curved geometry [10]. In $(2+1)$ dimension the vacuum represents a physical state of general relativity with a cosmological constant, and is closely connected to the Turaev-Viro state sum model. These different examples open up the questions of how many different quantum geometry realizations one is able to construct [22].

A very different approach, which avoids the selection of a vacuum state, is being developed in [11]. This framework replaces the inductive limit construction with a (dual) projective limit for the density functionals. But it is not clear yet, what kind of 'typical states' result from this framework.

3. Continuum Limit for the Dynamics of the Theory

We will now discuss a second — dynamical — understanding of the continuum limit. This would be the construction of the continuum physical Hilbert space of states satisfying the Hamiltonian and diffeomorphism,[b] constraints. Such physical

---

[a]Remarks about $(2+1)$ dimensions refer to the Riemannian signature because the gauge group in the Lorentzian signature is non-compact.
[b]Even if these constraints can be defined only a posteriori as discussed in Section 6.

states are expected not to be normalizable with respect to the kinematical Hilbert space. In fact we have now at our disposal several kinematical Hilbert spaces, all based on an inductive limit, but with different embedding maps.

We expect that also the physical Hilbert space can be organized in the form of an inductive limit Hilbert space. In this case the embedding maps $\iota_{\alpha,\alpha'}$ will again differ from the embedding maps for the kinematical Hilbert space. We will outline here a construction of such a physical Hilbert space, which would then represent the continuum physical Hilbert space.

A strategy [2, 12] to construct physical states, known as refined algebraic quantization, is by 'projecting' kinematical states via a so-called rigging map $\eta : \mathcal{D}_{kin} \to \mathcal{D}^*_{phys}$.[c] For totally constrained systems, where time evolution is a gauge transformation, one can formally write a 'projector' onto the space of solutions to quantum constraints as

$$\left(\prod_I \delta(\hat{C}_I)\,\psi\right)(X_{fin}) = \int \mathcal{D}N^I \exp\left(\frac{i}{\hbar}N^I\hat{C}_I\right)\psi(X_{fin})$$

$$= \int \mathcal{D}X_{ini} \int_{X_{ini}, X_{fin} \text{ fixed}} \mathcal{D}X \exp\left(\frac{i}{\hbar}S(X)\right)\psi_{ini}(X_{ini}). \quad (4)$$

In the second line we wrote the path integral over some set of configuration variables X with the corresponding action $S(X)$ for general relativity. Equation (4) states that this path integral serves as a (formal) projector onto states satisfying the Hamiltonian and diffeomorphism constraints [13].

The path integral is however only a formal object — so far, the only way to make it well defined is to turn to a discretization. This is one route to spin foam models [14] for which, as discussed in *Chapter 3*, the (boundary) variables X can be made to match those of loop quantum gravity.[d] However, a discretization comes with several drawbacks:

(a) A discretization typically breaks diffeomorphism symmetry for 4D gravity theories [16]. This prevents the discrete path integral to be a projector onto constraints, these are rather weakened to pseudo constraints [16–18].
(b) Related to the loss of diffeomorphism symmetry the path integral will in general depend on the choice of discrete structure, i.e. choice of (bulk) triangulation. This gives the triangulation an unwanted physical significance.
(c) There are many classical and quantum ambiguities in constructing the discrete amplitudes.
(d) The discrete path integral (4) can be defined on the Hilbert spaces $\mathcal{H}_\alpha$ associated to a given discretization α. However as an operator on the family of Hilbert spaces $\mathcal{H}_\alpha$ the path integral will in general not be cylindrically consistent and thus not be well defined on the continuum Hilbert space $\mathcal{H}$ [19–21].

---
[c] Here $\mathcal{D}_{kin}$ is a dense subspace of the kinematical Hilbert space, $\mathcal{D}_{kin} \subset \mathcal{H}$, whereas $\mathcal{D}^*_{phys}$ is given by the algebraic dual of a dense subspace $\mathcal{D}_{phys}$ in $\mathcal{H}_{phys}$. The Rigging map is called the 'group averaging map' in *Chapter 2*.
[d] More precisely the boundary Hilbert spaces match [15], at least on the discrete level.

(e) Finally a discrete path integral requires also an organization of the target (physical Hilbert) space as an inductive limit. The path integral as an operator should then also be cylindrically consistent with respect to dynamical embedding maps [20, 22] describing the physical Hilbert space.

We will argue that all these issues can be addressed by coarse graining the initial discrete path integral. As we will see this can also be interpreted as refining and amounts to the construction of the continuum limit for the discretized path integral.

To achieve the continuum limit for the dynamics of quantum gravity means in particular to turn the path integral into a cylindrical consistent operator, that is solve issues (d) and (e). We will describe here an iterative coarse graining process that aims at the construction of such a cylindrical consistent path integral.

This iterative process produces a coarse graining flow. Fixed points of such coarse graining flows often enjoy an enhanced symmetry. Several examples [23, 24] and arguments show that the diffeomorphism symmetry, in particular, is likely to be restored (see Section 6), which addresses problem (a). The same examples and the realization of diffeomorphism symmetry in the discrete — so called vertex displacements — show that diffeomorphism symmetry is equivalent with triangulation independence [23, 25], which resolves problem (b). Finally the coarse graining flow is considered on a space of models. Such a flow allows the characterization of relevant and irrelevant directions in this space of models, which addresses the issue (c). In particular, diffeomorphism invariance and triangulation independence are extremely strong requirements, thus one can hope that a discrete model satisfying these requirements (and leading to a suitable semi-classical limit) is, if it exists at all, unique.

We will furthermore argue that the issue (e) will lead to

(f) a notion of physical vacuum for quantum gravity.

This physical vacuum will be encoded into amplitude maps $\mathcal{A}_\alpha : \mathcal{H}_\alpha \to \mathbb{C}$. These maps define the amplitudes for the cylindrically consistent path integral and thus replace the 'bare' amplitudes of the initial discretization of the path integral. Here the label α stands for a discretization that can be obtained by refinement from an 'empty' discretization $\emptyset$ with $\mathcal{H}_\emptyset = \mathbb{C}$.

The amplitude map applied to a boundary wave function[e] $\psi_\alpha \in \mathcal{H}_\alpha$ gives the pairing of this wave function with the wave function $\mathbf{K}_{\emptyset\alpha}\psi_\emptyset$, resulting from a refining time evolution of the (kinematical no boundary) wave function $\psi_\emptyset$ to a wave function associated to the boundary α. This refining time evolution is given by a path integral and can therefore be understood to implement a rigging map, see equation (4). That

---

[e]The framework [26] introduces a generalization of Cauchy boundaries to boundaries of arbitrary regions, which is useful in this context.

is we consider

$$\mathcal{A}_\alpha(\psi_\alpha) := \int \mathcal{D}X \mathcal{D}X_\alpha \exp\left(-\frac{i}{\hbar}S(X, X_\alpha)\right) \psi_\alpha(X_\alpha)$$
$$= \langle \psi_\emptyset | (\mathbf{K}_{\emptyset\alpha})^\dagger | \psi_\alpha \rangle = \eta(\psi_\emptyset) \cdot \psi_\alpha =: \langle \psi_\emptyset | \psi_\alpha \rangle_{phys} \, . \quad (5)$$

Here we wrote the path integral[f] with bulk variables denoted by X and boundary variables denoted by X_α. We have to regularize this path integral on a discretization that fits in between the two boundaries $\emptyset$ and α. This discretization introduces of course the problems mentioned above, turning the expressions in the second line into not well defined ones. We will discuss below an iterative procedure to take the refinement limit of this discretization, that addresses these problems.

We pointed out in equation (4) that the time evolution operator in the form of the path integral should act as a projector onto physical states, which defines the rigging map η, here applied to the no-boundary (kinematical) wave function $\psi_\emptyset$. The last equation just displays the definition of a physical inner product in the refined algebraic quantization procedure [12]

$$\langle \psi_\alpha | \psi'_{\alpha'} \rangle_{phys} := \eta(\psi_\alpha) \cdot \psi'_{\alpha'} \quad (6)$$

between the projections of two kinematical states ψ_α and $\psi'_{\alpha'}$. (It suffices to apply the rigging map once, as it is given by a time evolution which acts as an usually improper projector.)

The amplitude maps encode the dynamics of the system [27] and will replace the 'bare' amplitudes of the initial discretized path integral. Note that such a discretized path integral is often built by associating amplitudes $\mathcal{A}_B$ to basic building blocks B. Indeed from the definition (5) the basic amplitudes $\mathcal{A}_B$ give the amplitude map in the coarsest triangulation possible. To this end we assume that one can refine the empty discretization $\emptyset$ to the one given by the boundary of B by gluing the building block B to $\emptyset$.

The iterative refinement process will replace these basic amplitudes with improved amplitudes $\mathcal{A}_\alpha$ by (i) refining the bulk discretization and (ii) also allowing a refining of the boundary discretization, that is, generalize from the boundary of B to finer boundary discretizations α. This generalization of the basic building blocks, that allows the incorporation of more boundary data, is important to convert non-local couplings, that inadvertently are produced by coarse graining to local (nearest neighbor) couplings of the improved amplitudes.

The end point of the construction should lead to a cylindrically consistent amplitude map satisfying cylindrical consistency

$$\mathcal{A}_{\alpha'}(\iota_{\alpha\alpha'}(\psi_\alpha)) = \mathcal{A}_\alpha(\psi_\alpha) \quad (7)$$

---

[f]We denoted the complex conjugated path integral amplitudes $\exp(-\frac{i}{\hbar}S(X, X_\alpha))$ to indicate that the complex conjugation of the wave function evolved from $\psi_\emptyset$. In spin foams (and in other approaches which incorporate in (4) an integration over positive and negative lapse) the sum over the basic variables includes a sum over orientations of space time. This leads to real amplitudes. This feature is important to obtain the projector property of the path integral.

with respect to certain embedding maps $\iota_{\alpha\alpha'}$. As we will argue below, it might be much easier to construct such cylindrical consistent amplitudes if we replace the kinematical embedding maps with dynamical ones. Such cylindrical consistent amplitude maps are then defined on a continuum Hilbert space $\mathcal{H}_{[\emptyset]}$ associated to the equivalence class of discretizations, that can be obtained by applying refinement operations to the empty discretization $\emptyset$.

This brings us to the second interpretation of the amplitude maps as representing the (dualized) physical vacuum. This interpretation is due to two points:

Firstly we defined the amplitude map via a refining time evolution starting from a 'no-boundary' discretization $\emptyset$. The resulting wave function can be seen as a Hartle Hawking no-boundary wave function [28].[g] This point is also strengthened as the amplitude map $\mathcal{A}_\alpha(\cdot) = \eta(\psi_\emptyset)\cdot$ results from applying the rigging map, to the kinematical vacuum $\psi_\emptyset \in \mathbb{C}$, which one would expect to carry the notion of having no excitations and leading to a homogeneous state, see also [29]. This concept of generating a vacuum state by refining time evolution comes also up in formulations incorporating evolving phase spaces [30] or Hilbert spaces [31], classical and quantum examples that support this interpretation can be found in [22]. In the formulation employed here, evolving Hilbert spaces are taken into account via the concept of inductive limit Hilbert spaces.

Secondly, we will use the amplitude maps to define dynamical embedding maps. That is the amplitude maps lead to an improved, and in the refinement limit, perfect discretization of the path integral. This path integral can be used to define a refining time evolution, interpolating between a boundary α and a refined boundary α'. However, as we discussed, there is no proper time evolution in diffeomorphism invariant systems, it rather acts as a projector onto physical states. In case the initial state ψ_α is physical, the resulting state $\psi_{\alpha'}$ should therefore be equivalent to ψ_α. This is realized if we assume an inductive limit structure for the physical Hilbert space and use the refining time evolution as (dynamical) embedding maps $\iota_{\alpha\alpha'} = \mathbf{K}_{\alpha\alpha'}$, as proposed in [22].

Note that such embedding maps have to satisfy the consistency conditions $\iota_{\alpha'\alpha''} \circ \iota_{\alpha\alpha'} = \iota_{\alpha\alpha''}$ for any triple $\alpha \prec \alpha' \prec \alpha''$, as discussed in Section 2. For a (refining) time evolution these conditions follow from Kuchar's requirement of a path independence of evolution [32], which is equivalent to the constraint algebra being consistent, that is first class, which itself signifies that diffeomorphism symmetry is correctly implemented. We can therefore expect this consistency condition to hold in the refinement limit, in which we hope to restore diffeomorphism symmetry.

Another aspect of path independence of evolution is a condition involving as an in-between state one that is finer than the final state:

$$\mathbf{K}_{\alpha''\alpha'} \circ \mathbf{K}_{\alpha\alpha''} = \mathbf{K}_{\alpha\alpha'} \tag{8}$$

---

[g] In the actual proposal [28] Wick rotates part of the time evolution. We do not assume such a Wick rotation here, which would indeed be hard to define in a completely background independent context.

for $\alpha \prec \alpha' \prec \alpha''$. If in addition we can identify $\mathbf{K}_{\alpha\alpha'} = (\mathbf{K}_{\alpha'\alpha})^\dagger$, which should hold due to the projector property of time evolution, it follows that the amplitude maps are cylindrically consistent for dynamical embedding maps $\iota_{\alpha\alpha'} = \mathbf{K}_{\alpha\alpha'}$:

$$\mathcal{A}_{\alpha'}(\iota_{\alpha\alpha'}\psi_\alpha) = \langle\psi_\emptyset|(\mathbf{K}_{\emptyset\alpha'})^\dagger|\mathbf{K}_{\alpha\alpha'}\psi_\alpha\rangle$$
$$\overset{(8)}{=} \langle\psi_\emptyset|(\mathbf{K}_{\emptyset\alpha})^\dagger|\psi_\alpha\rangle = \mathcal{A}_\alpha(\psi_\alpha) . \qquad (9)$$

This suggest to also change the embedding maps on the kinematical Hilbert space, as this simplifies the construction of a cylindrical consistent amplitude map.

Indeed we can take (9) as defining an iterative procedure to improve the amplitude maps, in particular regarding property (8). To this end we understand the term on the RHS of the first line in (9) as consisting of two steps. The first is the computation of $\langle\psi_\emptyset|(\mathbf{K}_{\emptyset\alpha'})^\dagger$, that is basically the amplitude functional $\mathcal{A}_{\alpha'}$ for a more refined boundary α. One would build such an amplitude functional from gluing amplitudes $\mathcal{A}_\alpha$ for less refined boundaries α.

As we want to define an iterative process that improves the amplitude maps $\mathcal{A}_\alpha$, we need to find a way to 'evolve back' the amplitudes $\mathcal{A}_{\alpha'}$ to the boundary Hilbert space $\mathcal{H}_\alpha$, which is done by using the dynamical embedding map $\iota_{\alpha\alpha'} = \mathbf{K}_{\alpha\alpha'}$. Thus one defines the improved amplitudes $\mathcal{A}_\alpha^{imp}$ as

$$\mathcal{A}_\alpha^{imp} = \langle\psi_\emptyset|(\mathbf{K}_{\emptyset\alpha'})^\dagger|\mathbf{K}_{\alpha\alpha'}\psi_\alpha\rangle . \qquad (10)$$

Here both $(\mathbf{K}_{\emptyset\alpha'})^\dagger$ and $\mathbf{K}_{\alpha\alpha'}$ are built from using the initial $\mathcal{A}_\alpha$ as basic amplitudes.

The process is repeated for the improved amplitudes $\mathcal{A}_\alpha^{imp}$ until the procedure converges to a fixed point $\mathcal{A}_\alpha^{fix}$. This fixed point amplitude can be used to proceed to a more refined pair of boundaries (α', α'') with $\alpha' \prec \alpha''$ to find the next fixed point amplitude $\mathcal{A}_{\alpha'}^{fix}$ and so on.

One can take this amplitude $\mathcal{A}_{\alpha'}^{fix}$ and aim to construct a dynamical embedding map $\iota_{\alpha\alpha'} = \mathbf{K}_{\alpha\alpha'}$ from a coarser boundary α to a finer one α'. This allows to consider the pull back $\mathcal{A}_\alpha^{fix,\alpha'} := \iota_{\alpha\alpha'}^* \mathcal{A}_{\alpha'}^{fix}$. This amplitude will differ from $\mathcal{A}_\alpha^{fix}$, the amplitude constructed taking less boundary data, namely the pair (α, α') into account. Because of this we see $\mathcal{A}_\alpha^{fix,\alpha'}$ as an improvement on $\mathcal{A}_\alpha^{fix}$. Iterating in this way one constructs amplitude maps that are satisfying the cylindrical consistency conditions for finer and finer boundaries.

Tensor network renormalization schemes make this procedure explicit, by specifying more in detail how to construct the refined amplitudes $\mathcal{A}_{\alpha'}$ and the embedding maps $\iota_{\alpha\alpha'} \sim \mathbf{K}_{\alpha\alpha'}$ from the amplitudes $\mathcal{A}_\alpha$ for coarser boundaries α. We will explain a tensor network algorithm in Section 5.

Once one has constructed amplitude maps that are cylindrically consistent (to a satisfying degree), one can use these amplitude maps to define an improved discretization of the path integral (4) and with it the rigging map. This is using the interpretation of the amplitude maps as giving the amplitudes of building blocks, which can now carry more boundary data.

Let us examine the gluing properties of these improved building blocks, in particular in which sense the amplitude for a given (finer) boundary can be obtained from gluing building blocks with coarser boundary. If this would be the case we would achieve independence from the chosen discretization, i.e. form the decomposition of a given region into building blocks.

For this consider a simplified situation with two manifolds of topology $\Sigma \times [0,1]$ glued along a common Σ hypersurface. The amplitude for the first manifold with boundaries α, α' is given by $\mathbf{K}_{\alpha\alpha'}$, for the second manifold we have $\mathbf{K}_{\alpha'\alpha''}$ so that the glued amplitude is

$$\mathbf{K}_{\alpha'\alpha''} \circ \mathbf{K}_{\alpha\alpha'} \stackrel{?}{=} \mathbf{K}_{\alpha\alpha''}. \qquad (11)$$

Thus discretization independence (here invariance under subdivision) would be realized if the equality in (11) holds. This equation can however not be true for arbitrary coarse in-between boundary α'. A very coarse α' would restrict the amount of information that can propagate from α to α''.[h] Thus α' should in general be finer than both α and α''. In this case equation (11) coincides with (8) (or its time reversal), and thus is expected to hold for cylindrically consistent amplitudes.

The situation is less clear-cut if we generalize to situations where only certain parts of the boundary are glued. However, as this is also used in the coarse graining procedure which builds such cylindrical consistent amplitudes one would expect that — depending on the coarseness of the outer boundaries not glued over — the gluing property is satisfied to better and better degree for finer and finer boundaries and in particular satisfied exactly if one takes for the boundary glued over the refinement limit. For subtleties that come up even in the continuum, see [26].

4. Renormalization Flow and Scale in Background Independent Theories

Here we want to discuss the relations and differences of the framework developed in Section 3, where the construction of cylindrical consistent amplitudes is central, to the understanding of renormalization flow in systems with a notion of background (scale) [21, 34, 35]. We will in particular provide an extension of aspects developed in the work [21] from the AL embedding maps to dynamical embedding maps.

Consider a system with discretization scale a', whose dynamics is defined by amplitudes $\mathcal{A}^{a'}(X')$ (e.g. $\exp(\frac{i}{\hbar}S(X'))$), depending on variables X' (defined at scale a'). The Wilsonian renormalization flow [33] defines effective amplitudes $\mathcal{A}^a$ at a larger scale a through the condition

$$\int_{B_X^{a',a}(X')=X} \mathcal{D}X' \, \mathcal{A}^{a'}(X') = \int \mathcal{D}X \, \mathcal{A}^a(X). \qquad (12)$$

Here we denote by $B_X^{a',a}$ a blocking function that determines how the microscopic degrees of freedom X' are coarse grained into the coarser variables X.

---

[h]The equality can hold however in topological theories which do not have local propagating degrees of freedom.

Repeating (12) at different pairs of scales will give a renormalization flow of the amplitudes $\mathcal{A}^a$ parametrized by the scale a. The amplitudes at coarser scales encode the 'effective' dynamics of the system and allow to determine the expectation values of sufficiently coarse observables (that can be expressed in the variables X of this scale):

$$\int \mathcal{O}(B_X^{a',a}(X')) \, \mathcal{A}^{a'}(X') \mathcal{D}X' = \int \mathcal{O}(X) \, \mathcal{A}^a(X) \mathcal{D}X \; . \tag{13}$$

In background independent systems we do not have a background scale available. Instead there are two entities, which replace the background scale: one is the discretization labels α characterizing the coarseness or fineness of a given boundary. The other is the geometry, which is part of the dynamical variables and thus determined by the boundary data or wave function.

The renormalization trajectory $\mathcal{A}^a$, parametrized by the scale a is replaced in background independent systems by the cylindrically consistent family of amplitude maps. Thus a cylindrically consistent family of amplitude maps defines a renormalization flow.

To see this consider the path integral over a certain region built from building blocks B or regions with a certain homogeneous boundary fineness α. Subdivide each of these building blocks into further building blocks $\{B'\}$. We then want to compare the path integral based on amplitude maps for building blocks B with the path integral based on amplitude maps for building blocks B'. Here the amplitudes $\mathcal{A}_B$ for a building block B are defined from the (cylindrical consistent) amplitude maps via

$$\mathcal{A}_{\alpha(B)}(\psi_{\alpha(B)}) = \int \overline{\mathcal{A}_B(X)} \, \psi_{\alpha(B)}(X) \tag{14}$$

where $\alpha(B)$ denotes the boundary of the building block B. Similarly we define a kernel $\iota_{\alpha,\alpha'}(X_{\alpha'}, X_\alpha)$ for the embedding maps $\iota_{\alpha,\alpha'}$ by

$$\iota_{\alpha,\alpha'}(\psi_\alpha)(X_{\alpha'}) = \int \mathcal{D}X_\alpha \; \iota_{\alpha,\alpha'}(X_{\alpha'}, X_\alpha) \, \psi_\alpha(X_\alpha) \tag{15}$$

where X_α denote the boundary variables of α and $X_{\alpha'}$ those of α' (assuming that the Hilbert space is $\mathcal{H}_\alpha = L_2(\mathcal{C}_\alpha, \mathcal{D}X_\alpha)$) with $\mathcal{C}_\alpha$ denoting the configuration space and $\mathcal{D}X_\alpha$ the measure.

To connect the amplitudes for B' and B we integrate over the shared boundary variables when gluing the building blocks $\{B'\}$ to B. This will however result into a finer boundary than for the original building block B. We thus need to use embedding maps $\iota_{\alpha,\alpha'}$ from the boundary $\alpha = \alpha(B)$ of B to the boundary $\alpha' = \alpha'(\{B'\})$ of the set of glued building blocks $\{B'\}$. These embedding maps are applied in the inverse direction, as these act indeed on the boundary wave function, with which the amplitude is paired.

We denote by α'' the boundary of a given building block B' and with $\cup \alpha''/\alpha'$ the inner (shared) boundaries in the gluing of the set $\{B'\}$ to B. The amplitude

$\tilde{\mathcal{A}}_B$ constructed from the $\mathcal{A}_{B'}$ is then given as

$$\tilde{\mathcal{A}}_B(X_\alpha) = \int \mathcal{D}X_{\alpha'} \left(\int \mathcal{D}X_{\cup \alpha''/\alpha'} \prod \mathcal{A}_{B'}(X'_\alpha) \right) \overline{\iota_{\alpha\alpha'}(X_\alpha, X_{\alpha'})} . \qquad (16)$$

The arguments from the previous section show that at least approximately we can expect $\mathcal{A}_B = \tilde{\mathcal{A}}_B$. Thus the $\mathcal{A}_B$ are indeed (also) effective amplitudes, that is they can be obtained by integrating out degrees of freedom starting from amplitudes $\mathcal{A}_{B'}$. A fixed point condition follows if we choose the building blocks such that $B = B'$.

Comparing with the definition of the Wilsonian renormalization flow (12) we can argue that the role of the pair of scales (a', a) there is taking over by (α', α). (As the original building blocks B' might have the same boundary as the effective building blocks B, that is $\alpha'' = \alpha$, we rather compare with the boundary of the set of glued building blocks $\{B'\}$.) The blocking functions $B^{a',a}$ are replaced by the embedding maps $\iota_{\alpha,\alpha'}$, which allow for more general constructions. A replacement of the (bulk) observable condition (13) can also be stated [21] and is equivalent to the cylindrical consistency condition for (boundary) observables (3).

Here we argued from the 'boundary' cylindrical consistency of the amplitude maps on the boundary Hilbert space towards a 'bulk' cylindrical consistency of the path integral measure (which we here understand to include the amplitudes $\mathcal{A}_B$). This last point is the starting point of [19, 21] for configuration spaces of connections and with the AL embedding maps. See also the discussion in [21] for a derivation of boundary cylindrical consistency [20] from bulk cylindrical consistency.

Thus we see that indeed the renormalization trajectory A^a is replaced by the cylindrical consistent set of amplitude maps $\mathcal{A}_\alpha$. Still one should avoid to equate the scale a with the boundary coarseness α. To consider amplitudes at a certain scale one would have to fix properties of the boundary wave function ψ_α or alternatively for the amplitude kernels $\mathcal{A}_B(X)$ consider variables X restricted to describe a certain scale.

The question whether a continuum (or refinement) limit of a quantum gravity model exist can be now reformulated as follows: *Does there exist a family of cylindrical consistent amplitude maps that would display the correct semi-classical limit, at least for boundary fields describing a slowly varying geometry or alternatively small curvature?* Assuming that slowly varying geometry can be described on a coarse boundary one would need in particular to check the semi-classical limit for simple building blocks $\mathcal{A}_B$ with a coarse boundary α. The semi-classical limit involves to consider a scaling of geometric variables so that these describe lengths much larger than Planck length $l_B(X_{\alpha(B)}) \gg l_{Planck}$. (Here l_B can be understood as the scale on which the boundary geometry described by $X_{\alpha(B)}$ can vary.) In this limit we expect

$$\mathcal{A}_B(X_{\alpha(B)}) \sim \cos(S_H(X_{\alpha(B)})) \qquad (17)$$

where S_H is Hamilton's principal function, i.e. the action evaluated on the solution determined by the boundary values $X_{\alpha(B)}$). Here we assumed that building blocks will contribute with both possible orientations, as is the case in spin foams. As we saw in *Chapter 3*, Condition (5) is indeed satisfied for spin foams [36], at least for the simplest building blocks, that is simplexes.

Thus the semi-classicality requirement for the amplitudes is at 'mesoscopic' scales $l_B \gg l_{Planck}$. Indeed we need to regularize the path integral via a discretization. Even classically (non-perfect) discretization are only reliable reproducing observables which are (much) coarser than the (coarseness) scale of the discretization. If we consider a fixed boundary geometry we can translate this statement into the discretization reproducing observables on scales (much) larger than the discretization scale.

As the cylindrical consistency conditions are very restricting, we can hope that the condition of cylindrical consistency leads to a unique family of amplitudes, that then define the theory at all scales (i.e. for all boundary wave functions). This philosophy is similar to the asymptotic safety scenario [37] where one hopes to extrapolate to the UV starting from the IR dynamics of a given theory. *Thus the question whether a refinement limit exist is similar to the asymptotic safety conjecture, namely the existence of an interacting UV fixed point.* The question whether we find a unique family of cylindrically consistent amplitudes is connected to the number of relevant couplings at this fixed point, which the asymptotic safety scenario conjectures to be finite.

This question — whether a family of cylindrically consistent amplitudes exist or not — will also determine the allowed matter couplings. The reconstruction of the renormalization flow in terms of the usual notion of scale, as discussed further below, should also reproduce the flow of the standard model matter couplings — as far as known. Thus including matter couplings would also mean to construct an UV completion of the corresponding quantum field theories — if such UV completions exist. One expects restrictions on the allowed matter content — as has been already shown in the asymptotic safety scenario to arise [38].

As laid out in the previous section, the cylindrical consistent family of amplitudes, that is the renormalization trajectory, can be constructed via an iterative coarse graining procedure. The initial amplitudes for this procedure, can be constructed by using a discretization — as is done in the spin foam approach described in *Chapter 3*. The iterative coarse graining procedure reconstructs the renormalization flow in a larger and larger space of 'couplings', that also include the parametrization of discretization ambiguities. With respect to the auxiliary coarse graining flow, that is used to construct the family of cylindrically consistent amplitudes, one can apply the usual notions of relevant/irrelevant couplings and universality. Thus discretization ambiguities (irrelevant couplings with respect to this flow) are taken care of, see [23] for an explicit example. This addresses the issue (c) in Section 3. Of course one would hope that the flow does not change the semi-classical property (5) of the initial amplitudes, i.e. that the integrated out quantum effects do not change the amplitudes at mesoscopic scales in the sense described above.

A notion of flow, nearer to the Wilsonian one based on scale, would require a reconstruction of this scale from the geometric boundary data. For this one needs to find a way to decompose the geometric variables into small and large scale ones and to correspondingly organize the amplitudes into families of effective ones by integrating out small scale degrees of freedom. This procedure would basically involve the continuum amplitudes encoded in the cylindrical consistent family. A problem is then to find a (preferably non-perturbative) notion of geometric scale. Of course also with respect to this flow one can classify relevant/irrelevant couplings, which are now nearer to the standard notion.

5. (Decorated) Tensor Network Renormalization for Spin Nets and Spin Foams

The construction of cylindrical consistent amplitude-maps is a highly demanding task — it basically requires to solve the theory for arbitrary complicated boundary data. One therefore hopes for an efficient approximation scheme. The parameter describing the approximations is naturally given by the coarseness α. We can think of this parameter as determining the complexity of boundary data. This approximation scheme is similar to the calculation of scattering amplitudes for more and more particles (at infinity). Similar to the expectation that for a scattering amplitude involving few particles at infinity in-between states with many particles are less relevant, one can hope that the coarser the boundary data the less relevant become in-between states involving very fine α. For this to hold true it is essential that the embedding maps — that determine the properties of excitations supported by the discrete structure α — are derived from the dynamics of the system.

Tensor network coarse graining schemes [39] implement a recursive improvement of the amplitudes as in (10) and (16). The name 'tensor network' indicates that the amplitudes are encoded in tensors associated to vertices (and dual to space time regions). The indices of a tensor at a vertex v are associated to edges attached to the vertex v. These edges are also dual to the boundary of the space time region (i.e. the edges cross the boundary). Gluing two space time regions is then equivalent to contracting two indices of two neighboring tensors.

The complexity of the boundary data, that is the coarseness parameter α translates here into the rank of the tensor and the index range, the so-called bond dimension χ (assuming finite dimensional Hilbert spaces, which are associated to the edges). Note that several edges (indices) of a tensor can be encoded into one edge (index) — thus the bond dimension might increase during the algorithm.

Let us explain an algorithm for a 2D model, in which the amplitudes are encoded into rank four tensors with bond dimension χ. Thus we discretize the partition function (or path integral) with a regular square lattice, where the squares are dual to the four-valent vertices.

One now glues four of such squares to a new square. This however also increases the number of edges, i.e. the amount of boundary data — the bond dimension is now χ^2. One needs to reduce these back to the original size χ (which can be chosen to be much larger than the index range of the original tensors).

$$A \vdots \boxed{M} {}^\alpha_\beta \boxed{M} \vdots B \qquad A \vdots \boxed{M} {}^\alpha_\beta \!\!\bullet\!\!{}_V\!\!\!-i$$

Fig. 1. Left: Two vertices in a tensor network, encoded in the matrices M, are sharing two edges with labels $\{\alpha, \beta\}$, which have a total range of χ^2. Right: From the singular value decomposition we can define the map V depicted as a three-valent vertex, where we restrict the label i of the singular values to be $\leq \chi$.

In the case of tensor network methods one chooses a truncation — via an embedding map as in (16) — that is chosen from the dynamics of the system. The idea is to approximate as well as possible the summation between two tensors. The situation is depicted in Figure 1. One organizes the indices of the tensors such that we can rewrite them into matrices M. We would like to replace the edges carrying an index pair $\{\alpha, \beta\}$ of size χ^2 with an effective edge carrying only a number χ of indices. An optimal truncation for the summation over the index pair $\{\alpha, \beta\}$, is given by the singular value decomposition of $M_{A\alpha\beta}$:

$$M_{A\alpha\beta} = \sum_{i=1}^{\chi^2} U_{Ai} \lambda_i V_{i\,\alpha\beta} \tag{18}$$

where $\lambda_1 \geq \lambda_2 \geq \ldots \geq \lambda_{\chi^2} \geq 0$ are positive, and U, V are unitary matrices. The truncation drops the smaller set of singular values λ_i with $i > \chi$. Pictorially $V_{i\alpha\beta}$ restricted to $i \leq \chi$ defines a three-valent vertex and we can use these three-valent vertices as in Figure 1 to arrive at a coarse grained region with less boundary data.

Applying the three-valent tensors to the square just glued, we obtain a new effective tensor, with the same bond dimension as before. This algorithm is applicable to systems with and without a background scale. For both cases one hopes that the truncation picks indeed the coarse (homogeneous) data.

This is supported by several examples; see the discussion in Ref. [22]. A better truncation could be reached by choosing the embedding maps to be more non-local (i.e. involving all boundary data and not only those associated to a pair of edges). Indeed in this case the truncation can be even made exact [22]. To see this consider a 'radial' evolution from a coarser to a finer boundary. The evolution operator only maps to a subspace of the target Hilbert space with dimension equal or smaller than the initial Hilbert space. Thus a singular value decomposition would turn out to have only as many non-vanishing singular values as we would take into account in the truncation. A certain notion of locality is however needed to be able to glue the new squares to each other. For a more non-local truncation scheme than the one described here see [40].

Such 2D algorithms have been successfully applied to spin net models [41], which are analogue models to spin foams, that can be also defined in 2D [42, 43]. The spin nets can also be interpreted as specific ('melonic') spin foams, see [43] and are conjectured to have similar statistical properties to spin foams. To be able to do numerical simulations the models considered so far are based on either finite groups or quantum groups $SU(2)_k$. The latter are conjectured to describe quantum gravity with a cosmological constant [44].

The group symmetry protecting variant of the algorithm developed in [42, 43] allows to keep track of the behavior of intertwiner degrees of freedom, which signify the status of the simplicity constraints — the ingredient of spin foams that distinguishes them from standard lattice gauge theories. The initial model differ in the choice of these simplicity constraints. This allows to scan an entire set of models for a reasonable continuum limit. To this end one needs to find a good parametrization of the initial phase space [43, 45, 46].

In fact the simplicity constraints lead to a large extension of the phase space of the latter. A very rich structure of topological fixed points (corresponding to phases in statistical model language) and phase transitions (candidates for interacting theories) has been found in [43], based on a parametrization of intertwiners developed in [46]. The 2D models also allow to study the concepts discussed in Section 3. In particular the notion of dynamical embedding maps and related vacua states describe condensation phenomena — in the 2D intertwiner models of anyons described by $SU(2)_k$ fusion modules [46, 47].

Recently, $SU(2)_k \times SU(2)_k$ spin net models which impose Barrett–Crane [48] simplicity constraints have been tested [49] and show also an interesting phase structure, which arises by only varying the so-called face weights of the model.

The richness of the phase structure found so far reinforces the hope that spin foams lead to a reasonable continuum limit. Of course one needs to confirm this hope by coarse graining actual spin foam models. These models are more general in their structure than tensor models, which are basically vertex models, with variables on edges and weight on vertices. In spin foam models variables do also appear on two-dimensional objects, i.e. plaquettes.

Decorated tensor networks [50] can deal with this issue in an effective way. Here one returns to representing the partition function as a gluing of building blocks. These building blocks carry boundary variables as prescribed by the initial model in question. A interesting feature of the procedure is that the type of these initial boundary variables is not changed. This allows a much more straightforward interpretation of the coarse graining flow by keeping track of the behavior of these variables. For spin foams these variables coincide with the intertwiner degrees of freedom so important for spin nets — which is one reason to expect similar behavior under coarse graining. The geometric interpretation of the (spin) variables in spin foams allows to access whether the coarse graining leads in fact to a geometric coarse graining of the system. This feature will in particular be encoded in the embedding maps.

In lowest order approximation the building blocks will carry (almost) the same amount of boundary data, as the initial model. As mentioned this allows for a straightforward interpretation of the coarse graining flow of these systems. Going to higher order truncations one incorporates more boundary data by associating a tensor to the building blocks which now introduces 'higher order' variables. The entire coarse graining procedure is similar to tensor network algorithms (i.e. also based on singular value decompositions), but 'decorated' by the original variables of the model.

Another feature of decorated tensor networks is that they may allow for (semi–) analytical calculations, see also [51]. This is important to be able to treat spin foam models based on Lie groups, where the issue of divergences arise [53, 54], see also the discussion in Section 6.

In tensor network algorithms the truncation is determined by the dynamics. This is so far the only way to find a reliable truncation, but makes the algorithms computationally very demanding. An alternative might be to use truncations, informed by some geometric intuition. E.g. [52] imposes a restriction to discretizations built out of cuboids, that describe geometries without curvature but with torsions. The truncations can be again imposed by an embedding map, that this time is however chosen by hand. The flatness makes the action contribution to the amplitudes vanish, thus the coarse graining flow tests only the measure terms. This flow does however indicate a restoration of (a remnant of) diffeomorphism symmetry, as we will explain in the next section.

6. Diffeomorphism Symmetry in the Discrete, Constraints and Divergences

In this section we will elaborate more on a notion of diffeomorphism symmetry in the discrete. This symmetry is very powerful [55, 56]. In fact its realization signifies that the continuum limit has been reached in the following way: although the physics is expressed on a discrete structure, the predictions for observables, which can be supported by this discrete structure coincide with those of the continuum model. Such a discretization is called perfect [24, 57] — it exactly mirrors continuum physics. Thus the refinement limit is necessary to reach diffeomorphism symmetry and thus a notion of physical states.

The notion of diffeomorphism symmetry under discussion also arises for discretizations which do not explicitly involve coordinates. For instance in Regge calculus [58] the variables are given by the lengths of the edges in a triangulation.[i] These geometric data of the discrete elements allow one to determine the relative position of the vertices with respect to each other. In fact if there is a symmetry of

---
[i]Alternatively one can use areas and angles [59], which is nearer to the variables used in spin foams.

the action[j] allowing for these relative vertex positions — expressed in the geometric data of the discretization — to change, we speak of a realization of diffeomorphism symmetry in the discrete. This symmetry is also referred to as vertex translations, as it coincides in the 3D BF formulation of gravity with the shift or translation symmetry of the triad fields [60].

Such a symmetry has been indeed identified for linearized Regge calculus [61] and a number of examples [23, 24, 56]. It is however broken if one considers a (Regge) solution of 4D gravity with curvature [16] or perturbative Regge gravity beyond linear order [18]. Here 'broken symmetry' means that the Hessian will display modes with very small eigenvalues (compared to the other eigenvalues), instead of null modes. This breaking has severe repercussions. It prevents the path integral — for the regularization of which we need to introduce the discretization — from acting as a projector onto physical states.

One can define a canonical discrete time formulation that is consistent with the covariant one, i.e., that reproduces the equations of the covariant framework [17, 18]. This formalism also consistently transfers the (broken) symmetries into (pseudo) constraints. Whereas constraints are given as equations of motions that only involve the canonical data of one time step, pseudo constraints will also involve, with a weak dependence, the data of a neighboring time step.

Thus, one reason to take the refinement limit is actually to restore the diffeomorphism symmetry [24, 62], as is also used in the perfect action program [57] for lattice QCD with regard to Lorentz symmetry. There are a number of arguments for such a restoration: one is that the pseudo gauge modes should have a small lattice correlation length and decouple in the continuum limit [62]. Another is that for Regge calculus with flat building blocks, for instance, the eigenvalues of the Hessian of the action associated to the pseudo gauge modes scale with the curvature per building block of the solution [16]. In the refinement limit this curvature goes to zero, thus leading to a restoration of the symmetries.

We described the symmetry as allowing displacements of vertices. This is basically the reason why this symmetry is so powerful and requires the continuum limit: For a system with such a symmetry it means that, given a solution, we can move the vertices around (i.e. change the associated geometric data) without changing the value of the action. Thus we can for instance move vertices on top of each other, reaching a coarser triangulation. This is basically the argument that diffeomorphism symmetry implies triangulation independence [23]. One can furthermore move the vertices such that one region appears very finely grained and another very coarse grained. (Again this is with respect to a solution, which provides a scale). Thus our model has to display continuum physics reliable on all scales and show no discretization artifacts, i.e. it has to be a perfect

---

[j]That is, the Hessian of the action evaluated on a solution needs to have null modes making the solutions non-unique [16, 25]. The action itself (away from solutions) will allow a huge class of invariant deformations, most of these trivialize however if restricted to solutions.

discretization. Such a perfect discretization avoids all problems (ambiguities, breaking of symmetries, triangulation dependence) of a 'typical' discretization.

For interacting systems one can hope only for non-local actions or amplitudes to display such a powerful symmetry. This is shown explicitly, with the non-existence of a local path integral measure for linear Regge calculus [63]. Non-local amplitudes are very difficult to deal with — in fact the framework described in Sections 3 and 5 avoids non-local couplings by introducing building blocks with more boundary data — since they are akin to introducing more fields in the continuum to absorb higher derivatives. Since diffeomorphism symmetry implies triangulation invariance we can also hope that coarse graining schemes on a regular lattice are sufficient to recover fully triangulation invariant models, which is indeed confirmed so far for spin net models [42, 43, 46].

As noted above (first class Hamiltonian and diffeomorphism) constraints can only appear if the discretization shows diffeomorphism symmetry and hence is perfect. Thus for 4D gravity one has to expect non-local constraints. Again, the framework developed in Section 3 could be of help here, as it might be possible to derive constraints on very coarse Hilbert spaces $\mathcal{H}_\alpha$ first and then going to finer and finer ones. This does not exclude graph-changing Hamiltonians [64], although one would expect that an inductive Hilbert space, based on dynamical embedding maps, allows for graph-non-changing ones. In fact for the simplest triangulations, leading to only flat bulk solutions, it is possible to find first class constraints [65]. Note that constraints which are derived from cylindrically consistent amplitudes, do also describe the flow of (matter) coupling constants. This can for example appear in the form of couplings, that depend on the geometric variables associated to building blocks. This information on the couplings of the running is dynamical information which, if the constraints are indeed derived from the consistent amplitudes, is obtained from the coarse graining process that led to these amplitudes. It seems impossible to construct consistent Hamiltonian constraints, without having such an explicit process that determines the running of the couplings.

One can also turn the argument around and say that if a refinement limit does not lead to a restoration of the symmetries (or first class property of the constraints), the system is inherently discrete [68]. The question of whether a refinement limit 'exist' or not might however depend on many details of how the system is constructed as well as how one attempts to construct[k] the continuum limit.

Let us also remark on the relation between divergences and diffeomorphism symmetry. As the gauge orbits of this symmetry are non-compact (with the exception of Euclidean gravity with positive cosmological constant) one has to expect that the partition function diverges in the case the symmetries are realized. (Vertex translations may also cross building blocks, reversing orientations of these [66], which

---

[k]Reference [68] makes a choice of (local) constraints, turns these into a master constraint [17, 67] and tests this master constraint on a certain class of semi-classical states. Each step involves a number of ambiguities.

allows for non-compact orbits.) Thus one would expect a divergence of Λ^{ND} for D space time dimension and N triangulation vertices. This is indeed confirmed for (topological) 3D spin foams with the link to the diffeomorphism symmetry made explicit [60]. The divergence structure of the 4D models is less clear [53, 54], as it also depends on a choice of path integral measure in the form of so-called edge and face weights [54, 69]. A correct divergence structure in itself would of course not be sufficient for a model to display diffeomorphism symmetry, as this structure can be easily tuned by only changing face and edge weights [54], but leaving the (discretized) action unaffected. Additionally the existence of degenerate configurations, which may display enhanced symmetries [54] and divergences complicate the issue.

As symmetries are typically broken one would expect the initial model to be finite. (As noted in [54] special configurations might actually exist, which show enhanced symmetries and might lead to divergences.) However, with the restoration of symmetries under coarse graining, the path integral becomes however more and more divergent. One could expect a problem here, however one can indeed deal with this successfully even in a numerical approach [23]. In fact the coarse graining procedure involves a rescaling of the amplitudes in each step. One would then expect that the (candidate) divergences lead to an enhancement of the terms in the amplitude that do lead to diffeomorphism symmetry and a suppression of the other terms. Thus diffeomorphism symmetry might enhance its own restoration in the refinement limit in this way.

7. Summary and Outlook

A refinement limit is inescapable for the construction of the full theory of (loop) quantum gravity. Only in this limit can we expect the realization of diffeomorphism symmetry, thus a notion of Hamiltonian and diffeomorphism constraints and finally physical states. Indeed as we explained here constructing the refinement limit means to construct physical states and a physical Hilbert space.

We presented a framework to formulate and construct the refinement limit using the essential structure of inductive limit Hilbert spaces and the concept of cylindrical consistent amplitudes, where the notion of cylindrical consistency is induced from the dynamics. The (tensor network) coarse graining procedures we discussed construct such amplitudes iteratively in a truncation scheme. The dynamics automatically determines this truncation, by introducing a notion of coarse states with few excitations and very fine states with many excitations. The excitations are with respect to a vacuum state that is also determined from the dynamics.

Although the construction of the refinement limit requires basically the solution of the model, it can be organized in a truncation scheme. The approximation improves with finer and finer discretizations that support more and more complicated boundary data. Some physical situations, such as cosmology, involve rather coarse data. Therefore, one might hope that a derivation of cosmology can be obtained at a low truncation order [70] (but sufficiently fine to determine the dynamical embedding maps essential for the understanding of the truncation).

We laid out the relation of the refinement scheme to renormalization involving a (background) scale. The scale is basically replaced by the coarseness of the discretizations — although one should be careful in equating the two. The notion of a complete renormalization trajectory is replaced by the notion of cylindrically consistent amplitudes, showing the correct semi-classical limit behavior (i.e. for large geometries). The crucial question is whether such cylindrically consistent amplitudes exist.

Renormalization comes up also in group field theories (GFT's) discussed in *Chapter 4*, again in order to regulate divergences [71, 72]. In this case one sums over triangulations and hopes to achieve a continuum limit by choosing weights such that configurations with infinitely many building blocks dominate [73]. The relation between renormalization in a GFT sense [74], which involves an explicit scale, and the coarse graining scheme presented here needs to be better understood, in particular since the divergences (may) correspond to gauge symmetries in the spin foam framework. (A GFT understanding of vertex translations leads to global symmetries [75].)

One can argue that, due to the restoration of diffeomorphism symmetry in the form of vertex displacement invariance, a given sufficiently fine lattice may simulate many coarser lattices. A variant of this argument is used in [76], to show that refinement and summing over triangulations should lead to the same result.[1] To inquire more about this relation, it is essential to clarify the relations between the Hilbert spaces involved, as the notion of cylindrical consistency is rather different [78]. Indeed, whether one prefers refining or summing over triangulations to obtain (bulk) triangulation independent amplitudes, in both cases we demand the amplitudes to be cylindrically consistent with respect to some choice of embedding maps. This latter notion specifies the relation between different boundary discretizations, and turns the amplitudes into well-defined maps on the continuum Hilbert space.

We will conclude by commenting on some possible future developments. Coarse graining results from spin net models hint at a rich phase space structure for spin foams. With an explicit coarse graining scheme for spin foams at hand [50], we can soon expect results that will allow deep insights into the dynamical mechanisms of spin foams and thus, hopefully, the workings of quantum spacetime.

Even the identification of topological field theories in the phase diagram of spin foams can give rise to exciting developments. Such topological field theories lead to cylindrical consistent embedding maps, thus to new inductive limit Hilbert spaces [9, 10, 22]. These can be used to construct further alternative vacua and representations for loop quantum gravity,[m] possibly with a notion of simplicity constraints and

---

[1] See also [71, 77] for a related discussion of this issue, namely whether the sum over triangulations leads to the path integral as projector.

[m] Note that the uniqueness results [79] pertaining to the AL representation do not apply in the more general situations in which the flux operators exist *only* in an exponentiated form.

in-between the AL representation [3] and the one based on BF theory developed in [9]. Different vacua and representations allow to expand the theories around different regimes and to thus organize the dynamics of the theory with respect to different notions of excitations. This opens new perspectives for loop quantum gravity and can lead to a large extension of the framework.

Acknowledgements

This research was carried out at the Perimeter Institute for Theoretical Physics. Research at Perimeter Institute is supported by the Government of Canada through Industry Canada and by the Province of Ontario through the Ministry of Research and Innovation.

References

[1] A. Ashtekar and J. Lewandowski, "Differential geometry on the space of connections using projective techniques," *J. Geom. Phys.* **17** (1995) 191 [hep-th/9412073].

[2] T. Thiemann, *Modern Canonical Quantum General Relativity* (Cambridge, UK: Cambridge Univ. Pr. 2007) [gr-qc/0110034].

[3] A. Ashtekar and C. J. Isham, "Representations of the holonomy algebras of gravity and nonAbelian gauge theories," *Class. Quant. Grav.* **9** (1992) 1433 [hep-th/9202053]. A. Ashtekar and J. Lewandowski, "Projective techniques and functional integration for gauge theories," *J. Math. Phys.* **36** (1995) 2170 [gr-qc/9411046].

[4] A. Ashtekar, "New variables for classical and quantum gravity," *Phys. Rev. Lett.* **57** (1986) 2244.

[5] J. F. Barbero, " A Real polynomial formulation of general relativity in terms of connections," *Phys. Rev. D* **49** (1994) 6935, arXiv:gr-qc/9311019 [gr-qc].

[6] T. Thiemann, "Gauge field theory coherent states (GCS): 1. General properties," *Class. Quant. Grav.* **18** (2001) 2025 [hep-th/0005233]. T. Thiemann, "Complexifier coherent states for quantum general relativity," *Class. Quant. Grav.* **23** (2006) 2063 [gr-qc/0206037]. A. Ashtekar, S. Fairhurst and J. L. Willis, "Quantum gravity, shadow states, and quantum mechanics," *Class. Quant. Grav.* **20** (2003) 1031 [gr-qc/0207106]. M. Varadarajan, "The Graviton vacuum as a distributional state in kinematic loop quantum gravity," *Class. Quant. Grav.* **22** (2005) 1207 [gr-qc/0410120]. B. Bahr and T. Thiemann, "Gauge-invariant coherent states for loop quantum gravity. II. Non-Abelian gauge groups," *Class. Quant. Grav.* **26** (2009) 045012 [arXiv:0709.4636 [gr-qc]]. D. Oriti, R. Pereira and L. Sindoni, "Coherent states in quantum gravity: a construction based on the flux representation of LQG," *J. Phys. A* **45** (2012) 244004 [arXiv:1110.5885 [gr-qc]].

[7] H. G. Diaz-Marin and J. A. Zapata, "Curvature function and coarse graining," *J. Math. Phys.* **51** (2010) 122307 [arXiv:1101.3818 [hep-th]]. E. R. Livine, "Deformation operators of spin networks and coarse-graining," arXiv:1310.3362 [gr-qc].

[8] T. A. Koslowski, "Dynamical quantum geometry (DQG programme)," (2007), arXiv:0709.3465 [gr-qc]. H. Sahlmann, "On loop quantum gravity kinematics with non–degenerate spatial background," *Class. Quant. Grav.* **27** (2010) 225007, arXiv:1006.0388 [gr-qc]. T. Koslowski and H. Sahlmann, "Loop quantum gravity vacuum with nondegenerate geometry," *Sigma* **8** (2012) 026, arXiv:1109.4688 [gr-qc]. M. Varadarajan, "The generator of spatial diffeomorphisms in the

Koslowski–Sahlmann representation," *Class. Quant. Grav.* **30** (2013) 175017, arXiv:1306.6126 [gr-qc].

[9] B. Dittrich and M. Geiller, "A new vacuum for loop quantum gravity," *Class. Quant. Grav.* **32** (2015) no. 11, 112001 doi:10.1088/0264-9381/32/11/112001 [arXiv:1401.6441 [gr-qc]]. B. Dittrich and M. Geiller, "Flux formulation of loop quantum gravity: Classical framework," "*Class. Quant. Grav.* **32** (2015) no. 13, 135016 doi:10.1088/0264-9381/32/13/135016 [arXiv:1412.3752 [gr-qc]]. B. Bahr, B. Dittrich and M. Geiller, "A new realization of quantum geometry," arXiv:1506.08571 [gr-qc].

[10] B. Dittrich and M. Geiller, "Quantum gravity kinematics from extended TQFTs," arXiv:1604.05195 [hep-th]. C. Delcamp and B. Dittrich, "From 3D TQFTs to 4D models with defects," arXiv:1606.02384 [hep-th].

[11] S. Lanry and T. Thiemann, "Projective loop quantum gravity I. State space," arXiv:1411.3592 [gr-qc]. S. Lanry and T. Thiemann, "Projective loop quantum gravity II. Searching for semi-classical states," arXiv:1510.01925 [gr-qc]. S. Lanéry, "Projective limits of state spaces: Quantum field theory without a vacuum," arXiv:1604.05629 [hep-th].

[12] A. Ashtekar and R. S. Tate, "An algebraic extension of Dirac quantization: Examples," *J. Math. Phys.* **35** (1994) 6434 [gr-qc/9405073].

[13] J. J. Halliwell and J. B. Hartle, "Wave functions constructed from an invariant sum over histories satisfy constraints," *Phys. Rev. D* **43** (1991) 1170. C. Rovelli, "The projector on physical states in loop quantum gravity," *Phys. Rev. D* **59** (1999) 104015 [arXiv:gr-qc/9806121]. K. Noui and A. Perez, "Three dimensional loop quantum gravity: Physical scalar product and spin foam models," *Class. Quant. Grav.* **22** (2005) 1739 [arXiv:gr-qc/0402110].

[14] M. P. Reisenberger and C. Rovelli, "'Sum over surfaces' form of loop quantum gravity," *Phys. Rev. D* **56** (1997) 3490 [gr-qc/9612035]. C. Rovelli, *Quantum Gravity* (Cambridge University Press, Cambridge 2004). A. Perez, "The spin foam approach to quantum gravity," *Living Rev. Rel.* **16** (2013) 3 [arXiv:1205.2019 [gr-qc]].

[15] Y. Ding and C. Rovelli, "Physical boundary Hilbert space and volume operator in the Lorentzian new spin-foam theory," *Class. Quant. Grav.* **27** (2010) 205003 [arXiv:1006.1294 [gr-qc]]. M. Dupuis and E. R. Livine, "Lifting SU(2) spin networks to projected spin networks," *Phys. Rev. D* **82** (2010) 064044 [arXiv:1008.4093 [gr-qc]]. W. Kaminski, M. Kisielowski and J. Lewandowski, "Spin-foams for all loop quantum gravity," *Class. Quant. Grav.* **27** (2010) 095006 [Erratum-ibid. **29** (2012) 049502] [arXiv:0909.0939 [gr-qc]]. B. Dittrich, F. Hellmann and W. Kaminski, "Holonomy spin foam models: Boundary Hilbert spaces and time evolution operators," *Class. Quant. Grav.* **30** (2013) 085005 [arXiv:1209.4539 [gr-qc]].

[16] B. Bahr and B. Dittrich, "(Broken) Gauge symmetries and constraints in Regge calculus," *Class. Quant. Grav.* **26** (2009) 225011 [arXiv:0905.1670 [gr-qc]]. B. Bahr and B. Dittrich, "Breaking and restoring of diffeomorphism symmetry in discrete gravity," *AIP Conference Proceedings* Volume 1196, ed. J. Kowalski-Glikman *et al.*, pp. 10-17 arXiv:0909.5688 [gr-qc].

[17] R. Gambini and J. Pullin, "Consistent discretizations as a road to quantum gravity," In *Approaches to Quantum Gravity*, ed. Oriti, D., pp. 378-392 [gr-qc/0512065]. M. Campiglia, C. Di Bartolo, R. Gambini and J. Pullin, "Uniform discretizations: A new approach for the quantization of totally constrained systems," *Phys. Rev. D* **74** (2006) 124012 [arXiv:gr-qc/0610023]. B. Bahr, R. Gambini and J. Pullin, "Discretisations, constraints and diffeomorphisms in quantum gravity," *SIGMA* **8** (2012) 002 [arXiv:1111.1879 [gr-qc]].

[18] B. Dittrich and P. A. Höhn, "From covariant to canonical formulations of discrete gravity," *Class. Quant. Grav.* **27** (2010) 155001 [arXiv:0912.1817 [gr-qc]].

[19] B. Bahr, "Operator spin foams: holonomy formulation and coarse graining," *J. Phys. Conf. Ser.* **360** (2012) 012042 [arXiv:1112.3567 [gr-qc]].
[20] B. Dittrich, "From the discrete to the continuous: Towards a cylindrically consistent dynamics," *New J. Phys.* **14** (2012) 123004 [arXiv:1205.6127 [gr-qc]].
[21] B. Bahr, "On background-independent renormalization of spin foam models," arXiv:1407.7746 [gr-qc].
[22] B. Dittrich and S. Steinhaus, "Time evolution as refining, coarse graining and entangling," *New J. Phys.* **16** (2014) 123041 doi:10.1088/1367-2630/16/12/123041 [arXiv:1311.7565 [gr-qc]].
[23] B. Bahr, B. Dittrich and S. Steinhaus, "Perfect discretization of reparametrization invariant path integrals," *Phys. Rev. D* **83** (2011) 105026 [arXiv:1101.4775 [gr-qc]].
[24] B. Bahr and B. Dittrich, "Improved and perfect actions in discrete gravity," *Phys. Rev. D* **80** (2009) 124030 [arXiv:0907.4323 [gr-qc]].
[25] B. Dittrich, "Diffeomorphism symmetry in quantum gravity models," *Adv. Sci. Lett.* **2** (2009) 151 [arXiv:0810.3594 [gr-qc]].
[26] R. Oeckl, "A 'General boundary' formulation for quantum mechanics and quantum gravity," *Phys. Lett. B* **575** (2003) 318 [hep-th/0306025].
[27] A. Perez and C. Rovelli, "Observables in quantum gravity," In '*Quanta of Maths*', eds. E. Blanchard, D. Ellwood, M. Khalkhali, M. Marcolli, H. Moscovici, S. Popa, pp. 501-518, American Mathematical Society (2011) [gr-qc/0104034]. C. Rovelli, "On the structure of a background independent quantum theory: Hamilton function, transition amplitudes, classical limit and continuous limit," arXiv:1108.0832 [gr-qc].
[28] J. Hartle and S. Hawking, "Wave function of the universe," *Phys. Rev. D* **28** (1983) 2960–2975.
[29] F. Conrady, L. Doplicher, R. Oeckl, C. Rovelli and M. Testa, "Minkowski vacuum in background independent quantum gravity," *Phys. Rev. D* **69** (2004) 064019 [gr-qc/0307118].
[30] B. Dittrich and P. A. Höhn, "Canonical simplicial gravity," *Class. Quant. Grav.* **29** (2012) 115009 [arXiv:1108.1974 [gr-qc]]. B. Dittrich and P. A. Höhn, "Constraint analysis for variational discrete systems," *J. Math. Phys.* **54** (2013) 093505 [arXiv:1303.4294 [math-ph]].
[31] P. A. Hoehn, "Quantization of systems with temporally varying discretization I: Evolving Hilbert spaces," arXiv:1401.6062 [gr-qc]. P. A. Hoehn, "Quantization of systems with temporally varying discretization II: Local evolution moves," arXiv:1401.7731 [gr-qc].
[32] K. Kuchar, "Geometry of Hyperspace. 1.," *J. Math. Phys.* **17** (1976) 777.
[33] K. G. Wilson and J. B. Kogut, "The Renormalization group and the epsilon expansion," *Phys. Rept.* **12** (1974) 75.
[34] J. A. Zapata, "Loop quantization from a lattice gauge theory perspective," *Class. Quant. Grav.* **21** (2004) L115 [gr-qc/0401109]. E. Manrique, R. Oeckl, A. Weber and J. A. Zapata, "Loop quantization as a continuum limit," *Class. Quant. Grav.* **23** (2006) 3393 [hep-th/0511222].
[35] R. Oeckl, "Renormalization of discrete models without background," *Nucl. Phys. B* **657** (2003) 107 [arXiv:gr-qc/0212047].
[36] J. W. Barrett and T. J. Foxon, "Semiclassical limits of simplicial quantum gravity," *Class. Quant. Grav.* **11** (1994) 543 [gr-qc/9310016]. J. W. Barrett, R. J. Dowdall, W. J. Fairbairn, H. Gomes and F. Hellmann, "Asymptotic analysis of the EPRL four-simplex amplitude," *J. Math. Phys.* **50** (2009) 112504 [arXiv:0902.1170 [gr-qc]]. F. Conrady and L. Freidel, "On the semiclassical limit of 4d spin foam models," *Phys. Rev. D* **78** (2008) 104023 [arXiv:0809.2280 [gr-qc]].

[37] O. Lauscher and M. Reuter, "Quantum Einstein gravity: Towards an asymptotically safe field theory of gravity," *Lect. Notes Phys.* **721** (2007) 265. M. Reuter and F. Saueressig, "Quantum Einstein gravity," *New J. Phys.* **14** (2012) 055022 [arXiv:1202.2274 [hep-th]].

[38] P. Donà, A. Eichhorn and R. Percacci, "Matter matters in asymptotically safe quantum gravity," *Phys. Rev. D* **89** (2014) no. 8, 084035 [arXiv:1311.2898 [hep-th]]. P. Don, A. Eichhorn, P. Labus and R. Percacci, "Asymptotic safety in an interacting system of gravity and scalar matter," *Phys. Rev. D* **93** (2016) no. 4, 044049 Erratum: [*Phys. Rev. D* **93** (2016) no. 12, 129904] [arXiv:1512.01589 [gr-qc]].

[39] M. Levin, C. P. Nave, "Tensor renormalization group approach to 2D classical lattice models," *Phys. Rev. Lett.* **99** (2007) 120601, [arXiv:cond-mat/0611687 [cond-mat.stat-mech]]. Z. -C. Gu and X. -G. Wen, "Tensor-entanglement-filtering renormalization approach and symmetry protected topological order," *Phys. Rev. B* **80** (2009) 155131 [arXiv:0903.1069 [cond-mat.str-el]].

[40] Z. Y. Xie, H. C. Jiang,Q. N. Chen,Z. Y. Weng,T. Xiang, "Second renormalization of tensor-network states" *Phys. Rev. Lett.* **103** (2009) 160601, [arXiv:0809.0182[cond-mat.str-el]].

[41] B. Bahr, B. Dittrich and J. P. Ryan, "Spin foam models with finite groups," *J. Grav.* **2013** (2013) 549824 [arXiv:1103.6264 [gr-qc]].

[42] B. Dittrich, F. C. Eckert and M. Martin-Benito, "Coarse graining methods for spin net and spin foam models," *New J. Phys.* **14** (2012) 035008 [arXiv:1109.4927 [gr-qc]]. B. Dittrich and F. C. Eckert, "Towards computational insights into the large-scale structure of spin foams," *J. Phys. Conf. Ser.* **360** (2012) 012004 [arXiv:1111.0967 [gr-qc]]. B. Dittrich, M. Martín-Benito and E. Schnetter, "Coarse graining of spin net models: dynamics of intertwiners," *New J. Phys.* **15** (2013) 103004 [arXiv:1306.2987 [gr-qc]].

[43] B. Dittrich, M. Martin-Benito and S. Steinhaus, "Quantum group spin nets: refinement limit and relation to spin foams," *Phys. Rev. D* **90** (2014) 024058 [arXiv:1312.0905 [gr-qc]].

[44] L. Smolin, "Linking topological quantum field theory and nonperturbative quantum gravity," *J. Math. Phys.* **36** (1995) 6417 [gr-qc/9505028]. S. Major and L. Smolin, "Quantum deformation of quantum gravity," *Nucl. Phys. B* **473** (1996) 267 [gr-qc/9512020]. W. J. Fairbairn and C. Meusburger, "Quantum deformation of two four-dimensional spin foam models," *J. Math. Phys.* **53** (2012) 022501 [arXiv:1012.4784 [gr-qc]]. M. Han, "4-dimensional spin-foam model with quantum Lorentz group," *J. Math. Phys.* **52** (2011) 072501 [arXiv:1012.4216 [gr-qc]]. M. Dupuis and F. Girelli, "Quantum hyperbolic geometry in loop quantum gravity with cosmological constant," *Phys. Rev. D* **87** (2013) 121502 [arXiv:1307.5461 [gr-qc]].

[45] B. Bahr, B. Dittrich, F. Hellmann and W. Kaminski, "Holonomy spin foam models: Definition and coarse graining," *Phys. Rev. D* **87** (2013) 044048 [arXiv:1208.3388 [gr-qc]].

[46] B. Dittrich and W. Kaminski, "Topological lattice field theories from intertwiner dynamics," arXiv:1311.1798 [gr-qc].

[47] F. A. Bais and J. K. Slingerland, "Condensate induced transitions between topologically ordered phases," *Phys. Rev. B* **79** (2009) 045316 [arXiv:0808.0627 [cond-mat.mes-hall]].

[48] J. W. Barrett and L. Crane, "Relativistic spin networks and quantum gravity," *J. Math. Phys.* **39**, 3296 (1998) [gr-qc/9709028].

[49] B. Dittrich, E. Schnetter, C. J. Seth and S. Steinhaus, "Coarse graining flow of spin foam intertwiners," arXiv:1609.02429 [gr-qc].

[50] B. Dittrich, S. Mizera and S. Steinhaus, "Decorated tensor network renormalization for lattice gauge theories and spin foam models," *New J. Phys.* **18** (2016) no. 5, 053009 doi:10.1088/1367-2630/18/5/053009 [arXiv:1409.2407 [gr-qc]]. C. Delcamp, B. Dittrich "Towards a phase diagram for spin foams," to appear.

[51] A. Banburski, L. Q. Chen, L. Freidel and J. Hnybida, "Pachner moves in a 4d Riemannian holomorphic Spin Foam model," *Phys. Rev. D* **92** (2015) no. 12, 124014 [arXiv:1412.8247 [gr-qc]].

[52] B. Bahr and S. Steinhaus, "Investigation of the Spinfoam Path integral with Quantum Cuboid Intertwiners," *Phys. Rev. D* **93** (2016) no. 10, 104029 [arXiv:1508.07961 [gr-qc]]. B. Bahr and S. Steinhaus, "Numerical evidence for a phase transition in 4d spin foam quantum gravity," *Phys. Rev. Lett.* **117** (2016) no. 14, 141302 [arXiv:1605.07649 [gr-qc]].

[53] C. Perini, C. Rovelli and S. Speziale, "Self-energy and vertex radiative corrections in LQG," *Phys. Lett. B* **682**, 78 (2009) [arXiv:0810.1714 [gr-qc]]. V. Bonzom and M. Smerlak, "Bubble divergences from cellular cohomology," *Lett. Math. Phys.* **93**, 295 (2010) [arXiv:1004.5196 [gr-qc]]. A. Riello, "Self-energy of the Lorentzian EPRL-FK spin foam model of quantum gravity," *Phys. Rev. D* **88** (2013) 024011 [arXiv:1302.1781 [gr-qc]].

[54] V. Bonzom and B. Dittrich, "Bubble divergences and gauge symmetries in spin foams," *Phys. Rev. D* **88** (2013) 124021 [arXiv:1304.6632 [gr-qc]].

[55] C. Rovelli, "Discretizing parametrized systems: the magic of Ditt-invariance," arXiv:1107.2310 [hep-lat].

[56] B. Dittrich, "How to construct diffeomorphism symmetry on the lattice," *PoS QGQGS* **2011** (2011) 012 [arXiv:1201.3840 [gr-qc]].

[57] P. Hasenfratz and F. Niedermayer, "Perfect Lattice Action For Asymptotically Free Theories," *Nucl. Phys. B* **414** (1994) 785 [arXiv:hep-lat/9308004].

[58] T. Regge, "General relativity without coordinates," *Nuovo Cim.* **19** (1961) 558.

[59] B. Dittrich and S. Speziale, "Area-angle variables for general relativity," *New J. Phys.* **10** (2008) 083006 [arXiv:0802.0864 [gr-qc]]. B. Bahr and D. Dittrich, "Regge calculus from a new angle," *New J. Phys.* **12** (2010) 033010 [arXiv:0907.4325 [gr-qc]].

[60] L. Freidel and D. Louapre, "Diffeomorphisms and spin foam models," *Nucl. Phys. B* **662** (2003) 279 [arXiv:gr-qc/0212001].

[61] M. Rocek and R. M. Williams, "Quantum Regge calculus," *Phys. Lett. B* **104** (1981) 31. B. Dittrich, L. Freidel and S. Speziale, "Linearized dynamics from the 4-simplex Regge action," *Phys. Rev. D* **76** (2007) 104020 [arXiv:0707.4513 [gr-qc]].

[62] M. Lehto, H. B. Nielsen and M. Ninomiya, "Diffeomorphism symmetry in simplicial quantum gravity," *Nucl. Phys. B* **272** (1986) 228.

[63] B. Dittrich and S. Steinhaus, "Path integral measure and triangulation independence in discrete gravity," *Phys. Rev. D* **85**, 044032 (2012) [arXiv:1110.6866 [gr-qc]]. B. Dittrich, W. Kaminski and S. Steinhaus, "Discretization independence implies non-locality in 4D discrete quantum gravity," arXiv:1404.5288 [gr-qc].

[64] T. Thiemann, "Anomaly-free formulation of non-perturbative, four-dimensional Lorentzian quantum gravity," *Phys. Lett. B* **380** (1996) 257 [arXiv:gr-qc/9606088]. "Quantum spin dynamics (QSD)," *Class. Quant. Grav.* **15** (1998) 839 [arXiv:gr-qc/9606089].

[65] B. Dittrich and J. P. Ryan, "Phase space descriptions for simplicial 4d geometries," *Class. Quant. Grav.* **28** (2011) 065006 [arXiv:0807.2806 [gr-qc]]. V. Bonzom and B. Dittrich, "Dirac's discrete hypersurface deformation algebras," *Class. Quant. Grav.* **30** (2013) 205013 [arXiv:1304.5983 [gr-qc]].

[66] M. Christodoulou, M. Langvik, A. Riello, C. Roken and C. Rovelli, "Divergences and

orientation in spinfoams," *Class. Quant. Grav.* **30** (2013) 055009 [arXiv:1207.5156 [gr-qc]].

[67] T. Thiemann, "The Phoenix project: Master constraint programme for loop quantum gravity," *Class. Quant. Grav.* **23** (2006) 2211 [arXiv:gr-qc/0305080]. B. Dittrich and T. Thiemann, "Testing the master constraint programme for loop quantum gravity. I: General framework," *Class. Quant. Grav.* **23** (2006) 1025 [arXiv:gr-qc/0411138].

[68] R. Gambini, J. Pullin and S. Rastgoo, "Quantum scalar field in quantum gravity: The vacuum in the tpherically symmetric case," *Class. Quant. Grav.* **26** (2009) 215011 [arXiv:0906.1774 [gr-qc]].

[69] M. Bojowald and A. Perez, "Spin foam quantization and anomalies," arXiv:gr-qc/0303026.

[70] M. Bojowald, "Loop quantum cosmology," *Living Rev. Rel.* **11** (2008) 4. E. Bianchi, C. Rovelli and F. Vidotto, "Towards Spinfoam Cosmology," *Phys. Rev. D* **82** (2010) 084035 [arXiv:1003.3483 [gr-qc]]. A. Ashtekar and P. Singh, "Loop quantum cosmology: A status report," *Class. Quant. Grav.* **28** (2011) 213001 [arXiv:1108.0893 [gr-qc]].

[71] L. Freidel, "Group field theory: An overview," *Int. J. Theor. Phys.* **44** (2005) 1769 [hep-th/0505016].

[72] D. Oriti, "The microscopic dynamics of quantum space as a group field theory," arXiv:1110.5606 [hep-th]. D. Oriti, "Group field theory and loop quantum gravity," arXiv:1408.7112 [gr-qc].

[73] A. Baratin, S. Carrozza, D. Oriti, J. P. Ryan and M. Smerlak, "Melonic phase transition in group field theory," arXiv:1307.5026 [hep-th].

[74] S. Carrozza, D. Oriti and V. Rivasseau, "Renormalization of tensorial group field theories: Abelian U(1) models in four dimensions," *Commun. Math. Phys.* **327** (2014) 603 [arXiv:1207.6734 [hep-th]]. S. Carrozza, "Discrete renormalization group for SU(2) tensorial group field theory," arXiv:1407.4615 [hep-th].

[75] A. Baratin, F. Girelli and D. Oriti, "Diffeomorphisms in group field theories," *Phys. Rev. D* **83** (2011) 104051 [arXiv:1101.0590 [hep-th]].

[76] C. Rovelli and M. Smerlak, "In quantum gravity, summing is refining," *Class. Quant. Grav.* **29** (2012) 055004 [arXiv:1010.5437 [gr-qc]].

[77] T. Thiemann and A. Zipfel, "Linking covariant and canonical LQG II: Spin foam projector," *Class. Quant. Grav.* **31** (2014) 125008 [arXiv:1307.5885 [gr-qc]].

[78] D. Oriti, "Group field theory as the 2nd quantization of loop quantum gravity," arXiv:1310.7786 [gr-qc].

[79] J. Lewandowski, A. Okolow, H. Sahlmann and T. Thiemann, "Uniqueness of diffeomorphism–invariant states on holonomy-flux algebras," *Commun. Math. Phys.* **267** 703 (2006), `arXiv:gr-qc/0504147`. C. Fleischhack, "Representations of the Weyl algebra in quantum geometry", *Commun. Math. Phys.* **285** (2009) 67, `arXiv:math-ph/0407006`.

Part 3
Applications

... One may not assume the validity of field equations at very high density of field and matter and one may not conclude that the beginning of the expansion should be a singularity in the mathematical sense.

– Albert Einstein (Meaning of Relativity)

Chapter 6

Loop Quantum Cosmology

Ivan Agullo and Parampreet Singh

Department of Physics and Astronomy,
Louisiana State University,
Baton Rouge, LA 70803-4001, USA

1. Introduction

The goal of this chapter is to apply the techniques of loop quantum gravity (LQG) to cosmological spacetimes. The resulting framework is known as loop quantum cosmology (LQC). This chapter has a two-fold motivation: to highlight various developments on the theoretical and conceptual issues in the last decade in the framework of loop quantum cosmology, and to demonstrate the way these developments open novel avenues for explorations of Planck scale physics and the resulting phenomenological implications.

From the theoretical viewpoint, cosmological spacetimes provide a very useful stage to make significant progress on many conceptual and technical problems in quantum gravity. These geometries have the advantage of being highly symmetric, since spatial homogeneity reduces the infinite number of degrees of freedom to a finite number, significantly simplifying the quantization of these spacetimes. Difficult challenges and mathematical complexities still remain, but they are easier to overcome than in more general situations. The program of canonical quantization of the gravitational degrees of freedom of cosmological spacetimes dates back to Wheeler and De Witt [1, 2]. In recent years, LQC has led to significant insights and progress in quantization of these mini-superspace cosmological models and fundamental questions have been addressed. These include: whether and how the classical singularities are avoided by quantum gravitational effects; how a smooth continuum spacetime emerges from the underlying quantum theory; how quantum gravitational effects modify the classical dynamical equations; the problem of time and inner product; quantum probabilities; etc. (see [3–8] for reviews in the subject). Spacetimes where detailed quantization has been performed include Friedmann-Lemaitre-Robertson-Walker (FLRW) [9–11, 13–23], Bianchi [24–30] and Gowdy models [31–34], the latter with an infinite number of degrees of freedom. A coherent picture of singularity resolution and Planck scale physics has emerged based on a rigorous mathematical framework, complemented with powerful numerical techniques. This new paradigm has provided remarkable insights on quantum gravity, and allowed a systematic exploration of the physics of the very early universe. On the other hand, simplifications also entail limitations. Since the formulation and the resulting physics is most rigorously studied in the mini-superspace setting, it is natural to question its robustness when infinite number of degrees of freedom are present, and whether the framework captures the implications from the full quantum theory. The problem of relating a model with more degrees of freedom to its symmetry reduced version is present even at the mini-superspace level. In this setting important insights have been gained on the relation between the loop quantization of Bianchi-I spacetime and spatially flat ($k = 0$) isotropic model, which provide useful lessons to relate quantization of spacetimes with different number of degrees of freedom [26]. Moreover, the Belinsky-Khalatnikov-Lifshitz (BKL) conjecture [35] — that the structure of the spacetime near the singularities is determined by the time derivatives and spatial derivatives become negligible, which is substantiated by rigorous mathematical and numerical results [36, 37], alleviates some of these concerns and provides a support to the quantum cosmology program. Finally, recently there has been some concrete progress on the relation between LQC and full LQG, discussed briefly in Section 6. From the phenomenological perspective, we are experiencing a fascinating time in cosmology. The observational results of WMAP [38] and PLANCK [39] satellites have provided strong evidence for a primordial origin of the CMB temperatures anisotropies. There is no doubt that the excitement in early universe cosmology is going to continue for several more years, providing a promising opportunity to test implications of quantum gravity in cosmological observations.

This chapter provides a review, including the most recent advances, of loop quantization of cosmological spacetimes and phenomenological consequences. It is organized as follows. Section 2 provides a summary of loop quantization of the spatially flat, isotropic and homogeneous model sourced with a massless scalar field. This model was the first example of the rigorous quantization of a cosmological spacetime in LQC [9–11]. Because the quantization strategy underlying this model has been implemented for spacetimes with spatial curvature, anisotropies and also in presence of inhomogeneities, we discuss it in more detail. After laying down the classical framework in Ashtekar variables, we discuss the kinematical and dynamical features of loop quantization and the way classical singularity is resolved and replaced by a bounce. This section also briefly discusses the effective continuum spacetime description which provides an excellent approximation to the underlying quantum dynamics for states which are sharply peaked. For a specific choice of lapse, equal to the volume, and for the case of a massless scalar field one obtains an exactly solvable model of LQC (sLQC) which yields important robustness results on the quantum bounce [12]. In Section 3, we briefly discuss the generalization of loop quantization and the resulting Planck scale physics to spacetimes with spatial curvature, Bianchi, and Gowdy models. Section 4 is devoted to cosmological perturbations. We review the formulation of a quantum gravity extension of the standard theory of gauge invariant cosmological perturbations in LQC. These techniques provide the theoretical arena to study the origin of matter and gravitational perturbations in the early universe. This is the goal of Section 5 where we summarize the LQC extension of the inflationary scenario and discuss the quantum gravity corrections to physical observables [41–43]. Due to space limitations, it is difficult to cover various topics and details in this chapter. These include the earlier developments in LQC [44–46], the path integral formulation of LQC [47], entropy bounds [48], consistent quantum probabilities [49–53], application to black hole interiors [54–59], and various mathematical [60–63] and numerical results [64–66] in LQC. Issues with inverse triad modifications [10, 11], limitations of the earlier quantizations in LQC and the role of fiducial scalings [11, 67, 68], and issues related to quantization ambiguities and the resulting physical effects [69, 70] are also not discussed. For a review of some of these developments and issues in LQC, we refer the reader to Ref. [3] and the above cited references. We are also unable to cover all the current approaches to studying LQC effects on cosmological perturbations; see [71–84] for different approaches to that problem. Further information can be found in *Chapter 8*, and in the review articles [3, 4, 8, 85–87]. Related to LQC, there have been developments in spin foams and group field theory, for which we refer the reader to Refs. [88, 89].

Our convention for the metric signature is $-+++$, we set $c = 1$ but keep G and $\hbar$ explicit in our expressions, to emphasize gravitational and quantum effects. When numerical values are shown, we use Planck units.

2. Loop Quantization of Spatially Flat Isotropic and Homogeneous Spacetime

In this section, we illustrate the key steps in loop quantization of homogeneous cosmological models using the example of spatially flat FLRW spacetime sourced with a massless scalar field ϕ. Though simple, this model is rich in physics and provides a blueprint for the quantization of models with spatial curvature, anisotropies and other matter fields. Loop quantization of this spacetime was first performed in Refs. [9–11] where a rigorous understanding of the quantum Hamiltonian constraint, the physical Hilbert space and the Dirac observables was obtained, and detailed physical predictions were extracted using numerical simulations. It was soon realized that this model can also be solved exactly [12]. This feature serves as an important tool to test the robustness of the physical predictions obtained using numerical simulations. In the following, in Section 2.1, we begin with the quantization of this cosmological model in the volume representation. We discuss the classical and the quantum framework, and the main features of the quantum dynamics. We also briefly discuss the effective spacetime description which captures the quantum dynamics in LQC for sharply peaked states to an excellent approximation and provides a very useful arena to understand various phenomenological implications. The exactly solvable model is discussed in Section 2.2.

2.1. *Loop Quantum Cosmology: k = 0 Model*

In the following, we outline the classical and the quantum framework of LQC in the spatially flat isotropic and homogeneous spacetime following the analysis of Refs. [9–11]. In literature this quantization is also known as '$\bar{\mu}$ quantization' or 'improved dynamics' [11]. In the first part we introduce the connection variables, establish their relationship with the metric variables, find the classical Hamiltonian constraint in the metric and the connection variables and obtain the singular classical trajectories in the relational dynamics expressing volume as a function of the internal time ϕ. This is followed by the quantum kinematics, properties of the quantum Hamiltonian constraint in the geometric (volume) representation, the physical Hilbert space and a summary of the physical predictions. A comparison with the Wheeler-DeWitt theory is also provided both at the kinematical and the dynamical level. An effective description of the quantization performed here, following the analysis of Refs. [90, 91] is discussed in Section 2.1.3.

2.1.1. *Classical framework*

The spatially flat homogeneous and isotropic spacetime is typically considered with a spatial topology $\mathbb{R}^3$ or of a 3-torus $\mathbb{T}^3$. For the non-compact spatial manifold extra care is needed to introduce the symplectic structure in the canonical framework because of the divergence of the spatial integrals. For the non-compact case one introduces a fiducial cell $\mathcal{V}$, which acts as an infra-red regulator [11]. Physical implications must be independent of the choice of this regulator, which is the case for the present analysis.[a] Such a cell is not required for the compact topology. The spacetime metric is given by

$$ds^2 = -dt^2 + a^2 \,\mathring{q}_{ab} dx^a dx^b \tag{1}$$

where t is the proper time, a denotes the scale factor of the universe and $\mathring{q}_{ab}$ denotes the fiducial metric on the spatial manifold. With the matter source as the massless scalar field which serves as a physical clock in our analysis, instead of proper time it is natural to introduce a harmonic time τ satisfying $\Box \tau = 0$ since ϕ satisfies the wave equation $\Box \phi = 0$. This corresponds to the choice of the lapse $N = a^3$. The spacetime metric then becomes

$$ds^2 = -a^6 d\tau^2 + a^2 \left(dx_1^2 + dx_2^2 + dx_3^2 \right). \tag{2}$$

In terms of the physical spatial metric $q_{ab} = a^2 \mathring{q}_{ab}$, the physical volume of the spatial manifold is $V = a^3 V_o$, where V_o is the comoving volume of the fiducial cell

---
[a]This is not true for the earlier quantization in LQC [10, 46], and the lattice refined models [92]. For a detailed discussion of these difficulties in other quantization prescriptions we refer the reader to Refs. [11, 67].

in case the topology is $\mathbb{R}^3$, or the comoving volume of $\mathbb{T}^3$ in case the topology is compact.

Due to the underlying symmetries of this spacetime, the spatial diffeomorphism constraint is satisfied and the only non-trivial constraint is the Hamiltonian constraint. Let us first obtain this constraint in the metric variables. In such a formulation, the canonical pair of gravitational phase space variables consists of the scale factor a and its conjugate $p_{(a)} = -a\dot{a}$, with 'dot' denoting derivative with respect to the proper time. These variables satisfy $\{a, p_{(a)}\} = 4\pi G/3V_o$. The matter phase space variables are ϕ and $p_{(\phi)} = V\dot{\phi}$, which satisfy $\{\phi, p_{(\phi)}\} = 1$. In terms of the metric variables, the Hamiltonian constraint is given by

$$\mathcal{C}_H = -\frac{3}{8\pi G}\frac{p_{(a)}^2 V}{a^4} + \frac{p_{(\phi)}^2}{2V} \approx 0, \qquad (3)$$

which yields the classical Friedman equation in terms of the energy density, $\rho = p_{(\phi)}^2/2V^2$, for the spatially flat FRW model:

$$\left(\frac{\dot{a}}{a}\right)^2 = \frac{8\pi G}{3}\rho. \qquad (4)$$

In order to obtain the classical Hamiltonian constraint in terms of the variables used in LQG: the Ashtekar-Barbero SU(2) connection A_a^i and the conjugate triad E_i^a, we first notice that due to the symmetries of the isotropic and homogeneous spacetime, the connection A_a^i and triad E_i^a can be written as [46]

$$A_a^i = c V_o^{-1/3} \mathring{\omega}_a^i, \quad E_i^a = p V_o^{-2/3} \sqrt{\mathring{q}}\, \mathring{e}_i^a, \qquad (5)$$

where c and p denote the isotropic connection and triad, and $\mathring{e}_i^a$ and $\mathring{\omega}_a^i$ are the fiducial triads and co-triads compatible with the fiducial metric $\mathring{q}_{ab}$. The canonically conjugate pair (c, p) satisfies $\{c, p\} = 8\pi G\gamma/3$, and is related to the metric variables as $|p| = V_o^{2/3} a^2$ and $c = \gamma V_o^{1/3} \dot{a}/N$, where γ is the Barbero-Immirzi parameter in LQG, whose value is set to $\gamma \approx 0.2375$ using black hole thermodynamics [93]. The modulus sign over the triad arises because of the two possible orientations, the choice of which does not affect physics in the absence of fermions. It is important to note that the above relation between the triad and the scale factor is true kinematically, whereas the relation between the isotropic connection and the time derivative of the scale factor is true only for the physical solutions of GR.

It turns out that in the quantum theory, it is more convenient to work with variables b and v which are defined in terms of c and p as [12]:

$$\mathrm{b} := \frac{c}{|p|^{\frac{1}{2}}}, \quad \mathrm{v} := \mathrm{sgn}(p)\frac{|p|^{\frac{3}{2}}}{2\pi G}, \qquad (6)$$

where sgn(p) is ± 1 depending on whether the physical and fiducial triads have the same orientation (+), or the opposite (-). The conjugate variables b and v satisfy $\{\mathrm{b}, \mathrm{v}\} = 2\gamma$, and in terms of which the classical Hamiltonian constraint becomes

$$\mathcal{C}_H = -\frac{3}{4\gamma^2}\mathrm{b}^2|\mathrm{v}| + \frac{p_{(\phi)}^2}{4\pi G|\mathrm{v}|} \approx 0. \qquad (7)$$

For a given value of $p_{(\phi)}$ and for a given triad orientation, Hamilton's equations yield an expanding and a contracting trajectory, given by

$$\phi = \pm \frac{1}{\sqrt{12\pi G}} \ln \frac{\mathrm{v}}{\mathrm{v}_c} + \phi_c \qquad (8)$$

where v_c and ϕ_c are integration constants. Both trajectories encounter a singularity. In the classical theory, the existence of a singularity either in past of the expanding branch or in the future of the contracting branch is thus inevitable.

2.1.2. *Quantum framework*

To pass to the quantum theory, the strategy is to promote the classical phase variables and the classical Hamiltonian constraint to their quantum operator analogs. For the metric variables, this strategy leads to the Wheeler-DeWitt quantum cosmology. Since, we wish to obtain a loop quantization of the cosmological spacetimes based on LQG we cannot use the same strategy for the connection-triad variables. In LQG, variables used for quantization are the holonomies of the connection A_a^i along edges, and the fluxes of the triads along 2-surfaces. (See *Chapter 1*.) For the homogeneous spacetimes, the latter turn out to be proportional to the triad [46]. The holonomy of the symmetry reduced connection A_a^i along a straight edge $\mathring{e}_k^a$ with fiducial length μ is,

$$h_k^{(\mu)} = \cos\left(\frac{\mu c}{2}\right) \mathbb{I} + 2\sin\left(\frac{\mu c}{2}\right) \tau_k \qquad (9)$$

where $\mathbb{I}$ is a unit 2×2 matrix and $\tau_k = -i\sigma_k/2$, where σ_k are the Pauli spin matrices. Due to the symmetries of the homogeneous spacetime, the holonomy and flux are thus captured by functions $N_\mu(c) := e^{i\mu c/2}$ of c, and the triads p respectively. Since μ can take arbitrary values, N_μ are *almost periodic functions* of the connection c. The next task is to find the appropriate representation of the abstract $\star$-algebra generated by almost periodic functions[b] of the connection c: $e^{i\mu c/2}$, and the triads p. It turns out that there exists a unique kinematical representation of algebra generated by these functions in LQC [95–97]. This result has parallels with existence of a unique irreducible representation of the holonomy-flux algebra in full LQG [98, 99]. The gravitational sector of the kinematical Hilbert space $\mathcal{H}_{\mathrm{kin}}$ underlying this representation in LQC is a space of square integrable functions on the Bohr compactification of the real line: $L^2(\mathbb{R}_{\mathrm{Bohr}}, \mathrm{d}\mu_{\mathrm{Bohr}})$ [46]. Use of holonomies in place of connections does not directly affect the matter sector. For this reason, the matter sector of the kinematical Hilbert space is obtained by following the methods in the Fock quantization.[c] It is important to note the difference between the gravitational part of $\mathcal{H}_{\mathrm{kin}}$, and the one obtained by following the Wheeler-DeWitt procedure

---

[b]A continuous function F of an unrestricted real variable x is almost periodic if $F(x + \tau) = F(x)$ holds to an arbitrary accuracy for infinitely many values of τ, such that translations τ are spread over the whole real line without arbitrarily large intervals [94].

[c]Polymer quantization of matter sector in a similar setting has been studied in some of the works, see for eg. [100–102].

where the gravitational part of the kinematical Hilbert space is $L^2(\mathbb{R}, \mathrm{d}c)$. In LQC, the normalizable states are the countable sum of N_μ, which satisfy: $\langle N_\mu | N'_\mu \rangle = \delta_{\mu\mu'}$, where $\delta_{\mu\mu'}$ is a Kronecker delta. This is in contrast to the Wheeler-DeWitt theory where one obtains a Dirac delta. Thus, the kinematical Hilbert space in LQC is fundamentally different from one in the Wheeler-DeWitt theory. The intersection between the kinematical Hilbert space in LQC and the Wheeler-DeWitt theory consists only of the zero function. Since the system has only a finite degree of freedom, one may wonder why the von-Neumann uniqueness theorem, which leads to a unique Schrödinger representation in quantum mechanics, does not hold. It turns out that for the theorem to be applicable in LQC, N_μ should be weakly continuous in μ. This condition is not met in LQC, and the von-Neumann theorem is bypassed. (For further details on this issue, we refer the reader to Ref. [103].)

The action of the operators $\hat{N}_\mu$ and $\hat{p}$ on states $\Psi(c)$ is by multiplication and differentiation respectively. On the states in the triad representation labelled by eigenvalues μ of $\hat{p}$, the action of $\hat{N}_\mu$ is translational:

$$\hat{N}_\zeta \Psi(\mu) = \Psi(\mu + \zeta), \tag{10}$$

where ζ is a constant,[d] and $\hat{p}$ acts as:

$$\hat{p}\, \Psi(\mu) = \frac{8\pi\gamma l_{\mathrm{Pl}}^2}{6} \mu \Psi(\mu). \tag{11}$$

Before we proceed to the quantum Hamiltonian constraint, we note that the change in the orientation of the triads which does not lead to any physical consequences in the absence of fermions corresponds to a large gauge transformation by a parity operator $\hat{\Pi}$ which acts on $\Psi(\mu)$ as: $\hat{\Pi}\Psi(\mu) = \Psi(-\mu)$. The physical states in the absence of fermions are therefore required to be symmetric, satisfying $\Psi(\mu) = \Psi(-\mu)$.

To obtain the dynamics in the quantum theory, we start with the Hamiltonian constraint in full LQG in terms of triads E_i^a and the field strength of the connection $F_{ab}{}^k$:[e]

$$C_{\mathrm{grav}} = -\gamma^{-2} \int_{\mathcal{C}} \mathrm{d}^3 x \left[N (\det q)^{-\frac{1}{2}} \epsilon^{ij}{}_k E_i^a E_j^b \right] F_{ab}{}^k \tag{12}$$

which in terms of the symmetry reduced triads and lapse $N = a^3$ becomes,

$$C_{\mathrm{grav}} = \gamma^{-2} V_o^{-1/3} \epsilon^i{}_{jk} \mathring{e}_i^a \mathring{e}_j^b |p|^2 F_{ab}{}^k. \tag{13}$$

The field strength F_{ab}^k is expressed in terms of the holonomies over a square plaquette $\square_{ij}$ with length $\bar{\mu} V_o^{1/3}$ in the $i-j$ plane spanned by fiducial triads:

$$F_{ab}{}^k = -2 \lim_{Ar\square \to 0} \mathrm{Tr}\left(\frac{h_{\square_{ij}} - \mathbb{I}}{Ar\square} \tau^k \right) \mathring{\omega}_a^i \mathring{\omega}_b^j. \tag{14}$$

---

[d]Note that we have used N_ζ instead of N_μ to avoid confusion with the argument of the wave function $\Psi(\mu)$.

[e]The Hamiltonian constraint consists of two terms proportional to $\epsilon_{ijk} F_{ab}^i E^{aj} E^{bk}$ and $K_{[a}^i K_{b]}^j E_i^a E_j^b$, where K_a^i capture the extrinsic curvature. These two terms turn out to be proportional to each other for the spatially flat homogeneous and isotropic model. Equation (12) captures the resulting total contribution.

Here Ar☐ denotes the area of the square plaquette, and $h_{☐_{ij}} = h_i^{(\bar\mu)} h_j^{(\bar\mu)} (h_i^{\bar\mu})^{-1} (h_j^{\bar\mu})^{-1}$, with $\bar\mu$ denoting the *physical* edge length of the plaquette in the given state. Note that due to the underlying quantum geometry, the limit Ar☐ $\to 0$ in (14) does not exist in the quantum theory. Instead one has to shrink the area of the loop to the minimum non-zero eigenvalue of the area operator in LQG. We denote this minimum area to be Δl_{Pl}^2 where $\Delta = 4\sqrt{3}\pi\gamma$ [26]. This results in the following functional dependence of $\bar\mu$ on the triad [11]

$$\bar\mu^2 = \frac{\Delta l_{\text{Pl}}^2}{|p|} , \qquad (15)$$

where we have used the expression for the physical area of the loop which equals $\bar\mu^2 |p|$. Due to this form of $\bar\mu$, the action of $N_{\bar\mu}$ on the triad eigenstates is not by a simple translation. However, switching to the volume representation gives the simple translation action, and therefore in the quantum theory it is more convenient to work with this representation in which the action of the conjugate operator $\widehat{\exp(i\lambda b)}$ (with $\lambda^2 = \Delta l_{\text{Pl}}^2$) and the volume operator is:

$$\widehat{\exp(i\lambda b)} |\nu\rangle = |\nu - 2\lambda\rangle, \quad \hat V |\nu\rangle = 2\pi\gamma l_{\text{Pl}}^2 |\nu| |\nu\rangle \qquad (16)$$

where $\nu = \text{v}/\gamma\hbar$. Using these operators, we can find the solutions to $\hat C_H \Psi(\nu,\phi) = \hat C_{\text{grav}} + 16\pi G \hat C_{\text{matt}} \Psi(\nu,\phi) = 0$. For the massless scalar field as the matter source, the quantum constraint equation results in the following:

$$\partial_\phi^2 \Psi(\nu,\phi) = 3\pi G \nu \frac{\sin\lambda b}{\lambda} \nu \frac{\sin\lambda b}{\lambda} \Psi(\nu,\phi) =: -\Theta\Psi(\nu,\phi) \qquad (17)$$

where Θ is a positive definite, second order difference operator:

$$\Theta\Psi(\nu,\phi) := -\frac{3\pi G}{4\lambda^2} \nu \left((\nu + 2\lambda)\Psi(\nu + 4\lambda) - 2\nu\Psi(\nu,\phi) + (\nu - 2\lambda)\Psi(\nu - 4\lambda) \right) . \qquad (18)$$

The form of the quantum constraint turns out to be very similar to the Klein-Gordon theory, where ϕ plays the role of time and Θ acts like a spatial Laplacian operator. As in the Klein-Gordon theory, the physical states can be either positive or the negative frequency solutions. Without any loss of generality we choose the physical states to be solutions of the 'positive frequency' square root of the quantum constraint:

$$-i\,\partial_\phi \Psi(\nu,\phi) = \sqrt{\Theta}\,\Psi(\nu,\phi) . \qquad (19)$$

The inner product for these physical states can be obtained using group averaging [104–106], and is given by

$$\langle \Psi_1 | \Psi_2 \rangle = \sum_\nu \bar\Psi_1(\nu,\phi_o) |\nu|^{-1} \Psi(\nu_2,\phi_o) . \qquad (20)$$

To extract physical predictions, we introduce Dirac observables which are self-adjoint with respect to the above inner product. One of the Dirac observables is

$\hat{p}_{(\phi)}$ which is a constant of motion. The other is $\hat{V}|_\phi$, the volume at internal time ϕ. On states $\Psi(\nu, \phi)$, the action of these observables is

$$\hat{V}|_{\phi_o}\Psi(\nu, \phi) = 2\pi\gamma l_{\mathrm{Pl}}^2 \, e^{i\sqrt{\Theta}(\phi-\phi_o)}|\nu|\Psi(\nu, \phi_o) \qquad (21)$$

and

$$\hat{p}_\phi \Psi(\nu, \phi) = -i\hbar\, \partial_\phi \Psi(\nu, \phi) = \hbar\sqrt{\Theta}\Psi(\nu, \phi) \, . \qquad (22)$$

Note that the Dirac observables preserve the positive and negative frequency subspaces. The symmetric wave functions which satisfy equation (18) have support on a lattice $\nu = \pm\epsilon + 4n\lambda$ with $\epsilon \in [0, 4\lambda)$. Any subspace spanned by the wave functions labelled by ϵ is preserved under evolution and the action of the Dirac observables. Therefore, there is a superselection and it suffices to consider states with a particular value of ϵ. Further, physical predictions are insensitive to the choice of the lattice parameter. In the following analysis we choose $\epsilon = 0$, since this choice of lattice parameter results in the possibility of the evolution encountering the classical singularity at the zero volume. Any other value of ϵ can also be chosen, say $\epsilon = 0.1$, however in such case zero volume does not lie on support of the eigenfunctions of the Θ operator.

Before we discuss some of the key features of the quantum Hamiltonian constraint in LQC and the resulting physics, we note that a similar analysis goes through for the Wheeler-DeWitt theory based on the metric variables. At a kinematical level, the Wheeler-DeWitt Hilbert space consists of wave functions $\underline{\Psi}(a, \phi)$ on which the scale factor and ϕ operators act multiplicatively, and the operators corresponding to their conjugate variables act as differential operators. The physical states are found by promoting $\mathcal{C}_H$ (3) to an operator and solving $\hat{\mathcal{C}}_H \Psi(a, \phi) = 0$. The resulting quantum constraint turns out to be a differential equation [11, 12]:

$$\partial_\phi^2 \underline{\Psi}(z, \phi) = 12\pi G\, \partial_z^2 \underline{\Psi}(z, \phi) =: -\underline{\Theta}\, \underline{\Psi}(z, \phi) \qquad (23)$$

where $z = \ln a^3$ and $\underline{\Theta}$ is the evolution operator in Wheeler-DeWitt theory. This brings out another fundamental difference between the Wheeler-DeWitt theory and LQC. Unlike the Wheeler-DeWitt theory, the quantum constraint in LQC is a discrete operator with discreteness determined by the underlying quantum geometry in LQG. For the scales where the spacetime curvature is very small compared to the Planck scale, which corresponds to the large volumes for the present model, the Θ operator in LQC approximates the $\underline{\Theta}$ operator in the Wheeler-DeWitt theory [10, 46]. Thus, the continuum differential geometry is recovered from the underlying discrete quantum geometry at the small spacetime curvature.

Quantum evolution of physical states can be studied numerically using the quantum constraint equation (18). One considers an initial state far away from the Planck regime, with large volumes peaked at a certain value of $p_{(\phi)}$ at a classical trajectory. Recall that in the classical theory, for a given value of $p_{(\phi)}$ there exists an expanding and a contracting trajectory which are disjoint and singular. In numerical simulations, the state can be either chosen such that it is peaked on the

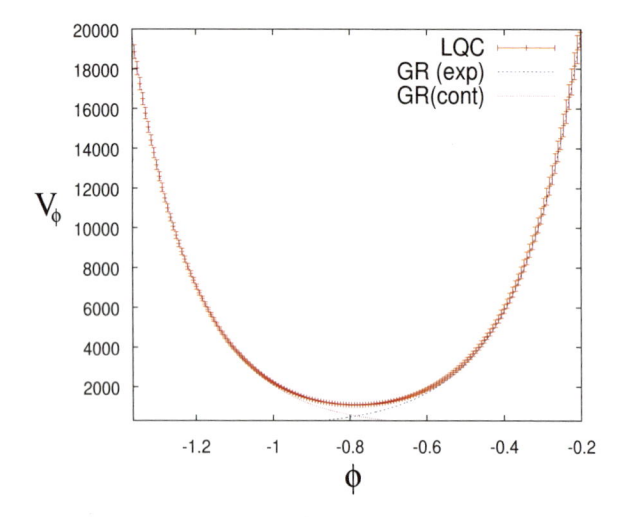

Fig. 1. A comparison of the quantum evolution in LQC for the volume observable (along with its dispersion) and the classical trajectories is shown. Unlike the general relativistic trajectories which lead to a singularity in the future evolution for the contracting branch and the past evolution for the expanding branch, the LQC trajectory is non-singular. The LQC trajectory bounces in the Planck regime and the loop quantum universe evolves in a non-singular way. The dispersions across the bounce are correlated, and their asymmetry depends on the method of initial state construction (see Ref. [10] for details of different methods). The state retains its peakedness properties in the above evolution, since the relative dispersion approaches a constant value at large volumes.

expanding trajectory at late times or on the contracting trajectory at early times Using ϕ as a clock, such a state, say chosen peaked on the expanding trajectory, is then numerically evolved towards the classical big bang singularity. The first numerical simulations were carried out using sharply peaked Gaussian states [10, 11]. Such states were shown to remain sharply peaked on the classical expanding trajectory for a long time in the backward evolution, till the spacetime curvature reaches approximately a percent of the Planck curvature. At the higher curvature scales, departures between the classical trajectory and quantum evolution become significant, and the loop quantum universe bounces when the energy density reaches a maximum value $\rho_{\max} \approx 0.41 \rho_{\text{Pl}}$ [11]. After the bounce, the quantum evolution is such that the state becomes sharply peaked on the classical contracting trajectory. Quantum gravitational effects thus bridge the two singular classical trajectories providing a non-singular evolution avoiding the classical singularity. A result of the variation of volume with respect to internal time from a typical simulation is illustrated in Figure 1 where the LQC evolution is also compared with the two classical trajectories in general relativity (GR). It is clearly seen that the quantum geometric effects play a role only near the bounce and quickly become negligible when spacetime curvature becomes small. These studies have been recently generalized for very widely spread states and highly squeezed and non-Gaussian states which capture the evolution of more quantum universes [107, 108], using high performance computing and faster algorithms [66]. The results of quantum bounce are found to be robust for all types of states. The existence of bounce does not require any fine tuning of the parameters or any special conditions. The quantum bounce is also found to be robust for slightly different quantization prescriptions in LQC [109, 110]. In contrast to the loop quantum evolution, the quantum evolution of Wheeler-DeWitt states yields a strikingly different picture. Initial states peaked on a classical trajectory remained peaked throughout the evolution and encounter the classical singularity. For the Wheeler-DeWitt states, the expectation values of the volume observable lie on the classical trajectory for all values of ϕ.

Thus, we find that unlike in the Wheeler-DeWitt theory, in LQC classical singularities are replaced by the bounce. The existence of bounce is tied to the underlying discrete quantum geometry — a feature which is absent in the Wheeler-DeWitt theory. The quantum evolution for various states in LQC illustrates the way classical GR is recovered in the low crvature regime. Thus, LQC not only provides a non-singular ultra-violet extension of the classical cosmological models, but also leads to the desired infra-red limit. Finally, we note that this feature provides an important criterion to single out the $\bar{\mu}$ quantization as performed in the above analysis of the various possible choices [67, 68]. In particular, it is useful to note that in the earlier quantization of LQC, called the μ_o scheme in literature, edge lengths of the loop over which holonomies were constructed did not take into account the physical geometry but were kept constant [10, 46]. It does not yield the correct infra-red limit and can lead to 'quantum gravitational effects' at arbitrarily small spacetime curvatures [10, 11, 67]. These difficulties are shared by the lattice refined models [92]. It is interesting to note that the conclusion that $\bar{\mu}$ quantization [11] is the only consistent quantization in LQC is not solely tied to the infra-red limit of the theory. This conclusion can also be reached by demanding that the physical predictions be invariant under the rescalings of the fiducial cell [67], by demanding the stability of the quantum difference equation [64, 65, 111, 112] and by demanding that the factor ordering ambiguities in gravitational part of the quantum constraint in LQC disappear in the limit where Wheeler-DeWitt theory is approached [113]. All these independent arguments provide a robust understanding of the viability of the $\bar{\mu}$ quantization of the cosmological spacetimes.

2.1.3. Effective spacetime description

In Section 2.1.2 we discussed the way underlying quantum geometry in LQG results in a quantum difference equation in LQC. Evolution of states with this difference equation predict a quantum bounce at the Planck scale. The question we are interested in now is whether it is possible to capture the key features of the quantum evolution, including the quantum bounce, in a continuum spacetime description at an effective level, allowing an $\hbar$-dependence in the metric coefficients. If so, is it possible to obtain a modified differential Friedmann and Raychaudhuri equations which incorporate the leading quantum gravitational effects? If such a set of reliable quantum-gravity-corrected equations exist, then the exploration of phenomenological implications would become much simpler numerically, within the approximations and caveats underlying these equations. Note however that, while physics obtained from such an effective spacetime description can provide important insights on the underlying quantum geometry, it is imperative to rigorously confirm the implications using full quantum dynamics in LQC wheereever possible. It turns out that for states which satisfy certain semi-classicality requirements and lead to a universe which is macroscopic at late times, an effective continuum spacetime description of the loop quantum dynamics can indeed be derived using a geometrical formulation of quantum mechanics [114, 115]. It provides an effective Hamiltonian from which a modified Friedmann equation can be obtained. The result is an effective dynamical trajectory which turns out to be in an excellent agreement with the quantum dynamics for the sharply peaked states [10, 11, 107]. In the following we briefly summarize the underlying method of deriving the effective dynamics following the analysis of Refs. [90, 91], obtain the modified Friedmann equation for the massless scalar field model, and discuss its main features. Various phenomenological implications of this modified Friedmann dynamics have been extensively discussed in the literature (see Ref. [3] for a review), a couple of which will be discussed briefly in Section 3.

In the geometrical formulation of quantum mechanics [114, 115], one treats the space of quantum states as an infinite dimensional phase space Γ_Q. The symplectic form (Ω_Q) on the phase space is given by the imaginary part of the Hermitian inner product on the Hilbert space. The real part of the inner product determines a Riemannian metric on Γ_Q. One then seeks a relation between the quantum phase space Γ_Q with symplectic structure Ω_Q and the classical phase space Γ with symplectic structure Ω. The relation is given by an embedding of the finite dimensional classical phase space onto $\bar\Gamma_Q \subset \Gamma_Q$. To capture the Hamiltonian flow on Γ_Q that generates full quantum dynamics, one must find an *astute* embedding such that the quantum Hamiltonian flow is tangential to $\bar\Gamma_Q$ to a high degree of approximation. This requirement is very non-trivial and there is no guarantee that such an embedding can be found. However, if such an embedding exists then the projection of the quantum Hamiltonian flow on $\bar\Gamma_Q$ provides the quantum corrected trajectories that capture the main quantum effects to a high degree of approximation. $\bar\Gamma_Q$ is by construction isomorphic with the classical phase space Γ and a point in Γ is labelled by $\xi^o \equiv (q_i^o, p_i^o)$. Therefore the quantum state corresponding to a point of $\bar\Gamma_Q$ is denoted by Ψ_{ξ^o}. A required embedding must satisfy $q_i = \langle \Psi_{\xi^o}, \hat q_i \Psi_{\xi^o} \rangle$, and $p_i = \langle \Psi_{\xi^o}, \hat p_i \Psi_{\xi^o} \rangle$. To find a suitable embedding, one makes careful choice of appropriate states Ψ_{ξ^o}, such as coherent states, by choosing appropriate parameters such as fluctuations. Once $\bar\Gamma_Q$ is found, the leading quantum corrections are well-captured in terms of the classical phase space variables. By carrying out this procedure, one thus obtains modified classical dynamical equations (the effective equations), which incorporate quantum gravitational corrections via a controlled approximation in terms of the parameters of the state. This approach to effective equations is called the 'embedding method'

Another approach is the 'truncation method' where one introduces a coordinate system on Γ_Q using the expectation values $(\bar{q}_i, \bar{p}_i)$, fluctuations *and* the higher order moments [117, 118]. The quantum Hamiltonian flow on the Hilbert space yields a set of coupled nonlinear differential equations which are infinite in number for all the moments. By suitably truncating this set up to a finite number of terms, one can then obtain classical dynamical equations with quantum corrections up to the truncated order. In comparison to the embedding approach where appropriate states and their parameters need to be chosen carefully to obtain approximately tangential Hamiltonian vector field, the truncation method is more systematic. However, it is difficult to understand the role played by the infinite number of moments which are truncated out, and the error associated with this truncation.

In LQC, effective equations have been derived using the embedding [90, 91, 116] as well the truncation method [117, 118]. However, since most of the numerical studies on confirming the validity of the effective dynamics and phenomenological implications have been performed for the embedding method, we will focus on this approach. For the massless scalar field, the effective Hamiltonian constraint, up to the approximation where terms proportional to the square of the quantum fluctuations of the state, using the embedding method is found to be [91]:

$$C_H^{(\text{eff})} = -\frac{3\hbar}{4\gamma\lambda^2}\nu\sin^2(\lambda b) + \frac{1}{4\pi\gamma l_{\text{Pl}}^2}\frac{p_{(\phi)}^2}{\nu}. \qquad (24)$$

Physical solutions satisfy $C_H^{(\text{eff})} \approx 0$, which yields

$$\frac{3}{8\pi G\gamma^2\lambda^2}V\sin^2(\lambda b) = \frac{p_{(\phi)}^2}{2V}. \qquad (25)$$

The modified Friedmann and Raychaudhuri equations can be found using Hamilton's equation for V and b respectively, which satisfy $\{b, V\} = 4\pi G\gamma$. As an example, Hamilton's equation for volume gives,

$$\dot{V} = \{V, C_H^{(\text{eff})}\} = -4\pi G\gamma\frac{\partial}{\partial b}C_H^{(\text{eff})} = \frac{3}{\gamma\lambda}\sin(\lambda b)\cos(\lambda b), \qquad (26)$$

from which it is straightforward to derive the modified Friedmann equation for the Hubble rate $H = \dot{V}/3V$ using equation (25):

$$H^2 = \frac{8\pi G}{3}\rho\left(1 - \frac{\rho}{\rho_{\text{max}}}\right) \quad \text{with} \quad \rho_{\text{max}} = \frac{3}{8\pi G\gamma^2\lambda^2}. \qquad (27)$$

The quantum gravitational correction thus appears as a ρ^2 modification to the classical Friedmann equation (4), with a negative sign.[f] The modified Raychaudhuri

---
[f]The ρ^2 modification albeit with a positive sign in front of ρ/ρ_{max} also appears in brane world scenarios in string cosmological models. For the modification to be negative one requires one of the extra dimensions to be time-like [119]. For a comparative analysis of the properties of the above modified Friedmann equation in LQC with the braneworld scenarios see Ref. [120, 121].

equation[g] can be similarly derived from the Hamilton's equation for $\dot b$, which yields

$$\frac{\ddot a}{a} = -\frac{4\pi G}{3}\rho\left(1 - 4\frac{\rho}{\rho_{\max}}\right) - 4\pi G P\left(1 - 2\frac{\rho}{\rho_{\max}}\right). \qquad (28)$$

where P denotes the pressure which is equal to the energy density for the massless scalar field model. The energy-matter conservation law remains unchanged from the classical theory.[h]

From equation (2.2) one concludes that the scale factor of the universe bounces when $\rho = \rho_{\max} \approx 0.41\rho_{\text{Pl}}$. Unlike the classical theory, the Hubble rate does not grow unboundedly through out the evolution, but is bounded above by a maximum value $|H|_{\max} = 1/(2\gamma\lambda)$ which occurs at $\rho = \rho_{\max}/2$. Note that the effective dynamics predicts the bounce at the same value of energy density which is found to be the supremum of the expectation values of the energy density observable ($\rho_{\sup}$) in exactly solvable LQC (as we shall see in equation (39)), and the value observed in various numerical simulations for the sharply peaked Gaussian states [11, 107]. For such states, the effective dynamical trajectory is in excellent agreement with the quantum evolution at all scales. This may seem surprising because the initial semi-classical state used to derive the effective dynamics is chosen in the regime where quantum gravitational effects are negligible and near the bounce some of the underlying assumptions on the parameters of the state can be suspect [91].[i] A careful analysis of the underlying assumptions in the derivation of effective dynamics shows that they are satisfied all the way up to the bounce for the sharply peaked states [116]. Thus, for such states effective dynamics provides a very reliable continuum spacetime description of this model. For states which have large relative fluctuations or have very large non-Gaussianity, numerical simulations find departures between the quantum evolution and the effective dynamics obtained from equations (2.2) and (28) [107, 108]. Interestingly, it turns out that the above effective dynamics always overestimates the energy density at the bounce. This observation is consistent with the result in sLQC that the maximum value of the expectation value of energy density $\rho_{\sup}$ is same as $\rho_{\max}$, and the bounce density for certain states in the quantum theory can be smaller [125, 126]. An insight from

---

[g]An interesting relation between this modified Raychaudhuri equation and the resulting structure of the canonical phase space has been explored as an inverse problem [121], without any a priori assumptions about the Hamiltonian framework. It has been suggested that existence of Raychaudhuri equation with such modifications, quadratic or higher order in energy density, requires holonomies of the connection as phase space variables.

[h]Strictly speaking this is true only when the Hamiltonian constraint does not contain any terms which include the inverse triad modifications which due to the choice of lapse are absent in our case. For models where such modifications can be consistently incorporated, such as in the $k = 1$ model, the conservation law is also modified [122, 123].

[i]It has been argued that for the non-compact topology, in the limit of removal of infra-red regulator quantum fluctuations do not affect the effective Hamiltonian [124]. However, this argument does not provide an answer to the puzzle about the validity of the effective dynamics at the bounce as discussed in the literature [91], since the apparently failing assumptions which are problematic are fluctuation independent [116].

sLQC is that for the case of a spatially flat model with a massless scalar field, the modified Friedman equation (2.2) can be generalized to arbitrary states in the physical Hilbert space with $\rho_{\max}$ replaced with the expectation value of the energy density observable at the bounce [127].

Effective dynamics has provided many important insights on the physics at the Planck scale in LQC. Using effective equations, a relationship of effective Hamiltonian in LQC with a covariant effective action containing infinite number of higher order curvature terms has been explored [128]. An extensive understanding is reached on genericity of singularity resolution and occurrence of inflation. For generic matter the above bound on the Hubble rate leads to the resolution of strong curvature singularities [129–132], and the bounce dynamics plays an important role to make the probability for inflation close to unity in LQC [133–136]. We discuss some of the applications of effective dynamics in Section 3. For a more complete discussion of various phenomenological implications we refer the reader to Ref. [3].

2.2. Solvable Loop Quantum Cosmology (sLQC)

The spatially flat loop quantum cosmological model with a massless scalar field can be solved exactly by passing to the b (the conjugate to volume) representation [12]. The exact solvability of this model proves extremely important to test the robustness of various physical implications obtained in Section 2.1. Another advantage of this analysis is that similarities and differences between LQC and the Wheeler-DeWitt theory become very transparent. In both frameworks, the underlying exactly soluble models are very similar, such as in the form of the quantum constraint and the action of momentum observable. But, there are also some important distinctions, in particular on the behavior of the expectation values of the volume. This difference is pivotal in proving some important results, including the genericness of quantum bounce in sLQC and the occurrence of the singularity in the Wheeler-DeWitt theory. Due to its simplicity and powerful features, sLQC has been widely applied in different settings to gain important insights on different problems in LQC, e.g., (i) to understand the growth of fluctuations for various states across the bounce [125, 126, 137, 138], (ii) to develop a path integral formulation of LQC to understand various conceptual issues and explore links with spin foam models [47, 140, 141], and, (iii) to understand the quantum probabilities for the occurrence of the bounce in LQC [50, 51] and singularities in Wheeler-DeWitt theory [49] using the consistent histories framework [142, 143]. Due to space limitations it is not possible for us to elaborate on these various interesting applications in detail, and refer the interested reader to the review [3].

In LQC, since the wave functions in the volume representation have support on a discrete interval $\nu = 4n\lambda$, the wave functions $\Psi(b, \phi)$ in the b representation have support on the continuous interval $(0, \pi/\lambda)$. In contrast, in the Wheeler-DeWitt theory $b \in (-\infty, \infty)$. In both the theories, the quantum Hamiltonian constraint in the b representation is a differential equation. Let us start with sLQC, where it is given by

$$\partial_\phi^2 \chi(b,\phi) = 12\pi G \left(\frac{\sin \lambda b}{\lambda} \partial_b\right)^2 \chi(b,\phi). \tag{29}$$

It is convenient to change the variable to x with $x \in (-\infty, \infty)$:

$$x = \frac{1}{\sqrt{12\pi G}} \ln\left(\tan\frac{\lambda b}{2}\right), \tag{30}$$

using which the quantum Hamiltonian constraint takes a very simple form of the wave equation,

$$\partial_\phi^2 \chi(x,\phi) = \partial_x^2 \chi(x,\phi) =: -\Theta \chi(x,\phi). \tag{31}$$

As in the case of the volume representation, the physical Hilbert space consists of the positive frequency solutions which satisfy $-i\partial_\phi \chi(x,\phi) = \sqrt{\Theta}\chi(x,\phi)$. Further, the requirement that the physics be invariant under the change of the orientation of the triads leads to $\chi(x,\phi) = -\chi(-x,\phi)$. Due to this antisymmetric condition, it turns out that every solution $\chi(x,\phi)$ can be expressed in terms of the right (x_-) and the left moving (x_+) parts $\chi(x,\phi) = \frac{1}{\sqrt{2}}(F(x_+) - F(x_-))$ where $x_\pm = \phi \pm x$. The physical inner product can be obtained in terms of the left moving (or the right moving) part as:

$$(\chi_1, \chi_2)_{\text{phys}} = -2i \int_{-\infty}^{\infty} dx \bar{F}_1(x_+) \partial_x F_2(x_+). \tag{32}$$

A similar construction can be carried out for the Wheeler-DeWitt theory, where, unlike sLQC, the resulting wave functions $\chi(y,\phi)$, where $y = (12\pi G)^{-1/2} \ln(b/b_o)$ with b_o an arbitrary constant, are not subject to the requirement that they must have support on the left and right moving sectors. In the Wheeler-DeWitt theory, these sectors decouple and one can choose wave functions composed solely of the left moving or the right moving solutions. Otherwise, the form of the quantum constraint, and the action of the momentum observable are identical. However, a crucial difference appears in the expectation value of the volume observable. For sLQC it turns out to be

$$(\chi, \hat{V}|_\phi \chi)_{\text{phys}} = 2\pi\gamma l_{\text{Pl}}^2 (\chi, |\hat{\nu}|_\phi \chi)_{\text{phy}} = V_+ e^{\sqrt{12\pi G}\phi} + V_- e^{-\sqrt{12\pi G}\phi}. \tag{33}$$

Here $V_\pm$ are positive constants determined by the initial state:

$$V_\pm = \frac{4\pi\gamma l_{\text{Pl}}^2 \lambda}{\sqrt{12\pi G}} \int_{-\infty}^{\infty} dx_+ \left|\frac{dF}{dx_+}\right|^2 e^{\mp\sqrt{12\pi G}x_+}. \tag{34}$$

In contrast, for the Wheeler-DeWitt theory the expectation value of $\hat{V}|_\phi$ for the left moving sector is

$$(\underline{\chi}_L, \hat{V}|_\phi \underline{\chi}_L)_{\text{phy}} = 2\pi\gamma l_{\text{Pl}}^2 (\underline{\chi}_L, |\hat{\nu}|_\phi \underline{\chi}_L)_{\text{phy}} = V_* e^{\sqrt{12\pi G}\phi} \tag{35}$$

with

$$V_* = \frac{8\pi\gamma l_{\text{Pl}}^2}{\sqrt{12\pi G} b_o} \int_{-\infty}^{\infty} dy_+ \left|\frac{d\underline{\chi}_L}{dy_+}\right|^2 e^{-\sqrt{12\pi G}y_+}. \tag{36}$$

In the Wheeler-DeWitt theory, the expectation values $\langle \hat{V}|_\phi \rangle$ approach zero as $\phi \to -\infty$. The left moving modes, which correspond to the expanding trajectory, thus encounter a big bang singularity in the past. Similarly, the right moving modes encounter a big crunch singularity in the future evolution. Note that this conclusion does not assume any profile of the initial state in this theory. An analysis of the quantum probabilities using consistent histories approach shows that the probability for a singularity to occur in this Wheeler-DeWitt model in asymptotic past of future is unity [49], even for the states composed of the arbitrary superpositions of the left and right moving sectors. On the other hand, in sLQC the expectation values of volume diverge both in asymptotic future and past ($\phi \to \pm\infty$). For any arbitrary state in the physical Hilbert space, $\langle \hat{V}|_\phi \rangle$ has a minimum value $V_{\min} = 2\sqrt{V_+ V_-}/||\chi||^2$ which is reached at the bounce time $\phi_B = (2\sqrt{12\pi G})^{-1} \ln(V_+/V_-)$. A consistent history analysis in sLQC yields the quantum probability for the bounce to be unity [51]. Unlike the Wheeler-DeWitt theory where big bang and big crunch singularities are inevitable, in sLQC these singularities are resolved for the generic states.

The fluctuations of the volume and the momentum observable can be computed in a similar way, which give important insights on the evolution of the states across the bounce. This issue is tied to understanding the way detailed properties of the universe in sLQC post-bounce branch are influenced by the initial state in the pre-bounce branch (or vice versa). Using sLQC constraints on the growth of the relative fluctuations have been obtained which show that a state which is semi-classical at very early times before the bounce retains its semi-classical properties after the bounce [125, 126, 137–139]. In particular, triangle inequalities relating the relative fluctuations of volume and momentum provide strong constraints on degree by which the fluctuations can change across the bounce [137]. These inequalities have been recently rigorously tested using extensive numerical simulations for semi-classical states as well as for states which are not sharply peaked [107, 108]. These inequalities are found to remain valid for all the numerical simulations performed till date.

Useful insights on the details of the singularity resolution in sLQC emerge on analyzing expectation values of the energy density of massless scalar field, whose corresponding observable is

$$\hat{\rho}|_\phi = \frac{1}{2}(\hat{A}|_\phi)^2 \quad \text{where} \quad \hat{A}|_\phi = (\hat{V}|_\phi)^{-1/2} \hat{p}_\phi (\hat{V}|_\phi)^{-1/2} \ . \tag{37}$$

The expectation values $\langle \hat{\rho}|_\phi \rangle$ computed at some $\phi = \phi_o$ is,

$$\langle \hat{\rho}|_{\phi_o} \rangle = \frac{3}{8\pi G \gamma^2} \frac{1}{\lambda^2} \frac{\left(\int_{-\infty}^\infty \mathrm{d}x |\partial_x F|^2\right)^2}{\left(\int_{-\infty}^\infty \mathrm{d}x |\partial_x F|^2 \cosh(\sqrt{12\pi G}x)\right)^2} \tag{38}$$

which is bounded above by $\rho_{\sup}$:

$$\rho_{\sup} = \frac{3}{8\pi\gamma^2 G \lambda^2} = \frac{\sqrt{3}}{32\pi^2 \gamma^3 G^2 \hbar} \approx 0.41 \rho_{\mathrm{Pl}} \ . \tag{39}$$

This value is in excellent agreement with the value of energy density at the bounce obtained using numerical simulations for the states which are sharply peaked at the classical trajectory at late times [11, 107]. As pointed out earlier, for the states which are widely spread, the value of energy density at the bounce in general turns out to be less than the above value [107]. The same conclusion holds true for the states which are squeezed [108, 126] or for states with more complex waveforms [108]. Extensive numerical simulations for various kinds of states have shown that the above supremum of the energy density always holds true [107, 108].

The generic bound on the energy density is a direct consequence of the quantum geometry which manifests through the area gap λ^2. It is also related to the ultra-violet cutoff for the eigenfunctions of the evolution operator which decay exponentially below the volume at which this energy density is reached. The evidence of this feature was first found numerically [10, 11], which has been recently rigorously confirmed using sLQC [144]. If the area gap is put to zero, the maximum of the energy density becomes infinity and the ultraviolet cutoff on the eigenfunctions disappears. Note that sLQC can be approximated to Wheeler-DeWitt theory for any given accuracy ϵ in a semi-infinite interval of time ϕ by appropriately shrinking the area gap. However, this is not possible if the entire infinite range of ϕ is considered. Then irrespective of the choice of a finite area gap, the differences between sLQC and Wheeler-DeWitt become arbitrarily large in some range of time ϕ. In this sense of global time evolution, the Wheeler-DeWitt theory is not a limiting case of sLQC. It turns out that sLQC is a fundamentally discrete theory and the limit $\lambda \to 0$ does not lead to a continuum theory. This feature of sLQC is not shared by the examples in polymer quantum mechanics where the continuum limit exists in the limit when the discreteness parameter vanishes [145, 146].

To summarize, sLQC has played an important role in proving the robustness of results on the bounce that were first observed in numerical simulations within LQC. The exact solvability of this model provides many insights on the supremum of the expectation value of the energy density observable, bounds on the growth of fluctuations across the bounce, and relation with the Wheeler-DeWitt theory.

3. LQC in More General Cosmological Spacetimes

In the previous section, we discussed the way quantum geometric effects in LQC resolve the classical big bang/big crunch singularity in the $k = 0$ isotropic and homogeneous spacetime, and result in a quantum bounce of the universe near the Planck scale. This result opens a new avenue to explore and develop novel non-singular paradigms in the very early universe, and to answer fundamental questions related to the structure of the spacetime near the singularities. For this it is necessary to extend the results on singularity resolution in LQC to more general settings. The goal of this section is to summarize the main developments in these directions. In Section 3.1, we start with a brief discussion on the generalization of the bounce results in different isotropic models — with spatial curvature, cosmological constant and an inflationary potential, focusing in particular on the properties of the quantum evolution operator and subtle features of the loop quantization.[j] This is followed by a discussion of two interesting applications in effective dynamics. Section 3.2 deals with the loop quantization of Bianchi models, where aspects of quantum theory and effective dynamics of Bianchi-I model are discussed in some detail. In Section 3.3, we discuss the application of LQC techniques to the Gowdy models which have provided useful insights on the singularity resolution in the presence of inhomogeneities.

3.1. *Quantization of Other Isotropic Models*

In the following, in Section 3.1.1 till Section 3.1.3, we summarize some of the main features of the isotropic models in LQC which have been quantized using the procedure outlined in Section 2.2. In all of these homogeneous models, one starts with the gravitational part of the classical Hamiltonian constraint in terms of the fluxes and the field strength of the connection, and express them in terms of the triads and the holonomies computed over a closed loop whose minimum area is given by $\lambda^2 = \Delta l_{\text{Pl}}^2$. The edge lengths of the holonomies $\bar{\mu}$ are functions of triads given by equation (15). The elements of holonomies form an algebra of almost periodic functions, and their action on the states in the volume representation is by uniform translations. As in the case of the $k = 0$ model, one obtains a quantum difference equation with uniform discreteness in volume. The scalar field plays the role of time in the quantum evolution, and one can introduce an inner product and a family of Dirac observables to extract physical predictions. Extensive numerical simulations confirm the existence of bounce which occurs at $\rho \approx 0.41 \rho_{\text{max}}$ for the sharply peaked states. Effective dynamics turns out to be in excellent agreement with the underlying loop quantum dynamics in all of these models. The last part of this section exhibits two applications of effective dynamics, where we discuss the way effective spacetime description provides important insights on the generic resolution of singularities and the naturalness of inflation in LQC.

3.1.1. *Spatially closed model*

The isotropic and homogeneous $k = 1$ model with a massless scalar field provides a very useful stage to carry out precise tests on the ultra-violet and infra-red limits in LQC. This is because in the classical theory, the scale factor in $k = 1$ model recollapses at a value determined by the momentum $p_{(\phi)}$. In the past of the classical evolution, the universe encounters a big bang singularity, and in the future it encounters a big crunch singularity. A non-singular quantum cosmological model should not only resolve both of the past and the future singularities, but must also lead to recollapse at the scales determined by the classical theory. Using the

---

[j]Loop quantization of isotropic model has also been performed in the presence of radiation. For details, we refer the reader to Ref. [23].

earlier quantization in LQC, Green and Unruh found that though the singularity is resolved, one is not able to obtain recollapse at the large scales predicted by the classical theory [147]. This limitation was tied to the unavailability of the inner product and detailed knowledge of the properties of the quantum evolution operator in the earlier works. These limitations were overcome in the loop quantization of the $k = 1$ model with a massless scalar field following the quantization procedure in the $k = 0$ model outlined in Section 2.2 [13, 14].[k] The resulting quantum Hamiltonian constraint, for lapse chosen to be $N = a^3$, takes the following form:

$$\partial_\phi^2 \Psi(\nu,\phi) = -\Theta_{(k=1)} \Psi(\nu,\phi)$$
$$= -\Theta\Psi(\nu,\phi) + \frac{3\pi G}{\lambda^2} \nu \left[\sin^2\left(\frac{\lambda}{\tilde{K}\nu^{1/3}} \frac{\ell_o}{2}\right) \nu \right.$$
$$\left. - (1+\gamma^2) \left(\frac{\lambda}{\tilde{K}} \frac{\ell_o}{2}\right)^2 \nu^{1/3} \right] \Psi(\nu,\phi) \qquad (40)$$

where $\Theta_{(k=1)}$ is a positive definite, self-adjoint operator. In contrast to the quantum evolution operator Θ of the $k = 0$ model, $\Theta_{(k=1)}$ has a discrete spectrum. This property is tied to the behavior of the eigenfunctions of $\Theta_{(k=1)}$ which decay exponentially for volumes greater than the recollapse volume in the classical theory, and also below a particular value of volume when the spacetime curvature reaches Planck scale. Numerical simulations with sharply peaked Gaussian states, and analysis of Dirac observables $\hat{p}_{(\phi)}$ and $\hat{V}|_\phi$, show that the $k = 1$ loop quantum universe bounces at the volume where the exponential decay of the eigenfunctions occurs in the Planck regime, with a maximum in the expectation values of energy density observable given by $0.41\rho_{\max}$. The loop quantized model also recollapses at the value in excellent agreement with the classical theory. States preserve their peakedness properties through bounces and recollapses, and the quantum evolution continues forever by avoiding the big bang and big crunch singularities, providing a non-singular cyclic model of the universe. The effective dynamics obtained from an effective Hamiltonian constraint with a form similar to (24) provides an excellent agreement to the loop quantum dynamics at all the scales. The loop quantization of $k = 1$ model successfully demonstrates that in LQC not only are the classical singularities resolved, but the theory also agrees with GR with an extraordinary precision at classical scales.

3.1.2. *Positive and negative cosmological constant*

The case of positive cosmological constant Λ, with a massless scalar field in the spatially flat isotropic and homogeneous spacetime is interesting due to various conceptual and phenomenological reasons. In the classical theory, the model has a big bang singularity in the past, and undergoes accelerated expansion in the future evolution. The universe expands to an infinite volume in an infinite proper time t, but at a finite value of the scalar field ϕ. Thus, in the relational dynamics with ϕ

---
[k]The model has been recently quantized in a different way following the same strategy [15].

as a clock, the Hamiltonian vector field on the phase space is incomplete. With an analytical extension, the dynamical trajectories in the classical theory start from a big bang singularity at $\phi = -\infty$, and encounter a big crunch at $\phi = \infty$ [21]. In terms of the time ϕ, the classical evolution thus turns out to be much richer. Following the strategy for the loop quantization of $k = 0$ FLRW model with a massless scalar, loop quantization can be performed rigorously, which leads to the following evolution equation [21]:

$$\partial_\phi^2 \Psi(\nu,\phi) = -\Theta_{\Lambda_+} \Psi(\nu,\phi) := -\Theta\, \Psi(\nu,\phi) + \frac{\pi G \gamma^2 \Lambda}{2} \nu^2\, \Psi(\nu,\phi), \qquad (41)$$

where $\Lambda > 0$. The operator Θ_{Λ_+} is not essentially self-adjoint, and one needs to find its self-adjoint extensions [20, 21]. For any choice of such an extension, the spectrum of Θ_{Λ_+} is discrete. It turns out that the details of the physics are independent of the choice of the extension for large eigenvalues of Θ_{Λ_+}. Numerical simulations with sharply peaked states show the existence of bounce when the energy density of massless scalar field and cosmological constant becomes approximately equal to $0.41\rho_{\mathrm{Pl}}$. Interestingly, for ϕ as a clock, infinite volume is reached in finite ϕ. In the quantum theory, evolution continues beyond this point in relational time ϕ and results in a contracting trajectory. In this way, the evolution in this model mimics the cyclic universe with ϕ as time.

The loop quantization of $k = 0$ model with a massless scalar field and a negative cosmological constant was first discussed briefly in Ref. [11], and studied in detail in Ref. [19]. As in the $k = 1$ model, the classical universe has a big bang singularity in the past, and a big crunch singularity in the future after the negative cosmological constant results in a recollapse in the expanding branch. In LQC, the quantum Hamiltonian constraint turns out to be,

$$\partial_\phi^2 \Psi(\nu,\phi) = -\Theta_{\Lambda_-} \Psi(\nu,\phi) := -\Theta\, \Psi(\nu,\phi) - \frac{\pi G \gamma^2 |\Lambda|}{2} \nu^2\, \Psi(\nu,\phi)\,. \qquad (42)$$

The operator Θ_{Λ_-} is essentially self-adjoint with a discrete spectrum. At large eigenvalues, the spacing between the eigenvalues is nearly uniform. Sharply peaked states constructed with such eigenvalues undergo nearly cyclic evolution in LQC, avoiding the big bang singularity in the past and the big crunch singularity in the future, with quantum bounces occurring at $\rho \approx 0.41\rho_{\mathrm{Pl}}$. As in the $k = 1$ model, the universe recollapses at the volume predicted by the classical theory.

3.1.3. *Inflationary potential*

In the classical theory, inflationary spacetimes are past incomplete [148]. A natural question is whether in LQC, one can construct non-singular inflationary models. The problem is challenging because of several reasons. Unlike the models considered so far, in the presence of a potential, $p_{(\phi)}$ is not a constant of motion and therefore ϕ does not serve as a global clock. However, ϕ can still be used as a local clock in portions of the pre-inflationary epoch where it has a monotonic behavior. For the

$\frac{1}{2}m^2\phi^2$ potential, the loop quantization leads to the following quantum evolution equation [22],

$$\partial_\phi^2 \Psi(\nu,\phi) = -\Theta_{(m)} \Psi(\nu,\phi) := -\left(\Theta - 4\pi G\gamma^2 m^2 \phi^2 \nu^2\right) \Psi(\nu,\phi). \quad (43)$$

Note that since $\Theta_{(m)}$ depends on time ϕ, obtaining the inner product becomes more subtle. The operator $\Theta_{(m)}$ is equivalent to $\Theta_{(\Lambda_+)}$ in (41) for any fixed value of ϕ, and hence fails to be essentially self-adjoint. For each value of ϕ, $\Theta_{(m)}$ admits self-adoint extensions. The physical Hilbert space can be obtained given a choice of these extensions. Numerical simulations with sharply peaked states show that the quantum evolution resolves the past singularity and results in a quantum bounce when the energy density of the inflaton field reaches $0.41\rho_{\rm Pl}$ [22]. Further, the classical GR trajectory is recovered when the spacetime curvature becomes much smaller than the Planck value. Thus, loop quantum gravitational effects make inflation past complete. A detailed understanding of the physics of this model and its relation to the choice of self-adjoint extensions is an open issue.

3.1.4. Some applications of the effective dynamics

We now discuss two applications of the effective dynamics in the isotropic model. The first example probes the question whether singularities are generically resolved in LQC, and under what conditions, and the second example deals with the naturalness of the inflationary scenario in LQC.

- *Generic resolution of strong singularities:* Apart from the resolution of the big bang/big crunch singularities in the quantum theory, the resolution of various other types of singularities such as the big rip singularity has been achieved in the effective spacetime description in LQC [129, 149–152]. An important question is whether quantum geometric effects resolve all the spacelike singularities, or are there certain types of singularities which are not resolved. Here it is to be noted that even in GR, not all singularities are harmful. Certain singularities even if characterized by divergences in the components of spacetime curvature, are harmless because the tidal forces turn out to be finite [153–155]. It turns out that such singularities — known as weak singularities, are not resolved in LQC [129, 156]. Such singularities are tied to the divergence in the spacetime curvature caused by the divergence in the pressure even when the energy density is bounded above. Note that for various LQC spacetimes, including isotropic, anisotropic and certain Gowdy models, expansion scalar and anisotropic shear are bounded [29, 68, 157–160]. However, it is straightforward to see that LQC allows divergence of curvature invariants. A straightforward computation of Ricci scalar R from the modified Friedman (2.2) and the Raychaudhuri equation (28), shows that if the equation of state of matter is such that $P(\rho) \to \pm\infty$ at a finite value of ρ, then the Ricci scalar diverges. However, geodesics can be extended beyond all such events [129]. On the other hand, strong singularities — the ones for which tidal forces are infinite, are generically resolved for matter with arbitrary equation of state in the $k = 0$ isotropic and homogeneous model in LQC [129]. The expansion scalar of the geodesics remains bounded in the effective spacetime which turns out to be geodesically complete. For the spatially curved model, a phenomenological analysis of effective dynamics confirms that strong singularities are resolved whereas weak singularities are ignored by the quantum geometric effects [150]. The results on generic resolution of strong singularities have also been generalized in the presence of anisotropies in the Bianchi-I model [130, 131] and Kantowski-Sachs spacetime [132]. These results provide a strong indication that resolution of singularities may be a very generic phenomena due to the quantum gravitational effects in LQC. A fundamental question is whether these results point towards a non-singularity theorem. Future work in this direction is expected to reveal an answer to this important question.

- *Probability for inflation:* The inflationary paradigm has been extremely successful in providing a description of the early universe in the FLRW model. It is natural to ask whether it can be successfully embedded in LQC. Various inflationary models have been considered in LQC using the modified Friedman dynamics, including single field inflation with $m^2\phi^2$ potential [133–135, 161, 162], multi-field inflationary models [163], tachyonic inflation [164], with non-minimally coupled scalar fields [165], and even in the presence of anisotropies [166]. An important question in this setting is, if we solve the spacetime dynamics in LQC starting from the evolution at the Planck era of the universe, does a phase of slow-roll inflation compatible with observations appear at some time in the future evolution? If so, does this happen for generic initial conditions or only for very specific, fine-tuned values of the initial data? In the context of GR, it has long been argued that inflationary trajectories are *attractors* in the space of solutions. Similar conclusions have been found in LQC [161], but existence of attractors does not tell us about the probabilities unless a suitable measure is defined. A detailed analysis of these questions was performed in Refs. [133, 134] in a spatially flat FLRW background. The authors of this reference proceed by computing the *fractional volume* in the space of solutions occupied by physical trajectories with the desired properties — an inflationary phase compatible with observation at some time in the evolution. But as they point out, the presence of gauge degrees of freedom makes the construction of the measure needed to compute those volumes quite subtle. The natural strategy, namely the projection of the natural Liouville measure to the reduced, or gauge fixed space of solutions, produces ambiguous results — it turns out, as pointed out in [135], that this fact is the origin of disparate results in the literature [167, 168]. The ambiguity, on the other hand, can be resolved by introducing a preferred moment in time in the evolution. In GR, there is no such preferred instant, but in LQC the existence of the bounce that *every* trajectory experiences provides the required structure. The authors of Refs. [133, 134] use this fact to show that, by assuming a flat probability distribution in the space of initial conditions for matter and geometry at the bounce time, a 99.9997% of the volume of that space corresponds to solutions that will encounter observationally favored inflation during the evolution. The conclusion is therefore that the inflationary attractor is also present in LQC. See Ref. [136] for further explanation of the underlying reason of this attractor mechanism.

3.2. Bianchi-I Model

The Bianchi-I model is one of the simplest settings to understand the way anisotropies in the spacetime play an important role on the physics near the classical singularities. Due to an interplay of the Ricci and Weyl components of the spacetime curvature, the structure of the singularities is very rich in comparison to the isotropic models. The spacetime metric of the Bianchi-I model is given by

$$ds^2 = -dt^2 + a_1^2 dx_1^2 + a_2^2 dx_2^2 + a_3^2 dx_3^2, \qquad (44)$$

where a_i denote the directional scale factors. Unlike the isotropic models, where the big bang singularity is characterized by the universe shrinking to a point of zero scale factor, the big bang singularity in the anisotropic models in GR can be of the shape of a cigar, a pancake, a barrel or a point depending on the behavior of the directional scale factors, captured by the Kasner exponents k_i, defined via $a_i \propto t^{k_i}$. It turns out that in general, unless one considers matter which has an equation of state, given by the ratio of the pressure and the energy density, to be greater than or equal to unity, the approach to the singularity is dictated by the anisotropic shear. In more general situations, such as in the Bianchi-IX models where the presence of spatial curvature leads to even richer dynamics, the approach to the singularity is oscillatory in the Kasner exponents and leads to the Mixmaster behavior [36]. According to the BKL conjecture, in the inhomogeneous spacetimes the approach to the singularity is such that the spatial derivatives can be ignored in comparison to the time derivatives, and each point of the space asymptotically behaves as in the Bianchi-IX model [35]. Thus, understanding the way singularity resolution occurs in Bianchi models is very important to understand singularity resolution in general.

In LQC, a quantization of Bianchi-I [26], Bianchi-II [27] and Bianchi-IX models [28, 29] has been performed which leads to a non-singular quantum constraint equation. However, physical implications in the quantum theory in terms of the expectation values of Dirac observables have only been studied for the Bianchi-I vacuum model [25] for an earlier quantization [24, 169].[1] Various interesting results on the physics of the Bianchi models have been obtained using effective dynamics, which include existence of Kasner transitions across the bounce in Bianchi-I model [138], existence of inflationary attractors [68], constructing non-singular cyclic models in the Ekpyrotic scenario [170], generic bounds on geometric scalars [68] and the resolution of strong curvature singularities [130, 131]. In the following, we outline

---

[1]This quantization has some limitations related to the dependence of physical predictions on the shape of the fiducial cell which is introduced to define symplectic structure [26, 68]. Nevertheless, it is consistent when the spatial topology is a 3-torus and provides important insights on the physics at the Planck scale and the nature of bounce in Bianchi-I model.

the quantization of the Bianchi-I model with a massless scalar fields as performed in Ref. [26], and discuss some of the features of the effective dynamics. For details of the loop quantization and the resulting physics of Bianchi-II and Bianchi-IX spacetimes, we refer the readers to the original works [27–29, 157, 159, 169, 171–173].

Utilizing the symmetries of the spatial manifold, which as in the $k = 0$ isotropic model can be of $\mathbb{R}^3$ or $\mathbb{T}^3$ topology, the Ashtekar-Barbero connection and the densitized triad in the Bianchi-I model can be written as

$$A_a^i = c^i (L^i)^{-1} \mathring{\omega}_a^i, \quad \text{and} \quad E_i^a = p_i L_i V_o^{-1} \sqrt{\mathring{q}} \, \mathring{e}_i^a \,, \qquad (45)$$

where c^i and p_i are the symmetry reduced connections and triads and L_i denote the coordinate lengths of the fiducial cell in the case of $\mathbb{R}^3$ topology. For the $\mathbb{T}^3$ topology, one does not need to introduce a fiducial cell and L_i can be set to 2π. Note that the coordinate volume $V_o = L_1 L_2 L_3$ with respect to the fiducial metric $\mathring{q}_{ab}$ changes if the individual L_i are rescaled. In the classical theory, physics is invariant under the change in the rescalings in L_i. This also turns out to be true for the loop quantization of the Bianchi-I model discussed here.[m] The triad components are related to the directional scale factors as

$$p_1 = \varepsilon_1 L_2 L_3 |a_2 a_3|, \quad p_2 = \varepsilon_2 L_1 L_3 |a_1 a_3|, \quad p_3 = \varepsilon_3 L_1 L_2 |a_1 a_2| \,, \qquad (46)$$

where $\varepsilon_i = \pm 1$ depending on the triad orientation. For the massless scalar field, the classical Hamiltonian constraint in terms of the symmetry reduced variables for lapse $N = a_1 a_2 a_3$, is given by

$$C_H = -\frac{1}{8\pi G \gamma^2}(c_1 p_1 \, c_2 p_2 + c_3 p_3 \, c_1 p_1 + c_2 p_2 \, c_3 p_3) + \frac{p_{(\phi)}^2}{2} \approx 0 \,. \qquad (47)$$

Using Hamilton's equations, one finds that

$$c_i p_i - c_j p_j = V(H_i - H_j) = \gamma \kappa_{ij} \,, \qquad (48)$$

where κ_{ij} is a constant antisymmetric matrix, and H_i denote the directional Hubble rates $H_i = \dot{a}_i/a_i$.

To understand the dynamical evolution, it is useful to introduce the mean Hubble rate and the shear scalar in this model. These are *kinematically* obtained from the trace and the symmetric tracefree parts of the expansion tensor defined as the covariant derivative of the timelike vector field tangential to the geodesics. The mean Hubble rate and the shear scalar in the Bianchi-I spacetime are:

$$H = \frac{1}{3}(H_1 + H_2 + H_3)\,, \quad \text{and} \quad \sigma^2 = \frac{1}{3}\left((H_1 - H_2)^2 + (H_2 - H_3)^2 + (H_3 - H_1)^2\right)\,, \qquad (49)$$

where $H = \dot{a}/a$ with $a = (a_1 a_2 a_3)^{1/3}$.

---

[m] In the earlier quantization prescription [24, 25], the resulting physics is not invariant under the change in shape of the fiducial cell if the topology is non-compact [26, 68].

In the classical theory, using equation (48), the shear $\Sigma^2 := \sigma^2 V^2/6$ turns out to be a constant.[n] The Hamilton's equations for the triads, yield the generalized Friedman equation:

$$H^2 = \frac{8\pi G}{3}\rho + \frac{\Sigma^2}{a^6} . \tag{50}$$

At the classical big bang singularity, the mean Hubble rate, energy density of the scalar field ρ and the shear scalar σ^2 diverge, and the geodesic evolution breaks down. Note that the shear scalar $\sigma^2 \propto a^{-6}$, and thus it diverges at the same rate as the massless scalar field energy density. In the presence of other matter sources such as dust and radiation, or an inflationary potential, since the energy density diverges slower than a^{-6}, the shear scalar dominates near the classical singularity and the singularity is necessarily anisotropic.

Let us now summarize the loop quantization of the Bianchi-I model. It is carried out with a similar procedure as in isotropic model, where one starts with a Hamiltonian constraint expressed in terms of triads E_i^a and the field strength of the holonomies $F_{ab}{}^i$. The holonomies yield an algebra of almost periodic functions of connections c_i, and the field strength can be computed by considering holonomies over a square loop $\Box_{ij}$. Due to the presence of anisotropies, the relation between the edge lengths $\bar\mu_i$ of the loops turns out to be: [26]

$$\bar\mu_1 = \lambda\sqrt{\frac{|p_1|}{|p_2 p_3|}}, \quad \bar\mu_2 = \lambda\sqrt{\frac{|p_2|}{|p_1 p_3|}}, \quad \text{and} \quad \bar\mu_3 = \lambda\sqrt{\frac{|p_3|}{|p_1 p_2|}} , \tag{51}$$

which reduces to the isotropic relation (15) when $p_1 = p_2 = p_3$. Due to the functional dependence on direction triads, the action of the elements of the holonomy algebra $\exp(i\bar\mu_i c^i)$ is very complicated on the states $\Psi(p_1, p_2, p_3)$. It is more convenient to work with states $\Psi(l_1, l_2, v)$ where l_i are defined via $p_i = (\text{sgn}\, l_i)\,(4\pi\gamma\lambda l_{\text{Pl}}^2)^{2/3}\, l_i^2$, and $v = 2(l_1 l_2 l_3)$. Computing the action of the field strength and the triad operators on these states, which also required to be symmetric under the change of the orientations of the triad, the quantum Hamiltonian constraint turns out to be

$$\partial_\phi^2 \Psi(l_1, l_2, v; \phi) = -\Theta_{\text{B-I}}\, \Psi(l_1, l_2, v; \phi) \tag{52}$$

where

$$\Theta_{\text{B-I}}\Psi(l_1,l_2,v;\phi) = \frac{\pi G \hbar^2}{8}\sqrt{v}\Big[(v+2)\sqrt{(v+4)}\Psi_4^+(l_1,l_2,v) - (v+2)\sqrt{v}\Psi_0^+(l_1,l_2,v;\phi)$$
$$- (v-2)\sqrt{v}\Psi_0^-(l_1,l_2,v;\phi) + (v-2)\sqrt{|v-4|}\Psi_4^-(l_1,l_2,v;\phi)\Big] . \tag{53}$$

---

[n]If matter has a non-vanishing anisotropic stress, as in the case of magnetic fields, Σ^2 is not constant in the classical theory. For a phenomenological investigation of the Bianchi-I model in LQC in such a situation, see Ref. [174].

Here $\Psi_4^\pm$ and $\Psi_0^\pm$ are defined as

$$\Psi_4^\pm(l_1, l_2, v; \phi) = \Psi\left(\frac{v\pm 4}{v\pm 2}l_1, \frac{v\pm 2}{v}l_2, v\pm 4\right) + \Psi\left(\frac{v\pm 4}{v\pm 2}l_1, l_2, v\pm 4\right)$$
$$+ \Psi\left(\frac{v\pm 2}{v}l_1, \frac{v\pm 4}{v\pm 2}l_2, v\pm 4\right) + \Psi\left(\frac{v\pm 2}{v}l_1, l_2, v\pm 4\right)$$
$$+ \Psi\left(l_1, \frac{v\pm 2}{v}l_2, v\pm 4\right) + \Psi\left(l_1, \frac{v\pm 4}{v\pm 2}l_2, v\pm 4\right), \quad (54)$$

and

$$\Psi_0^\pm(l_1, l_2, v; \phi) = \Psi\left(\frac{v\pm 2}{v}l_1, \frac{v}{v\pm 2}l_2, v\right) + \Psi\left(\frac{v\pm 2}{v}l_1, l_2, v\right)$$
$$+ \Psi\left(\frac{v}{v\pm 2}l_1, \frac{v\pm 2}{v}l_2, v\right) + \Psi\left(\frac{v}{v\pm 2}l_1, l_2, v\right)$$
$$+ \Psi\left(l_1, \frac{v}{v\pm 2}l_2, v\right) + \Psi\left(l_1, \frac{v\pm 2}{v}l_2, v\right). \quad (55)$$

An important property of the above quantum difference equation is the following. If one starts with an initial wave function which is peaked on a non-zero volume and which vanishes at $v = 0$, then in the quantum evolution it cannot have support on the zero volume. The classical singularity at $v = 0$ is decoupled from the evolution in the quantum theory [26]. This shows that for such wave functions, the spacetime curvature will remain bounded throughout the physical evolution. Further, it can be shown that the isotropic LQC for $k = 0$ model is recovered by integrating out anisotropies. To explore the physics in detail, numerical simulations on the lines of the isotropic models are required to compute expectation values of the Dirac observables, which in this case are $\hat{V}|_\phi, \hat{l}_1|_\phi$ and $\hat{l}_2|_\phi$. Since the form of the quantum evolution operator (52) is quite complicated in comparison to the isotropic quantum evolution operator (18), numerical simulations are technically difficult. However, insights have been gained on simplifying the quantum constraint to obtain physical solutions [175].

Useful insights on the Planck scale physics in the Bianchi-I model in LQC has been gained using the effective Hamiltonian constraint, which following the derivation in the isotropic model in LQC, is given by [26, 169]:

$$C_H^{\text{eff}} = -\frac{1}{8\pi G\gamma^2 (p_1 p_2 p_3)^{1/2}} \left(\frac{\sin(\bar\mu_1 c_1)}{\bar\mu_1} \frac{\sin(\bar\mu_2 c_2)}{\bar\mu_2} p_1 p_2 + \text{cyclic terms}\right) + \mathcal{H}_{\text{matt}} \quad (56)$$

where $\mathcal{H}_{\text{matt}}$ denotes the matter Hamiltonian. The expression for the energy density can be obtained from the vanishing of the effective Hamiltonian constraint, $C_H^{\text{eff}} \approx 0$, and it turns out to be

$$\rho = \frac{1}{8\pi G\gamma^2 \lambda^2} \left(\sin(\bar\mu_1 c_1)\sin(\bar\mu_2 c_2) + \text{cyclic terms}\right). \quad (57)$$

The energy density is bounded and has an absolute maximum $\rho_{\max} \approx 0.41 \rho_{\text{Pl}}$, same as in the isotropic $k = 0$ model. Using Hamilton's equations, one can compute the mean Hubble rate and the shear scalar which also turn out to be bounded [68, 157]: $H_{\max} = 1/(2\gamma\lambda)$ and $\sigma^2_{\max} = 10.125/(3\gamma^2\lambda^2)$. Similar bounds have been found in the presence of spatial curvature, in Bianchi-II and Bianchi-IX models [29, 157, 159]. It has been so far difficult to find a consistent modified generalized Friedmann equation in terms of the energy density ρ and anisotropic shear scalar σ^2, except when anisotropies are weak [169].º However, using Hamilton's equations the effective dynamics has been explored in a lot of detail. The effective dynamics of the Bianchi-I model turns out to be non-singular for various types of matter resulting in a bounce of the mean scale factor at the Planck scale. For perfect fluids with a vanishing anisotropic stress and with an arbitrary equation of state greater than -1, the curvature invariants are bounded and all strong curvature singularities are generically resolved [130, 131]. The approach to the bounce can be characterized using Kasner exponents and one finds that depending on the ratio of the initial anisotropy and energy density, the bounce can be associated to a barrel, cigar, pancake or a point like structure. Interestingly, the structures before and after the bounce can change in general and follow certain selection rules [176]. Depending on the anisotropic parameters, some transitions are also completely forbidden. Note that such Kasner transitions are absent in the Bianchi-I model in GR, and are so far known to arise only in LQC.

To summarize, the Bianchi-I model with a massless scalar field can be quantized in LQC in a similar way as the isotropic models, and the resulting quantum Hamiltonian constraint turns out to be a non-singular quantum difference equation. Unlike in the classical theory, the energy density, mean Hubble rate and anisotropies remain bounded throughout the evolution. The physics of the quantum Bianchi-I spacetime is considerably richer than the isotropic model and has provided a robust picture of the Planck scale physics in LQC. There are several interesting avenues to explore, including the way quantum geometry affects Mixmaster behavior which is expected to give important insights on the resolution of singularities in more general situations. Thanks to the formulation of the BKL conjecture in the connection variables [177], a stage is set to carry out a rigorous comparison between the quantum and the classical description of spacetimes in the Bianchi models, and to gain valuable insights on the generic resolution of singularities.

3.3. Gowdy Models and the Hybrid Quantization

So far we have discussed spacetimes with a finite number of degrees of freedom. A long standing issue in quantum cosmology is whether the physical implications obtained in the mini-superspace setting can be trusted for spacetimes with an infinite number of degrees of freedom. One of the directions which has been explored to go

---

ºOne can rearrange the Hamilton's equations to obtain an equation for the mean Hubble rate which mimics the classical generalized Friedmann equation exactly by defining a 'quantum shear' [178]. However, the limitation with such an approach is that the 'quantum shear' does not consistently capture the anisotropic shear in the effective spacetime, and one loses important information about the way anisotropies influence the Planck scale phenomena in LQC.

beyond the assumption of homogeneity is the Gowdy midi-superspace spacetimes which have been quantized using a hybrid method in which the homogeneous modes are loop quantized, and the inhomogeneous modes are Fock quantized [31–34]. The results of singularity resolution obtained in homogeneous models are found to be robust in the Gowdy models for vacuum [34] and also in presence of a massless scalar field [33]. In the following, we outline the main features of this approach for the vacuum case.

The Gowdy spacetimes which have been studied so far in LQC are the one with linear polarization. The spatial manifold is $\mathbb{T}^3$ coordinatized by (θ, σ, δ). The spacetime has two Killing fields, ∂_σ and ∂_δ, which are hypersurface orthogonal. The spatial dependence in the fields is captured completely by θ. Using the underlying symmetries, a partial gauge fixing can be performed and one is left with two global constraints: the diffeomorphism constraint C_θ and the Hamiltonian constraint C_H. The metric components are periodic in θ, and using this periodicity, one can perform a Fourier expansion of the metric fields and the reduced phase space can be decomposed into homogeneous (Γ_{hom}) and inhomogeneous sectors (Γ_{inhom}). The homogeneous sector is equivalent to the phase space of the compact vacuum Bianchi-I spacetime. For the inhomogeneous sector one can introduce the creation and annihilation variables, a_m and a_m^* respectively, using which the global diffeomorphism constraint can be written as

$$C_\theta = \sum_{n=1}^{\infty} n(a_n^\dagger a_n - a_{-n}^* a_{-n}) = 0 . \tag{58}$$

The Hamiltonian for the inhomogeneous modes consists of a free part H_o:

$$H_o = \frac{1}{8\pi G \gamma^2} \sum_{n \neq 0} |n| a_n^\star a_n \tag{59}$$

and an interaction part H_{int}:

$$H_{\text{int}} = \frac{1}{8\pi G \gamma^2} \sum_{n \neq 0} \frac{1}{2|n|} \left(2 a_n^\star a_n + a_n a_{-n} + a_n^\star a_{-n}^\star \right) \tag{60}$$

which mixes different inhomogeneous modes. The total Hamiltonian constraint for all the modes is given by

$$C_H = -\frac{2}{\gamma^2 V} \Bigg[(c_\theta p_\theta \, c_\sigma p_\sigma + c_\theta p_\theta \, c_\delta p_\delta + c_\sigma p_\sigma \, c_\sigma p_\sigma) \\ - G \left(32 \pi^2 \gamma^2 |p_\theta| H_o + \frac{(c_\sigma p_\sigma + c_\delta p_\delta)^2}{|p_\theta|} H_{\text{int}} \right) \Bigg] , \tag{61}$$

where the terms in the first parenthesis arise from the homogeneous part corresponding to the phase space of the Bianchi-I spacetime. This part of the Hamiltonian constraint is loop quantized following the technique elaborated in Section 3.2, but for the earlier quantization [24]. The inhomogeneous part of the constraint, expressed in terms of the annihilation and creation operators, is Fock quantized. The resulting quantum constraint operator is

$$\Theta_{\rm G} = \Theta_{\rm B-I} + \frac{1}{8\pi\gamma^2}\left[32\pi^2\gamma^2\widehat{|p_\theta|}\hat{H}_o + \left(\widehat{\frac{1}{|p_\theta|^{1/4}}}\right)^2 (\hat{\Theta}_\sigma + \hat{\Theta}_\delta)^2 \left(\widehat{\frac{1}{|p_\theta|^{1/4}}}\right)^2 \hat{H}_{\rm int}\right] \tag{62}$$

where $\Theta_{\rm B-I}$ denotes the quantum Hamiltonian constraint operator for the Bianchi-I spacetime and the inverse powers of $\hat{p}_\theta$ are computed by expressing them in terms of a Poisson bracket between the positive powers of p_θ and the holonomies in the classical theory, and promoting the latter to a commutator [179]. Once we have these quantum constraints, the inner product is obtained by using the same strategy as in the quantization of isotropic and anisotropic spacetimes. A complete set of Dirac observables are found which are self-adjoint with respect to the inner product. A feature of the above quantization, shared with the loop quantization of the Bianchi-I model, is the zero volume states in the homogeneous sector that are decoupled by the action of the Hamiltonian constraint. Thus the states corresponding to the classical singularity at zero volume are absent and in this sense the singularity is resolved.

The physics of the the Planck regime in these spacetimes is being explored using effective dynamics which confirms the existence of the bounce in the presence of inhomogeneities. Earlier work in the effective dynamics of the Gowdy model was based on using the effective Hamiltonian constraint for a slightly different quantization of the Bianchi-I spacetime [24], which has some undesirable features [26, 68, 130]. It was found that in the case where the bounce can be approximated by the dynamics of the vacuum Bianchi-I model, the statistical average of the inhomogeneities across the bounce is positive. On the other hand, when dynamics is dominated by inhomogeneities the statistical average of inhomogeneities across the bounce is preserved. More recently, the effective dynamics has been studied using the effective Hamiltonian constraint for the improved Bianchi-I quantization [26]. It turns out that in comparison to the homogeneous Bianchi-I spacetime in LQC, the inhomogeneities increase the volume at which the bounce occurs. These investigations provide a first glimpse of the bounce in the presence of Fock quantized inhomogeneities in LQC. It serves as a useful intermediate step towards the full loop quantization of Gowdy spacetimes.

4. Inhomogeneous Perturbations in LQC

The standard model of cosmology (see e.g. [180–183]) is based on classical general relativity, and therefore cannot describe the earliest epochs of cosmic expansion, when curvature invariants reach the Planck scale. The goal of this and next section is to use LQC to extend existing models to the Planck era. This extension opens the possibility of connecting Planck scale physics with cosmological observations, providing a new avenue to test some of the fundamental ideas on which the theory rests. The extension will also provide new physical mechanisms to account for the intriguing large scale anomalies observed in the CMB.

But to carry out this task, the theoretical framework summarized in previous sections is insufficient. A key ingredient in existing theories of the early universe, such as inflation, is the physics of first order cosmological *perturbations*, the so-called scalar and tensor modes, that propagate on a classical Friedmann-Lemaître-Robertson-Walker (FLRW) spacetime. We can observe features of these perturbations in the cosmic microwave background (CMB), allowing us to confront different models with observations. Therefore, if our goal is to incorporate LQC ideas in the description of the early universe, we need first to extend our theoretical framework to incorporate inhomogeneous cosmological perturbations propagating on the *quantum* cosmological spacetime, described in the previous section. Since cosmological observations [39] have confirmed that the early universe is very well described by a homogeneous and isotropic FLRW line element with spatial curvature compatible with zero, the rest of this section will focus on quantum spacetimes with these features, that were described in Section 2.

Incorporation of inhomogeneous perturbations on quantum spacetimes poses an interesting challenge. Far from the Planck regime, when the quantum aspects of gravity can be neglected, inhomogeneous first order perturbations are accurately described as quantum fields propagating in a classical expanding universe. The theory of quantum fields in curved spacetimes [184–186], well established since 1970's, provides the suitable theoretical arena. But in the Planck regime the background geometry is fully quantum. Since quantum states $\Psi(\nu,\phi)$ introduced in previous sections provide only probabilistic amplitudes for the occurrence of various metrics, a priori we do not have a classical FLRW geometry in the background. How do inhomogeneous perturbations propagate on theses quantum geometries $\Psi(\nu,\phi)$? This section will show that the answer to this question becomes tractable under the assumption that first order perturbations produce negligible back-reaction on the background quantum spacetime on which they propagate, i.e. when they can be considered as *test fields*. This assumption plays also a key role in the standard cosmological model, as well as in alternatives to inflation [188, 189]. In LQC, once the test field approximation is made, it becomes possible to obtain a well-defined *quantum field theory of cosmological perturbations on quantum FLRW spacetimes* [190]. The reader is referred to [42, 190–194] for further details. (See also Refs. [195] for application of similar techniques to spherically symmetric spacetimes.)

We begin by briefly summarizing the well-known classical theory of cosmological, gauge invariant, first order perturbations on spatially flat FLRW spacetimes (see e.g. [196] for details). This will establish notation and provide the arena for quantization. Then we construct the quantum theory upon it. Finally, we will use these results to include Planck scale physics in the description of the early universe, and to describe mechanisms to connect Planck scale physics with cosmological observations.

4.1. *Cosmological Perturbations in Classical FLRW Spacetimes*

Most explorations of the early universe rest on the assumption that the energy-momentum budget during the first stages of cosmic expansion was dominated by a scalar field ϕ commonly called the inflaton, that is subject to an effective potential $\mathrm{v}(\phi)$. (An example was discussed in Section 3.1.3, with $\mathrm{v}(\phi) = \frac{1}{2}m^2\phi^2$.) This scalar field can be thought either as a fundamental or an effective degree of freedom. Then, motivated by CMB observations which show that the early universe was extraordinarily homogeneous and isotropic, one looks for solutions of Einstein equations given by a FLRW metric $g_{ab}(t)$ with a homogeneous and isotropic scalar field $\phi(t)$ as source, *together with inhomogeneous first order perturbations* $\delta g_{ab}(\vec{x},t)$, $\delta\phi(\vec{x},t)$. The gravitational field by itself contains two physical degrees of freedom — so most of the metric components are purely coordinate (gauge) dependent functions — and adding the scalar field our system has three physical degrees of freedom in total. In order to avoid gauge artifacts, it is convenient to re-write the perturbation fields $\delta g_{ab}(\vec{x},t)$, $\delta\phi(\vec{x},t)$ in terms of these gauge invariant degrees of freedoms: they are made of the so-called scalar perturbation $\mathcal{Q}(x)$ — known as the Mukhanov-Sasaki variable — and two tensor perturbations $\mathcal{T}^{(1)}(x)$ and $\mathcal{T}^{(2)}(x)$. Physically one can think of $\mathcal{Q}$ as representing perturbations of the scalar field, and tensor modes as representing the two polarizations of a gravity wave. These variables, together with their conjugate momenta and the homogeneous and isotropic degrees of freedom, span the physical phase space $\Gamma_{\mathrm{phys}} = \Gamma_{\mathrm{hom}} \times \Gamma_{\mathrm{pert}}$, where Γ_{hom} is made of two pairs of canonically conjugate variables $(a, \pi_{(a)}; \phi, p_{(\phi)})$, with a the standard scale factor of the FLRW metric and Γ_{pert} is the phase space of scalar and tensor modes. Since the two tensor modes behave identically, from now on we will denote them collectively by $\mathcal{T}$.

We now discuss dynamics in Γ_{phys}. First, if we restrict to the homogeneous sector, Γ_{hom}, dynamical trajectories are generated by the restriction to FLRW of the Hamiltonian constraint of general relativity (see Section 2)

$$\mathcal{C}_H[N] = N\left[-\frac{3V_0}{8\pi G}\frac{p_{(a)}^2}{a} + \frac{1}{2}\frac{p_{(\phi)}^2}{a^3 V_0} + a^3 V_0\, \mathrm{v}(\phi)\right], \tag{63}$$

where $\kappa = 8\pi G$. The lapse function N indicates the time coordinate one is using: $N = 1$ corresponds to standard cosmic or proper time t, $N = a$ to conformal time η, and $N = V_0^3 a^3 / p_{(\phi)} := N_\phi$ to choosing the scalar field ϕ as a time variable, which turns out to be the most appropriate choice in the LQC.[p] The evolution generated by $\mathcal{C}_H[N]$ takes place entirely in Γ_{hom}; it does not involve inhomogeneous perturbations.

Dynamics in Γ_{pert} is generated by a true Hamiltonian $\mathcal{C}_2$, which is obtained from the second order piece of the scalar constraint of general relativity by keeping only terms which are quadratic in first order perturbations. This Hamiltonian has the

---
[p]As mentioned in Section 3.1.3, in general we only have a 'local clock' since ϕ is only a good time variable in patches of dynamical trajectories along which ϕ is monotonic.

form $\mathcal{C}_2 = \mathcal{C}_2^{(\mathcal{Q})} + \mathcal{C}_2^{(\mathcal{T}^{(1)})} + \mathcal{C}_2^{(\mathcal{T}^{(2)})}$, where in Fourier space

$$\mathcal{C}_2^{(\mathcal{T})}[N] = \frac{N}{2(2\pi)^3} \int d^3k \left(\frac{4\kappa}{a^3} |\mathfrak{p}_{\vec{k}}^{(\mathcal{T})}|^2 + \frac{a\,k^2}{4\kappa} |\mathcal{T}_{\vec{k}}|^2 \right), \qquad (64)$$

$$\mathcal{C}_2^{(\mathcal{Q})}[N] = \frac{N}{2(2\pi)^3} \int d^3k \left(\frac{1}{a^3} |\mathfrak{p}_{\vec{k}}^{(\mathcal{Q})}|^2 + a\,(k^2 + \mathcal{U}) |\mathcal{Q}_{\vec{k}}|^2 \right). \qquad (65)$$

Here $\mathcal{U} = [\mathrm{v}(\phi)\,r - 2\mathrm{v}_\phi(\phi)\sqrt{r} + \mathrm{v}_{\phi\phi}(\phi)]a^2$, with $r = (3\kappa p_{(\phi)}^2)/((1/2)p_{(\phi)}^2 + V_0^2 a^6 \mathrm{v}(\phi))$, and $\mathrm{v}_\phi(\phi)$ and $\mathrm{v}_{\phi\phi}(\phi)$, the first and second derivatives of the inflaton potential $\mathrm{v}(\phi)$ with respect to ϕ. Scalar and tensor modes evolve independently of each other.

The equations of motion for tensor and scalar perturbations generated by the Hamiltonian (64) and (65) take the form, in conformal time η

$$\mathcal{T}_{\vec{k}}'' + 2\frac{a'}{a}\mathcal{T}_{\vec{k}}' + k^2 \mathcal{T}_{\vec{k}} = 0; \qquad \mathcal{Q}_{\vec{k}}'' + 2\frac{a'}{a}\mathcal{Q}_{\vec{k}}' + (k^2 + \mathcal{U}(\eta))\mathcal{Q}_{\vec{k}} = 0. \qquad (66)$$

In physical space these equations are

$$\Box \mathcal{T}(x) = 0 \qquad \Box \mathcal{Q}(x) - \mathcal{U}\,\mathcal{Q}(x) = 0, \qquad (67)$$

where $\Box = \nabla_a \nabla^a$ is the D'Alembertian of g_{ab}. Therefore, tensor modes satisfy the same equation as a massless scalar field in FLRW, and similarly for scalar perturbations, except for the presence of the external potential $\mathcal{U}$. As expected, the dynamics in Γ_{pert} knows about Γ_{hom}: scalar and tensor perturbations satisfy linear differential equations (66) which contain coefficients involving background variables. Dynamics then is obtained by *first* solving the evolution for $a(\eta)$ and $\phi(\eta)$ using only $\mathcal{C}_H$, and *then* 'lifting' the resulting dynamical trajectory to Γ_{pert} using (66). This is a consequence of the main approximation underlying this construction: perturbations produce negligible back-reaction on the background metric.

In the inflationary scenario, the next step is to quantize the perturbation fields $\mathcal{Q}$ and $\mathcal{T}$, but keeping the homogeneous degrees of freedom as classical. This is justified because curvature invariants are well below the Planck scale at all times during and after inflation, and hence quantum effects of the gravitational background are expected to be negligible. Therefore, one keeps the homogeneous phase space Γ_{hom} unmodified, but replaces the classical phase space of perturbations Γ_{pert} by a Hilbert space $\mathcal{H}_{\mathrm{pert}}$ in which the perturbation fields $\hat{\mathcal{Q}}$ and $\hat{\mathcal{T}}$ are represented as quantum operators. This is a quantum field theory in a classical FLRW spacetime, and therefore well established techniques [180–183] are available to extract physical predictions from this system.

As in the classical theory, this semiclassical framework rests on the test field approximation. A necessary condition for its validity is that the stress-energy of perturbations must be subdominant compared to the background contribution. Since energy and momentum are quadratic in the basic variables, one needs to introduce renormalization techniques to obtain well-defined expressions. The ambiguities in the process of renormalization in curved spacetimes add difficulties in testing the validity of this semiclassical theory.

In the next subsection we describe the framework in which *both* the homogeneous as well as the inhomogeneous degrees of freedom are treated quantum-mechanically.

4.2. Quantum Theory of Cosmological Perturbations on a Quantum FLRW

The description of perturbations in a quantum cosmological background is more complicated than that in the classical FLRW case, although it will follow the same logical steps. In particular, the construction relies upon the assumption that perturbations produce negligible effects on the background. In the classical theory, the absence of back-reaction is reflected on the fact that dynamics of background fields $a(\eta)$ and $\phi(\eta)$ is completely independent of perturbations. In the quantum theory, the test field approximation is incorporated by assuming that the total wave function Ψ has the form of a product

$$\Psi(a, \phi, \mathcal{Q}_{\vec{k}}, \mathcal{T}_{\vec{k}}) = \Psi_{\text{hom}}(a, \phi) \otimes \Psi_{\text{pert}}(a, \phi, \mathcal{Q}_{\vec{k}}, \mathcal{T}_{\vec{k}}). \tag{68}$$

This structure implies the absence of correlations between background and inhomogeneous degrees of freedom initially, which is then maintained during evolution as long as the test field approximation holds. Our task now is, first, to construct the quantum theory describing Ψ_{hom}, and then study the evolution of Ψ_{pert} on the background geometry Ψ_{hom}.

The evolution of Ψ_{hom} follows exactly the steps described in Section 2 except that, as noted in Section 3.1.3, in presence of an inflationary potential $\text{v}(\phi)$, the operator $\sqrt{\Theta}$ appearing in the evolution equation (19) now must be replaced by $\left|\Theta - \text{v}(\phi)\nu^2 \frac{\pi G \gamma^2}{2}\right|^{1/2}$. The presence of $\text{v}(\phi)$ poses new challenges, because the resulting operator fails to be essentially self-adjoint for a generic potential $\text{v}(\phi)$. However, as we will see in the next subsection, we will be only interested in situation in which $\langle\Theta\rangle \gg \langle\text{v}(\phi)\nu^2 \frac{\pi G \gamma^2}{2}\rangle$ in the Planck era. Under these circumstances, the potential can be treated as a perturbation to Θ, and one can show it produces a negligible contribution to the evolution in the quantum gravity regime. More precisely, as discussed in [197], in situations of physical interest (see next subsection), the evolution with and without potential produce results for observable quantities with differences several orders of magnitude smaller than observational error bars. One can therefore obtain reliable predictions without including the potential in the quantum gravity regime. Then, we can directly import the result of Section 2 for the evolution of Ψ_{hom}. Nevertheless the mathematical subtleties appearing in the inclusion of the inflaton potential constitute an important open issue, although not of direct relevance for the phenomenological considerations. These issues have been studied for a constant potential, i.e. a cosmological constant Λ, in [21] where it was found that, although the operator $\left|\Theta - \Lambda\nu^2 \frac{\pi G \gamma^2}{2}\right|^{1/2}$ fails to be essentially self-adjoint, the quantum evolution is surprisingly insensitive to the choice of the self-adjoint extension.

The next task is to construct the theory of perturbations Ψ_{pert} propagating on the quantum spacetime Ψ_{hom}. Dynamics in this theory is extracted from the constraint equation (19) in the presence of perturbations. For briefly, we will write down some of the intermediate steps of the quantization for tensor modes, and simply provide the result for scalar perturbations at the end.

Evolution for the entire system will be obtained from the constraint equation (19), that now reads

$$-i\hbar \partial_\phi (\Psi_{\text{hom}} \otimes \Psi_{\text{pert}}) = \left| \hbar^2 \Theta - 2V_0\, \mathcal{C}_2^{(\mathcal{T})} \right|^{\frac{1}{2}} (\Psi_{\text{hom}} \otimes \Psi_{\text{pert}}) \qquad (69)$$

The test field approximation allows us to use perturbation theory to solve this equation. We will treat Θ as the Hamiltonian of the 'heavy' degree of freedom and $\mathcal{C}_2^{(\mathcal{T})}$ as the Hamiltonian of the light one. Then, the previous equation can be approximated by (see [190] for details)

$$(-i\hbar \partial_\phi \Psi_{\text{hom}}) \otimes \Psi_{\text{pert}} + \Psi_{\text{hom}} \otimes (-i\hbar \partial_\phi \Psi_{\text{pert}})$$
$$= \hbar \sqrt{\Theta} \Psi_{\text{hom}} \otimes \Psi_{\text{pert}} - \mathcal{C}_2^{(\mathcal{T})}[N_\phi](\Psi_{\text{hom}} \otimes \Psi_{\text{pert}}) \qquad (70)$$

where we have used that, while the Θ operator acts on Ψ_{hom} but not on Ψ_{pert}, $\mathcal{C}_2^{(\mathcal{T})}[N_\phi]$ in contrast acts on both states, since it contains background as well as perturbation operators. The first term in each side of the previous equality cancel out by virtue of the evolution of the background state (19), and we are left with

$$\Psi_{\text{hom}} \otimes (i\hbar \partial_\phi \Psi_{\text{pert}}) = \mathcal{C}_2^{(\mathcal{T})}[N_\phi](\Psi_{\text{hom}} \otimes \Psi_{\text{pert}}) \qquad (71)$$

This equation tells us that in the test field approximation the right-hand side is proportional to Ψ_{hom}, and we therefore can take the inner product of this equation with Ψ_{hom} without losing information. The last equation then reduces to

$$i\hbar \partial_\phi \Psi_{\text{pert}} = \langle \mathcal{C}_2^{(\mathcal{T})}[N_\phi] \rangle \Psi_{\text{pert}}, \qquad (72)$$

where the expectation value is taken in the physical Hilbert space of the homogeneous sector. In other words, as long as the test field approximation holds, the evolution of perturbation is obtained from $\mathcal{C}_2^{(\mathcal{T})}[N_\phi]$ where the *background operators* are replaced by their expectation value in the state Ψ_{hom}.

This theory is conceptually different from quantum field theory (QFT) in classical spacetimes. As previously mentioned, the spacetime geometry is not described by a classical metric, but it is rather characterized by a wave function Ψ_{hom} that contains the quantum fluctuations of the geometry. Equation (72) tell us that, indeed, the propagation of perturbations *is* sensitive to those fluctuations, and not only to the mean-value trajectory of the scale factor $\langle \hat{a} \rangle$.

An interesting aspect of the previous evolution equation (72) is that, by simple manipulation, it can be written as

$$\hat{\mathcal{T}}_{\vec{k}}'' + 2\frac{\tilde{a}'}{\tilde{a}} \hat{\mathcal{T}}_{\vec{k}}' + k^2 \hat{\mathcal{T}}_{\vec{k}} = 0, \qquad (73)$$

where we have defined

$$\tilde{a}^4 := \frac{\langle \hat{\Theta}^{-\frac{1}{4}} \hat{a}^4(\phi) \hat{\Theta}^{-\frac{1}{4}} \rangle}{\langle \hat{\Theta}^{-\frac{1}{2}} \rangle}, \quad (74)$$

and the quantum conformal time $\tilde{\eta}$ is related to the internal time by

$$\mathrm{d}\tilde{\eta} := \tilde{a}^2(\phi) \langle \hat{\Theta}^{-\frac{1}{2}} \rangle \, \mathrm{d}\phi. \quad (75)$$

But interestingly, (73) has the same form as the equation of the field $\hat{\mathcal{T}}$ propagating on a smooth FLRW geometry (see equation (66)). More explicitly, the evolution (73) is *mathematically indistinguishable* from a QFT on a *smooth FLRW metric* $\tilde{g}_{ab}$

$$\tilde{g}_{ab} \, \mathrm{d}x^a \mathrm{d}x^b := \tilde{a}^2(\tilde{\eta}) \left(-\mathrm{d}\tilde{\eta}^2 + \mathrm{d}\vec{x}^2\right) \quad (76)$$

In position space, the equation for $\hat{\mathcal{T}}$ reads $\Box \hat{\mathcal{T}}(x) = 0$, where $\Box$ is the d'Alembertian of the metric $\tilde{g}_{ab}$. In a similar way the scalar perturbations satisfy the second order differential equation

$$\hat{\mathcal{Q}}''_{\vec{k}} + 2\frac{\tilde{a}'}{\tilde{a}} \hat{\mathcal{Q}}'_{\vec{k}} + (k^2 + \tilde{\mathcal{U}}(\tilde{\eta}))\hat{\mathcal{Q}}_{\vec{k}} = 0. \quad (77)$$

Scalar perturbations propagate in the *same* metric $\tilde{g}_{ab}$ as tensor modes but, additionally, they feel a potential given by

$$\tilde{\mathcal{U}} = \frac{\langle \hat{\Theta}^{-\frac{1}{4}} \hat{a}^2 \hat{\mathcal{U}} \hat{a}^2 \hat{\Theta}^{-\frac{1}{4}} \rangle}{\langle \hat{\Theta}^{-\frac{1}{4}} \hat{a}^4 \hat{\Theta}^{-\frac{1}{4}} \rangle}, \quad (78)$$

where $\hat{\mathcal{U}}$ is the operator associated to the classical external potential $\mathcal{U}$ written after equation (65).

Therefore, at the practical level, in order to evolve test fields on a quantum geometry Ψ_{hom} one only needs to compute the components of $\tilde{g}_{ab}$ from Ψ_{hom}, and them proceed as in standard quantum field theory in curved spacetimes.

The following remarks are in order:

(i) No further assumptions beyond the test field approximation have been made to obtain this result. In particular, the state Ψ_{hom} is not assumed to have small quantum dispersion in the configuration or momentum variables [197].

(ii) The metric $\tilde{g}_{ab}$ does not satisfy Einstein's equations. This is obvious from its definition: its coefficients are obtained as expectation values of background operators in the homogeneous and isotropic quantum geometry Ψ_{hom}, and therefore they depend on $\hbar$. Rather, $\tilde{g}_{ab}$ is a mathematical object that neatly captures the information in Ψ_{hom} that is relevant for the propagation of tensor and scalar modes. It is remarkable that, from the rich information contained in Ψ_{hom}, test fields only 'feel' a few of its moments, namely (74)(75)(78), and furthermore, that these moments can be codified in a smooth metric $\tilde{g}_{ab}$! This metric is called in the literature the *effective dressed metric*. [190]. Effective because it contains all the information in Ψ_{hom} that is relevant for perturbations; and dressed because it depends not only on the mean value of Ψ_{hom} but also on some of its quantum fluctuations.

(iii) One should not think about $\tilde{g}_{ab}$ as approximating the physical background geometry in any way. The spacetime geometry $\Psi_{\rm hom}$ is quantum, and in general cannot be approximated in any reasonable sense by a smooth metric tensor. Rather, $\tilde{g}_{ab}$ encodes only the information in $\Psi_{\rm hom}$ *that test fields care about*. In other words, if we probe the quantum geometry $\Psi_{\rm hom}$ using *only* tensor and scalar modes, we will be unable to distinguish it from a smooth geometry characterized by $\tilde{g}_{ab}$. But if other observables are used, e.g. powers of curvature invariants, we would easily realize that the background gravitational field has additional information not captured by $\tilde{g}_{ab}$.

(iv) If the state $\Psi_{\rm hom}$ is chosen to have very small quantum dispersion in the volume ν (equivalently in the scale factor a), then the dressed metric $\tilde{g}_{ab}$ is indistinguishable from the effective metric discussed in Section 2.1.3.

Now, because the theory of tensor and scalar test fields propagating in $\Psi_{\rm hom}$ has been written as a QFT in curved spacetime $\tilde{g}_{ab}$, we can import the well-known theoretical machinery developed in that context and construct a Fock-type quantization of the test fields. Then, we end up with a hybrid quantization approach, following the ideas introduced in Ref. [31] to study Gowdy models, in which homogeneous degrees of freedom are quantized using LQG techniques and inhomogeneous fields using standard Fock techniques. This strategy is mathematically consistent and physically attractive, and it allows to make contact with the treatment of test fields in standard cosmology, e.g. in the inflationary scenario.

As in the semi-classical theory, testing the validity of the test field approximation is not a simple task. In Refs. [42, 43], this has been done following the same steps as in the semi-classical theory, namely by comparing the expectation value of the renormalized energy and pressure of perturbations with the background contributions. The test field approximation may be a real limitation in situations of physical interest, and current efforts are focused on extending the range of applicability of framework summarized above [187].

4.3. *Other Approaches*

The LQC literature is vast, and other approaches to the quantization of first order perturbations exist (see [4, 6, 8, 71, 75, 84–87, 198]). The 'hybrid quantization approach' [71–73, 84, 199–201], originally suggested in Ref. [31] for Gowdy cosmologies, has similarities with the 'dressed metric approach' presented above, particularly the fact that test fields are quantized using standard Fock techniques, while the background FLRW geometry follows the non-perturbative methods of loop quantum gravity. The two methods however differ in the way in which constraints for perturbations are imposed, already at the classical level. But these conceptual difference have little impact in observable quantities, at least in the set of solutions that have been explored so far. Hence, the predictions for the primordial spectrum of perturbations are very similar in both approaches, adding robustness to the program. We focus on the 'dressed metric approach' on this chapter primarily because so far the most detailed calculations leading to phenomenological predictions have been carried out in this framework.

The 'separate universe' approach in LQC [75, 198] has the conceptual advantage of quantizing the homogeneous and inhomogeneous degrees of freedom 'in tandem', using only loop techniques, at the expenses of being only applicable to very long wavelengths.

The 'anomaly-free quantization approach' summarized in Refs. [6, 86, 87, 202] uses the algebra of gravitational constraints as guiding principle, and develops an effective approach in which quantum gravity corrections are incorporated. The expressions of these quantum corrections are guided by imposing that the constraint algebra closes at the desired order in perturbations. This program produces a physical picture of the very early universe which is very different from those in other LQC approaches. Its phenomenological consequences have been explored in [203–205], where it is claimed that some of the predictions are in sharp disagreement with CMB observations. (For further discussion, see *Chapter 8*.)

5. Application: LQC Extension of the Inflationary Scenario

Now that we have a quantization for FLRW spacetimes and a theory of scalar and tensor perturbations propagating thereon, we are ready to apply this theoretical framework to the early universe. Among the existing models, the inflationary scenario is perhaps the most accepted one, and this section summarizes an approach to extend it to the quantum gravity regime [41, 43]. See [206–208] for interesting work in LQC in the context of the "matter bounce scenario".

5.1. *The Strategy*

In the inflationary scenario one starts by assuming that tensor and scalar modes are in the Bunch–Davies vacuum at some early time when inflation begins. This state is then evolved until the end of inflation and relevant quantities for observations, e.g. the power spectra of tensor and scalar perturbations, are computed. The choice of the vacuum, however, rests on the implicit assumption that the evolution of the universe *before* inflation is unimportant in what the tensor and scalar modes respect; otherwise tensor and scalar modes would reach the onset of inflation in an excited state relative to the inflationary vacuum. But to show whether this is a reasonable assumption one needs a model for the pre-inflationary universe. Our strategy is to use LQC for this purpose, and *compute*, rather than postulate, the state of inhomogeneous perturbations at the onset of the slow-roll phase. In the resulting picture the universe contracts for an infinite amount of time, bounces at the Planck scale, and then inflation starts at some time after the bounce.[q] The strategy will

---

[q]In concrete simulations inflation starts around $10^{-35}s$ after the bounce, and lasts for a similar interval of time. However, the universe only expands for around fifteen e-folds from the bounce to the onset of inflation, while it expands for more than 60 e-folds during inflation. On the contrary,

then be to start with vacuum initial conditions for tensor and scalar modes at early times, possibly prior to the bounce, evolve them using the LQC equations, and show that the resulting state at the onset of inflation coincides with the Bunch-Davies vacuum to a good approximation. If the initial state can be specified in a compelling fashion and if it does not evolve to a state that is sufficiently close to the Bunch-Davies vacuum at the onset of inflation, the viability of the framework would be jeopardized. If, on the other hand, the evolved state turns out to be close to the Bunch-Davies vacuum but with appreciable deviations, there would be new observable effects.

Therefore, the effect of LQC in observable quantities that we are looking for does *not* come from quantum gravity corrections generated *during inflation*. There, the energy density and curvature of the universe are around eleven orders of magnitude below the Planck scale, and quantum gravity corrections are suppressed by a similar factor. On the contrary, the effects we are looking for are generated *before* inflation, when quantum gravity effects dominate. The important fact is that scalar and tensor perturbations keep memory of these effects [209, 210] which then can be imprinted in the CMB temperature anisotropies — if the amount of inflationary expansion is not much larger than 70 e-folds for these corrections not to be redshifted to super-horizon scales.

The exploration of phenomenological consequences in LQC follows these steps:
(i) Choose an inflationary potential $v(\phi)$.
(ii) Specify the quantum state for the FLRW geometry $\Psi_{\rm hom}(\nu, \phi)$.
(iii) Specify the quantum state for tensor and scalar modes $\Psi_{\rm pert}$.
(iv) Evolve the background and perturbations using the theoretical framework spelled out in previous sections.
(v) Compute observable quantities.

We now discuss these steps in more detail:

(i) Choose an inflationary potential $v(\phi)$. Although it would be desirable to derive the potential from first principles, at the present time there is no compelling candidate within LQC. One could expect though that $v(\phi)$ originates from a theory of particle physics, rather than from LQC which is a purely gravitational theory — although there also exist the exciting possibility that the inflaton field and its potential have a purely gravitational origin [211, 212]. Therefore, the strategy so far in LQC is the same as in standard inflation, namely to use different phenomenologically viable potentials and contrast the results with observations. Two choices have been explored in great detail in LQC: the quadratic[r] and the so-called Starobinsky

---

while the spacetime scalar curvature during inflation is almost constant, it decreases about eleven orders of magnitude from the bounce to the onset of slow-roll. These numbers are obtained using initial conditions that lead to interesting observable effects in the CMB, and they can vary for other choices.

[r]Upper bounds on the tensor-to-scalar ratio recently obtained from the Planck satellite observations [39] slightly disfavor the quadratic potential. However, LQC corrections can alleviate these constraints [215], and therefore this potential may still be of some interest.

potential [213, 214]:

$$\text{v}(\phi) = \frac{1}{2}m^2\phi^2 \quad \text{and} \quad \text{v}(\phi) = \frac{3m^2}{32\pi}\left(1 - \exp(-\sqrt{16\pi G/3}\,\phi)\right)^2 \tag{79}$$

The free parameter m can be fixed by CMB observations although, as analyzed in detail in [215], in LQC there is some extra freedom. It is also important to keep in mind that LQC effects have a purely quantum gravity origin and are largely independent of the choice of potential. Therefore, predictions obtained from different choices of v(ϕ) are all quite similar.

(ii) Quantum state for the FLRW geometry $\Psi_{\text{hom}}(\nu, \phi)$. As discussed at the beginning of Section 4.2, the states Ψ_{hom} of interest differ from the ones described in Section 2 in that now the dynamics contains an inflationary potential v(ϕ). But detailed analysis [43, 214, 215] have showed that LQC effects can be imprinted in the CMB only if the contribution from the potential v(ϕ) is subdominant around the bounce time, as compared to the kinetic energy of ϕ. This is because potential dominated bounces lead to very long inflationary phases, and LQC effects will then be red-shifted to super-Hubble scales. But even if the potential is subdominant around the bounce time, it gains relevance later in the evolution, and eventually dominates during inflation. However, it turns out that the potential dominated regime occurs well after the quantum gravity era, at a time when the energy density and curvature invariants are all well below the Planck scale, and general relativity becomes an excellent approximation. Therefore, the regime of physical interest for investigations of LQC phenomenology is such that the potential v(ϕ) can be treated as a perturbation during the quantum gravity era. Furthermore, as shown in Ref. [197], the relative effect of the potential in observable quantities under these circumstances is several orders of magnitude below observational sensitivity. Therefore, one can choose to simply ignore the potential in the quantum gravity era without affecting the accuracy of observational predictions. In this situation one is led to work with the states described in Section 2.

Among all the states for the homogeneous and isotropic gravitational field described in Section 2, the simplest choice is to work with states $\Psi_{\text{hom}}(\nu, \phi)$ that have very small quantum dispersions in volume ν around the bounce time. These states remain 'sharply peaked' during the entire evolution. More importantly, the resulting geometry can be accurately described by the effective metric presented in Section 2.1.3. For these states the energy density at the time of the bounce saturates the supremum in the entire physical Hilbert space, ρ_{max}. The only freedom in $\Psi_{\text{hom}}(\nu, \phi)$ is then the way this energy density is divided between potential and kinetic energy of the inflation field at the bounce. Different choices produce solutions that accumulate different amounts of expansion between the bounce and the end of inflation, $N_{\text{B}} := \ln\left(a_{\text{end}}/a_{\text{bounce}}\right)$; therefore, the freedom can be parameterized by the value of N_{B}.

One can also choose states $\Psi_{\text{hom}}(\nu,\phi)$ which have large dispersion in ν, and therefore are not sharply peaked in any of the variables. The computation of the scalar power spectrum is numerically more challenging in this case, and has been recently performed in [197] for states containing relative quantum dispersion in ν as large as 168% in the Planck epoch. The main lessons from this analysis are: (i) observational quantities *are* sensitive to the quantum dispersion in $\Psi_{\text{hom}}(\nu,\phi)$; (ii) however, the effects are quite simple and, within observational error bars, predicted observable quantities cannot distinguish between a widely spread state which produces a certain amount of expansion N_{B}, and a sharply peaked state with a slightly different valued of N_{B} (see [197] for details). Therefore, if one is only interested in observational predictions, there is no loss of generality in restricting to sharply peaked states, as long as different values of N_{B} are considered. This will be the strategy that we will follow in the rest of this section.

(iii) Quantum state for tensor and scalar modes. The specification of the initial state for inhomogeneous perturbations is an important question. Two main strategies have been followed in the LQC-literature: the state is specified by choosing vacuum initial conditions at [41, 43, 201, 215] or before the bounce [203, 204, 215]. It could seem at first that the far past is the natural place to set up initial data for perturbations. One cannot disregard, however, the possibility that quantum gravity effects make the pre-bounce evolution unimportant. Note that in the presence of an inflationary phase the radius of the observable universe is of the order of 10 Planck lengths at the time of the bounce. At these scales quantum gravity effects are very efficient, and there are indication that they could produce a diluting effect that makes scalar and tensor modes to forget about features acquired in the contracting phase. This is an interesting possibility, and in the absence of conclusive arguments the best strategy is to keep working with the two options and contrast their predictions with observations.

But even after making a choice between these two possibilities, one still has to face the inherent ambiguity in the definition of vacuum in quantum field theory in an expanding universe. If all wavelengths of interest are much smaller that the curvature radius, the adiabatic approach provides a useful criteria to reduce the ambiguity; if we are not in this situation, other arguments are needed. Different proposals for initial vacuum have been used in LQC. In [216], in a more general context, a covariant criterion was introduced to specify the notion of vacuum at a given time, by demanding that the expectation value of the adiabatically renormalized energy-momentum tensor vanishes at that time. In many situations of practical interest this condition provides a unique state. Reference [201], on the other hand, has fixed the freedom in the vacuum by demanding that certain time variations of the mode functions characterizing the quantum state in Fourier space are minimized during the evolution. Finally, in Refs. [217, 218] a quantum generalization of the Penrose's Weyl curvature hypothesis has been used to single out a vacuum state using considerations that bridge the Planck regime around the bounce and the end of inflation.

(iv) Evolution of the background and perturbations can be obtained by using the theoretical framework spelled out in previous sections.

(v) Computation of observable quantities. The quantities of interest are the power spectrum of tensor and scalar perturbations, and their spectral indices, which are defined in the standard way (see e.g. [183]).

5.2. *Results*

Chapter 8 in this volume focuses on the phenomenology of loop quantum gravity. In particular, Section 3 in that chapter is devoted to LQC, and includes results for CMB anisotropies obtained from the theoretical framework we have described in this chapter. Therefore, this section will be brief and its aim is to complement the analysis in the mentioned chapter.

Figure 2, extracted from Ref. [215], shows the scalar and tensor power spectrum for a given choice of parameters and initial state. The detailed analysis of these results was presented in [41, 43], and further analyzed in [215]. The main new feature in the power spectra is the appearance of a new scale, k_{LQC}, which is directly related to the value of the spacetime scalar curvature at the time of the bounce, $k_{LQC}/a(t_B) := \sqrt{R(t_B)/6}$. LQC corrections appear for Fourier modes with $k \lesssim k_{LQC}$. Modes with larger k are essentially insensitive to the bounce; they reach the onset of inflation in the Bunch-Davies vacuum and, as a consequence, their power spectrum is almost scale invariant.

The observable effects of LQC in the CMB therefore depend on the value that the new scale, which is of the order of the Planck scale at the time of bounce, has at *the present time* t_0, i.e. the value of $k_{LQC}/a(t_0)$, which in turn is dictated by the number of e-folds from the bounce to the end of inflation N_B (the expansion accumulated from the end of inflation until the present time is fixed by the standard model of cosmology). If $k_{LQC}/a(t_0)$ is larger than $k_*/a(t_0) := 0.002\,\text{Mpc}^{-1}$, the observed power spectrum is predicted to deviate significantly from scale invariance. Available data from the CMB temperature anisotropies constrain $k_{LQC}/a(t_0)$ to be of the same order or smaller than $k_*/a(t_0)$, so k_{LQC} is constrained to be smaller than k_*. On the other hand, if $k_{LQC}/a(t_0)$ is smaller than approximately $0.1(k_*/a(t_0))$, then LQC corrections are swept to length scales larger than our observable universe. Therefore, if LQC correction were to appear in CMB, we must have $k_{LQC} \in [0.1k_*, k_*]$. This is equivalent of saying that the amount of expansion from the bounce to the present time is such that a wavelength of size twice the Planck length at the bounce is red-shifted to approximately 3000Mpc at the present time. Although there are no mechanisms based on precise arguments to explain why such coincidence should happen, Ref. [217] has provided concrete physical principles that lead to this situation. Furthermore, observations have detected deviations from

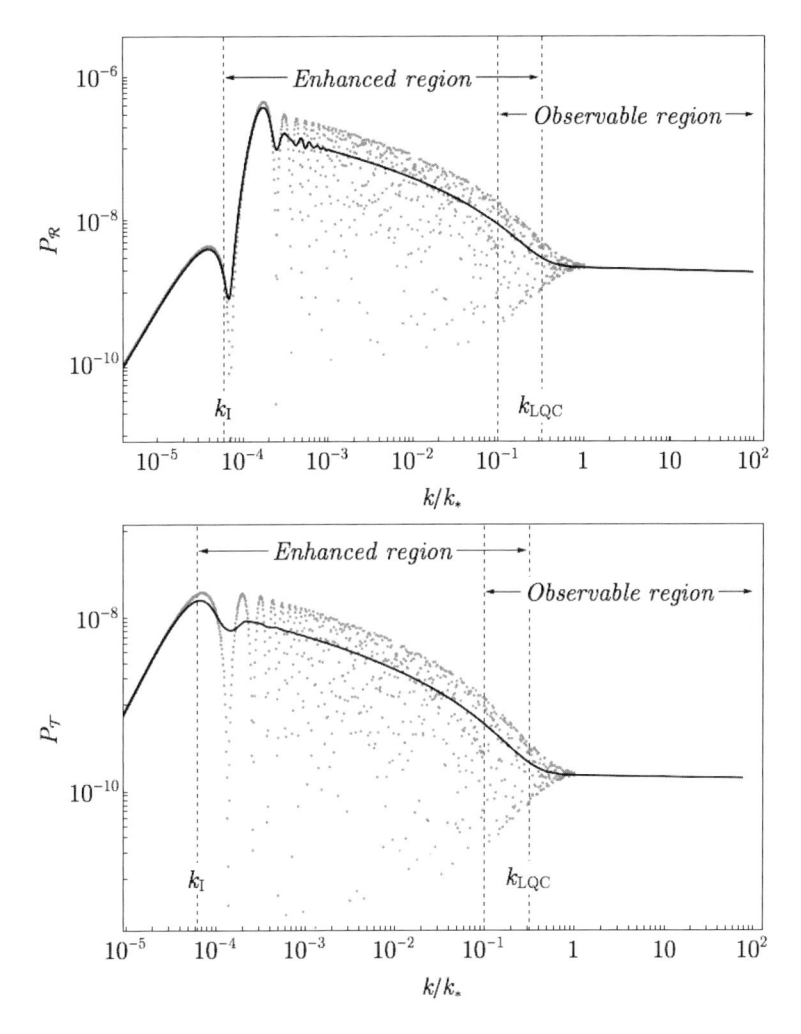

Fig. 2. The LQC scalar (top) and tensor (bottom) power spectrum as a function of k/k_*, where k_* is the pivot comoving scale corresponding to $0.002\,\text{Mpc}^{-1}$ at the present time. These plots are obtained for parameter values $m = 1.3 \times 10^{-6}$, preferred instantaneous vacuum [216] initial data for perturbations at initial time $t = -50000$, and value of the inflaton field at the bounce time $\phi_B = 1$, all quantities in Planck units. The numerically evolved spectrum, shown in gray, is rapidly oscillatory; its average, shown in black, has an amplitude which is enhanced with respect to the standard predictions of slow-roll inflation for modes $k_I \lesssim k \lesssim k_{LQC}$ but agrees with them for $k \ll k_{LQC}$. The region of Fourier modes that are observable in the CMB for this choice of parameters is also shown; smaller k's correspond to super-Hubble scales at the present time.

the standard featureless scale invariant spectrum for the low k region of the power spectrum, indicating that new physics may be needed to account for the observed anomalies [219]. Although the associated statistical significance of these anomalies is inconclusive, it is quite tempting to think that they may be visible traces of new physics, as indeed emphasized in [219].

A natural question is then whether the LQC bounce preceding inflation provides a suitable mechanism to quantitatively account for the observed anomalies. The Planck team has paid particular attention to two anomalous features in the CMB [219], namely: (i) a dipolar asymmetry arising from the fact that the averaged power spectrum is larger in a given hemisphere of the CMB than in the other, and; (ii) a power suppression at large scale, corresponding to a deficit of correlations at angular multipoles $\ell \lesssim 20$ as compared to the predictions of a scale invariant spectrum. We now briefly summarize existing ideas related to these anomalies in LQC.

A primordial dipolar asymmetry requires *correlations* between different wavenumbers k in the power spectrum. Such correlations do not arise at leading order in models for which the background is homogeneous, as the scenario discussed in the last two sections: the two-point function in Fourier space is diagonal. This motivated the authors of Ref [220] to go beyond leading order and discuss the corrections the primordial spectrum acquires from the three-point function (i.e. corrections from non-Gaussianitiy). As first pointed out in [221], non-Gaussian effects in the two-point function could indeed be responsible for the observed dipolar modulation in the CMB. In Ref. [220] this idea was implemented in LQC. The non-Gaussianity that inflation generates as a consequence of the pre-inflationary LQC bounce were computed and its effect on the primordial power spectrum were obtained. The result is that there exist values of the free parameters — the value of the inflaton field at the bounce ϕ_B (or equivalently, N_B) and its mass m — that make the non-Gaussian modulation of the power spectrum to induce a scale dependent dipolar modulation in the CMB that agrees with the observed anomaly. Furthermore, this mechanism also offers the possibility to account for the power suppression, since a *monopolar* modulation appears, in addition to the dipole, at large angular scales, which could reverse the enhancement of power shown in Figure 2. The analysis in [220] included the non-Gaussianity generated during inflation, but a contribution to the three-point function from the bounce is also expected. However, this contribution to non-Gaussianity is significantly more challenging to compute, even numerically, because of the absence of the slow-roll approximation normally used to simplify the computations in inflation. Work is in progress [222] to complete this computation with the goal of establishing that the non-Gaussian modulation in LQC is a viable mechanism to simultaneously account for the two observed anomalies.

Other ideas have also recently appeared to account for the power suppression at large scales [201, 217]. They are related to the choice of initial state for scalar perturbation at the time of the bounce mentioned above in this section. The statement in these works is that one can find physical criteria to select a preferred notion of ground state at the bounce which, when evolved until the end of inflation, produce a power spectrum which is suppressed compared to the standard scale invariant result for low values of k.

6. Discussion

LQC provides a remarkable example of successful quantization of the sector of classical GR spacetimes with symmetries observed at cosmological scales. It is based on a precise mathematical framework, supplemented with sophisticated state of the art numerical techniques. One starts by showing that the requirement of background independence is strong enough to uniquely fix the quantum representation, just as the Poincaré symmetry fixes the representation of the observable algebra in the standard quantum theory of free fields. One then uses this preferred representation. This procedure was first applied to a spatially flat FLRW background and the resulting quantum geometry was analyzed in detail. As described in this chapter, the final picture realizes many of the intuition that physicists, starting from Wheeler, have had about non-perturbative quantum gravity. Furthermore, interesting questions can now be answered in a precise fashion in LQC. Of particular interest is the way in which quantum effects are able to overwhelm the gravitational attraction and resolve the big bang singularity. While the LQC non-perturbative corrections dominate the evolution in the Planck regime and remove the big bang singularity, they disappear at low energies restoring agreement with the classical description. This is a non trivial result. The analysis has been extended to more complicated models containing spacial curvature, anisotropies, and even models with infinitely many degrees of freedom such as the Gowdy spacetime, adding significant robustness to the emergent physical picture. Using effective spacetime description of LQC, the problem of singularities in general has been addressed, which provides important insights on the generic resolution of strong curvature singularities.

One can further extend the regime of applicability of LQC by including cosmological perturbations. In standard cosmology one describes scalar and tensor curvature perturbations by quantum fields propagating in a classical FLRW spacetime. This is the theoretical framework — QFT in classical spacetimes — on which the phenomenological explorations of the early universe rely, e.g. in the inflationary scenario. In this chapter we have reviewed how such a framework can be generalized by replacing the classical spacetime by the quantum geometry provided by LQC. This framework provides a rich environment to analyze many interesting questions both conceptually and at the phenomenological level. It offers the theoretical arena to explore the evolution of scalar and tensor perturbations in the early universe, and to provide a self-consistent quantum gravity completion of the standard cosmological scenarios. It is our view that the level of detail and mathematical rigor attained in LQC is uncommon in quantum cosmology. The new framework has become a fertile arena to obtain new mechanisms that could explain some of the anomalous features observed in the CMB, which indicate that physics beyond inflation is required to understand the large scale correlations in the CMB [219].

Since LQC is a quantization of classical spacetimes with symmetries that are appropriate to cosmology, the theoretical framework shares the limitations of the symmetry reduced quantization strategy. Symmetry reduction often entails a drastic simplification, and therefore one may lose important features of the theory by restricting the symmetry prior to quantization. This is an important issue which has attracted efforts from different fronts. First let us recall that the BKL conjecture further supports the idea that quantum cosmological models are very useful in capturing the dynamics of spacetime near the singularities. Within LQC itself, the concern was initially alleviated by checking that models with larger complexity, such as anisotropic Bianchi I model, correctly reproduced the FLRW quantization previously obtained, when the anisotropies are 'frozen' at the quantum level. This test is even more remarkable when applied to models that have infinitely many degrees of freedom to begin with, as it is the case of the Gowdy model. More generally, there are interesting recent results on establishing a connection between LQC and LQG [224–226]. These include *quantum-reduced* loop quantum gravity [227], where the main idea is to capture symmetry reduction at the quantum level in LQG and then pass to the cosmological sector, and group field theory cosmology [89]. Promising results have been obtained in these approaches. As examples, improved dynamics as the one used in isotropic LQC has been found in quantum-reduced loop quantum gravity [228], and evidence of LQC like evolution and bounce have been reported in group field theory cosmology [229]. It is rather encouraging that results from different directions seem to yield a consistent picture of the Planck scale physics as has been extensively found in LQC

Another important ingredient in LQC is the process of de-parameterization. In the absence of a fundamental time variable in quantum gravity, in LQC one follows a relational-time approach in which one of the dynamical variables plays the role of time, and one studies the evolution of other degrees of freedom with respect to it. As explained in Section 2, in most of the LQC literature one uses a massless scalar field as time variable. An important question is how the physical results depend on the variable chosen as a time, i.e. if quantum theories constructed from different relational times are unitarily related. This is an age old question in quantum cosmology, but so far has not been systematically addressed.

It is worth commenting on some of the directions where significant progress has been made in LQC, in contrast to the earlier works in quantum cosmology. The first one deals with a rigorous treatment of fundamental questions in quantum cosmology about the probability of events — such as the probability for encountering a singularity or a bounce. These are hard questions whose answers had been elusive due to the lack of sufficient control over the physical Hilbert space structure, including properties of observables and a notion of time to define histories. Thanks to the quantization of isotropic and homogeneous spacetimes using a scalar field ϕ as a clock, a consistent history formulation can be completed both in the Wheeler-DeWitt theory and LQC [49–52]. A covariant generalization of these results has also been pursued [53]. Using exactly soluble model of sLQC computation of class operators, decoherence functional and probability amplitudes can be performed. It turns out that in the Wheeler-DeWitt theory the probability for bounce turns out to be zero even if one considers an arbitrary superposition of expanding and contracting states. The probability of bounce turns out to be unity in LQC. These developments show that not only LQC has been successful in overcoming problem of singularities which plague Wheeler-DeWitt theory, it has also established an analytical structure which has been used to answer foundational questions both in LQC and the Wheeler-DeWitt theory.

The second direction where developments in LQC are expected to have an impact beyond LQG are in the development of sophisticated numerical algorithms to understand the evolution in deep Planck regime for a wide variety of initial states, including with very large spreads [66, 107, 108]. Some of these techniques have been exported from traditional numerical relativity ideas which are modified and applied in the quantum geometric setting. Using high performance computing, these methods promise to yield a detailed picture of the physics of the Planck scale. These techniques can be replicated in a straightforward way for other quantum gravity approaches. More importantly they provide a platform to understand the structure of quantum spacetime analogous to the numerical works in classical gravity [36, 37]. A deeper understanding of how quantum gravitational effects modify the BKL conjecture and change our understanding of approach to singularity in the classical theory is a promising arena. Interesting results in this direction have started appearing, including on singularity resolution in Bianchi models [25, 230] and quantum Kasner transitions across bounces and selection rules on possible structures near the classical singularities [176].

Finally, we note that sometimes the limitations of LQC have been used to shed doubts on its results. These arguments, mainly articulated by the authors of [202], claim that a fully covariant approach with validity beyond symmetry reduced scenarios produces physical results inequivalent to those obtained from LQC. In particular, it is argued that, in the presence of inhomogeneities, there is an unavoidable change of signature, from Lorentzian to Euclidean, in an effective theory. The authors of this chapter disagree with the conclusions reached in [202] and subsequent papers along these lines. In our view, although the conceptual points raised by those authors are indeed interesting, their analysis relies on a series of assumptions and approximations that make their results far from being conclusive. Furthermore, recent results on limitations of the effective theories show that care must be taken in generalizing certain conclusions from the effective description to the full quantum theory [107, 108].

Acknowledgments

We are grateful to Abhay Ashtekar, Aurelien Barau, Chris Beetle, B. Bolliet, B. Bonga, Alejandro Corichi, David Craig, Peter Diener, Jonathan Engel, Rodolfo Gambini, J. Grain, Brajesh Gupt, Anton Joe, Wojciech Kaminski, Alok Laddha, Jerzy Lewandowski, E. Mato, Miguel Megevand, Jose Navarro-Salas, William Nelson, Javier Olmedo, Leonard Parker, Tomasz Pawlowski, Jorge Pullin, Sahil Saini,

David Sloan, Victor Taveras, Kevin Vandersloot, Madhavan Varadarajan, S. Vijayakumar, and Edward Wilson-Ewing for many stimulating discussions and insights. This work is supported in part by NSF grants PHY1068743, PHY1403943, PHY1404240, PHY1454832 and PHY-1552603. This work is also supported by a grant from John Templeton Foundation. The opinions expressed in this publication are those of authors and do not necessarily reflect the views of John Templeton Foundation.

References

[1] B. S. DeWitt, "Quantum theory of gravity I. The canonical theory," *Phys. Rev. D* **160** (1967) 1113.
[2] J. A. Wheeler, "Superspace and the nature of quantum geometrodynamics," in *Battelle Recontres*, eds. C. M. DeWitt and J. A. Wheeler, pp. 242 (Benjamin, New York (1968)).
[3] A. Ashtekar and P. Singh, "Loop quantum cosmology: A status report," *Class. Quant. Grav.* **28** (2011) 213001.
[4] K. Banerjee, G. Calcagni and M. Martin-Benito, "Introduction to loop quantum cosmology," *SIGMA* **8** (2012) 016.
[5] I. Agullo and A. Corichi, "Loop quantum cosmology," To appear in *Springer Handbook of Spacetime* eds. A. Ashtekar and V. Petkov, arXiv:1302.3833.
[6] J. Grain, "The perturbed universe in the deformed algebra approach of Loop Quantum Cosmology," *Int. J. Mod. Phys. D* **25** (2016) 1642003.
[7] A. Ashtekar, and A. Barrau, "Loop quantum cosmology: From pre-inflationary dynamics to observations," *Class. Quant. Grav.* **32** (2015) 234001.
[8] G. A. Mena Marugan, "Loop quantum cosmology: A cosmological theory with a view," *J. Phys. Conf. Ser.* **314** (2011) 012012; *A Brief Introduction to Loop Quantum Cosmology*, AIP Conf. Proc. 1130 (2009) 89.
[9] A. Ashtekar, T. Pawlowski and P. Singh, "Quantum nature of the big bang," *Phys. Rev. Lett.* **96** (2006) 141301.
[10] A. Ashtekar, T. Pawlowski and P. Singh, "Quantum nature of the big bang: An analytical and numerical investigation," *Phys. Rev. D* **73** (2006) 124038.
[11] A. Ashtekar, T. Pawlowski and P. Singh, "Quantum nature of the big bang: Improved dynamics," *Phys. Rev. D* **74** (2006) 084003.
[12] A. Ashtekar, A. Corichi and P. Singh, "Robustness of predictions of loop quantum cosmology," *Phys. Rev. D* **77** (2008) 024046.
[13] A. Ashtekar, T. Pawlowski, P. Singh and K. Vandersloot, "Loop quantum cosmology of k=1 FRW models," *Phys. Rev. D* **75** (2006) 0240035.
[14] L. Szulc, W. Kaminski and J. Lewandowski, "Closed FRW model in loop quantum cosmology," *Class. Quant. Grav.* **24** (2007) 2621.
[15] A. Corichi and A. Karami, "Loop quantum cosmology of k=1 FRW: A tale of two bounces," *Phys. Rev. D* **84** (2011) 044003.
[16] A. Corichi and A. Karami, "Loop quantum cosmology of k = 1 FLRW: Effects of inverse volume corrections," *Class. Quant. Grav.* **31** 035008 (2014).
[17] K. Vandersloot, "Loop quantum cosmology and the k = - 1 RW model," *Phys. Rev. D* **75** (2007) 023523.
[18] L. Szulc, "Open FRW model in loop quantum cosmology," *Class. Quant. Grav.* **24** (2007) 6191.

[19] E. Bentivegna and T. Pawlowski, "Anti-deSitter universe dynamics in LQC," *Phys. Rev. D* **77** (2008) 124025.

[20] W. Kaminski and T. Pawlowski, "The LQC evolution operator of FRW universe with positive cosmological constant," *Phys. Rev. D* **81** (2010) 024014.

[21] A. Ashtekar and T. Pawlowski, "Loop quantum cosmology with a positive cosmological constant," *Phys. Rev.* **85** (2012) 064001.

[22] A. Ashtekar, T. Pawlowski and P. Singh, "Loop quantum cosmology in the pre-inflationary epoch," (in preparation).

[23] T. Pawlowski, R. Pierini and E. Wilson-Ewing, "Loop quantum cosmology of a radiation-dominated flat FLRW universe," arXiv:1404.4036 [gr-qc].

[24] D. Chiou, "Loop quantum cosmology in Bianchi Type I models: Analytical investigation," *Phys. Rev. D* **75** (2007) 024029.

[25] M. Martin-Benito, G. A. Mena Marugan and T. Pawlowski, "Loop quantization of vacuum Bianchi I cosmology," *Phys. Rev. D* **78** (2008) 064008; "Physical evolution in loop quantum cosmology: The example of vacuum Bianchi I," *Phys. Rev. D* **80** (2009) 084038.

[26] A. Ashtekar and E. Wilson-Ewing, "Loop quantum cosmology of Bianchi type I models," *Phys. Rev. D* **79** (2009) 083535.

[27] A. Ashtekar and E. Wilson-Ewing, "Loop quantum cosmology of Bianchi type II models," *Phys. Rev. D* **80** (2009) 123532.

[28] E. Wilson-Ewing, "Loop quantum cosmology of Bianchi type IX models," *Phys. Rev. D* **82** (2010) 043508.

[29] P. Singh and E. Wilson-Ewing, "Quantization ambiguities and bounds on geometric scalars in anisotropic loop quantum cosmology," *Class. Quant. Grav.* **31** (2014) 035010.

[30] A. Corichi and A. Karami, "Loop quantum cosmology of Bianchi IX: Inclusion of inverse triad corrections," *Int. J. Mod. Phys. D* **25** (2016) 1642011.

[31] M. Martin-Benito, L. J. Garay and G. A. Mena Marugan, "Hybrid quantum Gowdy cosmology: Combining loop and Fock quantizations," *Phys. Rev. D* **78** (2008) 083516; L. J. Garay, M. Martin-Benito and G. A. Mena Marugan, "Inhomogeneous loop quantum cosmology: Hybrid quantization of the Gowdy model," *Phys. Rev. D* **82** (2010) 044048.

[32] D. Brizuela, G. A. Mena Marugan and T. Pawlowski, "Big bounce and inhomogeneities," *Class. Quant. Grav.* **27** (2010) 052001.

[33] M. Martin-Benito, D. Martin-de Blas and G. A. Mena Marugan, "Matter in inhomogeneous loop quantum cosmology: The Gowdy T^3 model," `arXiv:1012.2324`.

[34] M. Martin-Benito, G. A. Mena Marugan and E. Wilson-Ewing, "Hybrid quantization: From Bianchi I to the Gowdy model," *Phys. Rev. D* **82** (2010) 084012.

[35] V. A. Belinskii, I. M. Khalatnikov, I. M. and E. M. Lifshitz, "Oscillatory approach to a singular point in the relativistic cosmology," *Adv. Phys.* **19** (1970) 525.

[36] B. K. Berger, "Numerical Approaches to Spacetime Singularities," *Living Rev. Rel.* **5** (2002).

[37] D. Garfinkle, "Numerical simulations of general gravitational singularities," *Class. Quant. Grav.* **24** (2007) 295.

[38] Komatsu, E. *et al.*, "Seven-year Wilkinson microwave anisotropy probe observation: Cosmological interpretation," *Astrophys. J. Suppl.* **192** (2011) 18.

[39] Planck Collaboration, *Planck 2013 Results. XXII. Constraints on Inflation*, arXiv:1303.5082.

[40] BICEP2 Collaboration, *BICEP2 I: Detection Of B-mode Polarization at Degree Angular Scales*, *Phys. Rev. Lett.* **112** (2014) 241101.

[41] I. Agullo, A. Ashtekar and W. Nelson, "A quantum gravity extension of the inflationary scenario," *Phys. Rev. Lett.* **109** (2012) 251301.

[42] I. Agullo, A. Ashtekar and W. Nelson, "An extension of the quantum theory of cosmological perturbations to the Planck era," *Phys. Rev. D* **87** (2013) 043507.

[43] I. Agullo, A. Ashtekar and W. Nelson, "The pre-inflationary dynamics of loop quantum cosmology: Confronting quantum gravity with observations," *Class. Quant. Grav.* **30** (2013) 085014.

[44] M. Bojowald, "Absence of singularity in loop quantum cosmology," *Phys. Rev. Lett.* **86** (2001) 5227-5230.

[45] M. Bojowald, "Homogeneous loop quantum cosmology," *Class. Quant. Grav.* **20** (2003) 2595.

[46] A. Ashtekar, M. Bojowald and J. Lewandowski, "Mathematical structure of loop quantum cosmology," *Adv. Theo. Math. Phys.* **7** (2003) 233–268.

[47] A. Ashtekar, M. Campiglia and A. Henderson, "Casting loop quantum cosmology in the spin foam paradigm," *Class. Quant. Grav.* **27** (2010) 135020; "Path integrals and the WKB approximation in loop quantum cosmology," *Phys. Rev. D* **82** (2010) 124043.

[48] A. Ashtekar and E. Wilson-Ewing, "The Covariant entropy bound and loop quantum cosmology," *Phys. Rev. D* **78** (2008) 064047.

[49] D. A. Craig and P. Singh, "Consistent probabilities in Wheeler-DeWitt quantum cosmology," *Phys. Rev. D* **82** (2010) 123526.

[50] "Consistent histories in quantum cosmology," *Found. Phys.* **41** (2011) 371.

[51] D. A. Craig and P. Singh, "Consistent probabilities in loop quantum cosmology," *Class. Quant. Grav.* **30** (2013) 205008.

[52] D. A. Craig, "The consistent histories approach to loop quantum cosmology," *Int. J. Mod. Phys. D* **25** (2016) 1642009.

[53] D. Craig and P. Singh, "The vertex expansion in the consistent histories formulation of spin foam loop quantum cosmology," arXiv:1603.09671 [gr-qc].

[54] A. Ashtekar and M. Bojowald, "Quantum geometry and the Schwarzschild singularity," *Class. Quant. Grav.* **23** (2006) 391.

[55] L. Modesto, "Loop quantum black hole," *Class. Quant. Grav.* **23** (2006) 5587.

[56] M. Campiglia, R. Gambini and J. Pullin, "Loop quantization of spherically symmetric midi-superspaces: The interior problem," *AIP Conf. Proc.* **977** (2008) 52.

[57] C. G. Boehmer and K. Vandersloot, "Loop quantum dynamics of the Schwarzschild interior," *Phys. Rev. D* **76** (2007) 104030.

[58] N. Dadhich, A. Joe and P. Singh, "Emergence of the product of constant curvature spaces in loop quantum cosmology," *Class. Quant. Grav.* **32** (2015) 185006.

[59] A. Corichi and P. Singh, "Loop quantization of the Schwarzschild interior revisited," *Class. Quant. Grav.* **33** (2016) 055006.

[60] W. Kaminski, J. Lewandowski and L. Szulc, "The status of quantum geometry in the dynamical sector of loop quantum cosmology," *Class. Quant. Grav.* **25** (2008) 055003.

[61] W. Kaminski and J. Lewandowski, "The flat FRW model in LQC: The self-adjointness," *Class. Quant. Grav.* **25** (2008) 035001.

[62] W. Kaminski, J. Lewandowski and T. Pawlowski, "Physical time and other conceptual issues of QG on the example of LQC," *Class. Quant. Grav.* **26** (2009) 035012.

[63] W. Kaminski, J. Lewandowski and T. Pawlowski, "Quantum constraints, Dirac observables and evolution: Group averaging versus Schroedinger picture in LQC," *Class. Quant. Grav.* **26** (2009) 245016.

[64] D. Brizuela, D. Cartin and G. Khanna, "Numerical techniques in loop quantum cosmology," *SIGMA* **8** (2012) 001.

[65] P. Singh, "Numerical loop quantum cosmology: An overview," *Class. Quant. Grav.* **29** (2012) 244002.

[66] P. Diener, B. Gupt and P. Singh, "Chimera: A hybrid approach to numerical loop quantum cosmology," *Class. Quant. Grav.* **31** (2014) 025013.

[67] A. Corichi and P. Singh, "Is loop quantization in cosmology unique?" *Phys. Rev. D* **78** (2008) 024034.

[68] A. Corichi and P. Singh, "A geometric perspective on singularity resolution and uniqueness in loop quantum cosmology," *Phys. Rev. D* **80** (2009) 044024.

[69] P. Singh, "Is classical flat Kasner spacetime flat in quantum gravity?" *Int. J. Mod. Phys. D* **25** (2016) 1642001.

[70] J. L. Dupuy and P. Singh, "Implications of quantum ambiguities in k=1 loop quantum cosmology: Distinct quantum turnarounds and the super-Planckian regime," arXiv:1608.07772.

[71] M. Fernandez-Mendez, G. A. Mena Marugan and J. Olmedo, "Hybrid quantization of an inflationary universe," *Phys. Rev. D* **86** (2012) 024003.

[72] M. Fernandez-Mendez, G. A. Mena Marugan and J. Olmedo, "Hybrid quantization of an inflationary model: The flat case," *Phys. Rev. D* **88** (2013) 044013.

[73] M. Fernandez-Mendez, G. A. Mena Marugan and J. Olmedo, "Effective dynamics of scalar perturbations in a flat Friedmann-Robertson-Walker spacetime in loop quantum cosmology," *Phys. Rev. D* **89** (2014) 044041.

[74] E. Wilson-Ewing, "Holonomy corrections in the effective equations for scalar mode perturbations in loop quantum cosmology," *Class. Quant. Grav.* **29** (2012) 085005.

[75] E. Wilson-Ewing, "Lattice loop quantum cosmology: Scalar perturbations," *Class. Quant. Grav.* **29** 215013 (2012); "The matter bounce scenario in loop quantum cosmology," `arXiv:1211.6269`.

[76] M. Bojowald and G. M. Hossain, "Loop quantum gravity corrections to gravitational wave dispersion," *Phys. Rev. D* **77** (2008) 023508.

[77] W. Nelson and M. Sakellariadou, "Lattice refining loop quantum cosmology and inflation," *Phys. Rev. D* **76** (2007) 044015.

[78] J. Grain and A. Barrau, "Cosmological footprints of loop quantum gravity," *Phys. Rev. Lett.* **102** (2009) 081301.

[79] J. Grain, T. Cailleteau, A. Barrau and A. Gorecki, "Fully loop-quantum-cosmology-corrected propagation of gravitational waves during slow-roll inflation," *Phys. Rev. D* **81** (2010) 024040.

[80] J. Mielczarek, T. Cailleteau, J. Grain and A. Barrau, "Inflation in loop quantum cosmology: Dynamics and spectrum of gravitational waves," *Phys. Rev. D* **81** (2010) 104049.

[81] J. Grain, A. Barrau, T. Cailleteau and J. Mielczarek, "Observing the big bounce with tensor modes in the cosmic microwave background: Phenomenology and fundamental LQC parameters," *Phys. Rev. D* **82** (2010) 123520.

[82] M. Bojowald, G. Calcagni and S. Tsujikawa, "Observational test of inflation in loop quantum cosmology," *JCAP* **1111** (2011) 046.

[83] T. Cailleteau, J. Mielczarek, A. Barrau and J. Grain, "Anomaly-free scalar perturbations with holonomy corrections in loop quantum cosmology," *Class. Quant. Grav.* **29** (2012) 095010.

[84] L. Castello Gomar, M. Fernandez-Mendez, G. Mena Marugan and J. Olmedo, "Cosmological perturbations in hybrid loop quantum cosmology: Mukhanov-Sasaki variables," arXiv:1407.0998.

[85] G. Calcagni, "Observational effects from quantum cosmology," *Annalen Phys.* **525** (2013) 323-338, Erratum-ibid. 525 10-11, A165 (2013).
[86] A. Barrau, T. Cailleteau, J. Grain and J. Mielczarek. "Observational issues in loop quantum cosmology," *Class. Quant. Grav.* **31** (2014) 053001.
[87] A. Barrau, M. Bojowald, G. Calcagni, J. Grain and M. Kagan, "Anomaly-free cosmological perturbations in effective canonical quantum gravity," arXiv:1404.1018.
[88] C. Rovelli and F. Vidotto, "On the spinfoam expansion in cosmology," *Class. Quant. Grav.* **27** (2010) 145005; E. Bianchi, C. Rovelli and F. Vidotto, "Towards spinfoam cosmology," *Phys. Rev. D* **82** (2010) 084035.
[89] S. Gielen, D. Oriti and L. Sindoni, Cosmology from group field theory formalism for quantum gravity," *Phys. Rev. Lett.* **111** (2013) 031301.
[90] J. Willis, "On the low energy ramifications and a mathematical extension of loop quantum gravity." Ph.D. Dissertation, The Pennsylvania State University (2004).
[91] V. Taveras, "LQC corrections to the Friedmann equations for a universe with a free scalar field," *Phys. Rev. D* **78** (2008) 064072.
[92] M. Bojowald, "Loop quantum cosmology and inhomogeneities," *Gen. Rel. Grav.* **38** (2006) 1771.
[93] K. A. Meissner, "Black hole entropy in loop quantum gravity," *Class. Quant. Grav.* **21** (2004) 5245.
[94] H. Bohr, "Zur theorie der fastperiodischen funktionen," *Acta. Math.* **45** (1924) 29.
[95] A. Ashtekar and M. Campiglia, On the uniqueness of kinematics of loop quantum cosmology," *Class. Quant. Grav.* **29** (2012) 242001.
[96] J. Engle and M. Hanusch, "Kinematical uniqueness of homogeneous isotropic LQC," arXiv:1604.08199 [gr-qc].
[97] J. Engle, M. Hanusch and T. Thiemann, "Uniqueness of the representation in homogeneous isotropic LQC," arXiv:1609.03548.
[98] J. Lewandowski, A. Okolow, H. Sahlmann and T. Thiemann, "Uniqueness of diffeomorphism invariant states on holonomy flux algebras," *Comm. Math. Phys.* **267** (2006) 703-733.
[99] C. Fleischchack, "Representations of the Weyl algebra in quantum geometry," *Commun. Math. Phys.* **285** (2009) 67-140.
[100] G. M. Hossain, V. Husain and S. S. Seahra, "Non-singular inflationary universe from polymer matter," *Phys. Rev. D* **81** (2010) 024005.
[101] A. Kreienbuehl and T. Pawlowski, "Singularity resolution from polymer quantum matter," *Phys. Rev. D* **88** (2013) 043504.
[102] S. M. Hassan, V. Husain and S. S. Seahra, "Polymer inflation," *Phys. Rev. D* **91** (2015) 065006.
[103] A. Ashtekar, S. Fairhurst and J. Willis, "Quantum gravity, shadow states, and quantum mechanics," *Class. Quant. Grav.* **20** (2003) 1031.
[104] D. Marolf, "Refined algebraic quantization: Systems with a single constraint," `arXiv: gr-qc/9508015`; "Quantum observables and recollapsing dynamics," *Class. Quant. Grav.* **12** (1994) 1199–1220.
[105] A. Ashtekar, J. Lewandowski, D. Marolf, J. Mourão and T. Thiemann, "Quantization of diffeomorphism invariant theories of connections with local degrees of freedom," *J. Math. Phys.* **36** (1995) 6456–6493.
[106] A. Ashtekar, L. Bombelli and A. Corichi, "Semiclassical states for constrained systems," *Phys. Rev. D* **72** (2005) 025008.
[107] P. Diener, B. Gupt and P. Singh, "Numerical simulations of a loop quantum cosmos: Robustness of the quantum bounce and the validity of effective dynamics," *Class. Quant. Grav.* **31** (2014) 105015.

[108] P. Diener, B. Gupt, M. Megevand and P. Singh, "Numerical evolution of squeezed and non-Gaussian states in loop quantum cosmology," *Class. Quant. Grav.* **31** (2014) 165006.

[109] M. Martin-Benito, G. A. M. Marugan and J. Olmedo, "Further improvements in the understanding of isotropic loop quantum cosmology," *Phys. Rev. D* **80** (2009) 104015.

[110] G. A. Mena Marugan, J. Olmedo and T. Pawlowski, "Prescriptions in loop quantum cosmology: A comparative analysis," *Phys. Rev. D* **84** (2011) 064012.

[111] D. Cartin and G. Khanna, "Matrix methods in loop quantum cosmology," *Proc. Quantum Gravity in Americas III*, Penn State (2006). http://igpg.gravity.psu.edu/events/conferences/QuantumGravityIII/proceedings.shtm.

[112] T. Tanaka, F. Amemiya, M. Shimano, T. Harada and T. Tamaki, "Discretisation parameter and operator ordering in loop quantum cosmology with the cosmological constant," *Phys. Rev. D* **83** (2011) 104049.

[113] W. Nelson and M. Sakellariadou, "Unique factor ordering in the continuum limit of LQC," *Phys. Rev. D* **78** (2008) 024006.

[114] T. Schilling, "Geomtery of quantum mechanics," Ph. D Dissertation, The Pennsylvania State University (1996).

[115] A. Ashtekar and T. A. Schilling, "Geometrical formulation of quantum mechanics," In: *On Einstein's Path: Essays in Honor of Engelbert Schücking*, ed. Harvey, A. (Springer, New York (1999)), pp. 23–65, arXiv:gr-qc/9706069.

[116] P. Singh and V. Taveras, "A note on the effective equations in loop quantum cosmology," (To appear).

[117] M. Bojowald and A. Skirzewski, "Effective theory for the cosmological generation of structure, *Rev. Math. Phys.* **18** (2006) 713.

[118] M. Bojowald, B. Sandhoefer, A. Skirzewski and A. Tsobanjan, "Effective constraints for quantum systems," *Rev. Math. Phys.* **21** (2009) 111.

[119] Y. Shtanov and V. Sahni, "Bouncing brane worlds," *Phys. Lett. B* **557** (2003) 1.

[120] P. Singh, "Loop cosmological dynamics and dualities with Randall-Sundrum braneworlds," *Phys. Rev. D* **73** (2006) 063508.

[121] P. Singh and S. K. Soni, "On the relationship between modifications to the Raychaudhuri equation and the canonical Hamiltonian structures," *Class. Quant. Grav.* **33** (2016) 125001.

[122] P. Singh, "Effective state metamorphosis in semi-classical loop quantum cosmology," *Class. Quant. Grav.* **22** (2005) 4203.

[123] J. Magueijo and P. Singh, "Thermal fluctuations in loop cosmology," *Phys. Rev. D* **76** (2007) 023510.

[124] C. Rovelli and E. Wilson-Ewing, "Why are the effective equations of loop quantum cosmology so accurate?" *Phys. Rev. D* **90** (2014) 023538.

[125] A. Corichi and E. Montoya, "On the semiclassical limit of loop quantum cosmology," *Int. J. Mod. Phys. D* **21** (2012) 1250076.

[126] A. Corichi and E. Montoya, "Coherent semiclassical states for loop quantum cosmology," *Phys. Rev. D* **84** (2011) 044021.

[127] A. Ashtekar and B. Gupt, "Generalized effective description of loop quantum cosmology," *Phys. Rev. D* **92** (2015) 084060.

[128] G. J. Olmo and P. Singh, "Effective action for loop quantum cosmology a la Palatini, *JCAP* **0901** (2009) 030.

[129] P. Singh, "Are loop quantum cosmologies never singular?" *Class. Quant. Grav.* **26** (2009) 125005.

[130] P. Singh, "Curvature invariants, geodesics and the strength of singularities in Bianchi-I loop quantum cosmology," arXiv:1112.6391.
[131] P. Singh, *Bull. Astron. Soc. India* **42** (2014) 121.
[132] S. Saini and P. Singh, "Geodesic completeness and the lack of strong singularities in effective loop quantum Kantowski-Sachs spacetime," arXiv:1606.04932 [gr-qc] (To appear in Class. Quant. Grav.).
[133] A. Ashtekar and D. Sloan, "Loop quantum cosmology and slow roll inflation," *Phys. Lett. B* **694** (2010) 108-112.
[134] A. Ashtekar and D. Sloan, "Probability of inflation in loop quantum cosmology," *Gen. Rel. Grav.* **43** (2011) 3619-3656.
[135] A. Corichi and A. Karami, "On the measure problem in slow roll inflation and loop quantum cosmology," *Phys. Rev. D* **83** (2011) 104006.
[136] A. Corichi and D. Sloan, "Inflationary attractors and their measures," *Class. Quant. Grav.* **31** (2014) 062001.
[137] W. Kaminski and T. Pawlowski, "Cosmic recall and the scattering picture of loop quantum cosmology," *Phys. Rev. D* **81** (2010) 084027.
[138] A. Corichi and P. Singh, "Quantum bounce and cosmic recall," *Phys. Rev. Lett.* **100** (2008) 209002.
[139] T. Pawlowski, "Universes memory and spontaneous coherence in loop quantum cosmology," *Int. J. Mod. Phys. D* **25** (2016) 1642013.
[140] A. Ashtekar, M. Campiglia and A. Henderson, "Path integrals and the WKB approximation in loop quantum cosmology," *Phys. Rev. D* **82** (2010) 124043.
[141] A. Ashtekar, M. Campiglia and A. Henderson, "Loop quantum cosmology and spin foams," *Phys. Lett. B* **681** (2009) 347-352.
[142] R. B. Griffiths, "Consistent histories and the interpretation of quantum mechanics," *J. Stat. Phys.* **36** (1984) 219; *Consistent Quantum Theory* (Cambridge University Press, Cambridge (2008)).
[143] J. B. Hartle, "Spacetime quantum mechanics and the quantum mechanics of spacetime," *Gravitation and Quantizations: Proc. 1992 Les Houches Summer School*, eds. B. Julia and J. Zinc-Justin (North Holland, Amsterdam (1995)).
[144] D. A. Craig, "Dynamical eigenfunctions and critical density in loop quantum cosmology," *Class. Quant. Grav.* **30** (2013) 035010.
[145] A. Ashtekar, S. Fairhurst, J. L. Willis, "Quantum gravity, shadow states, and quantum mechanics," *Class. Quant. Grav.* **20** (2003) 1031.
[146] A. Corichi, T. Vukasinac, J. A. Zapata, "Polymer quantum mechanics and its continuum limit," *Phys. Rev. D* **76** (2007) 044016.
[147] D. Green and W. Unruh, "Difficulties with recollapsing models in closed isotropic loop quantum cosmology," *Phys. Rev. D* **70** (2004) 103502.
[148] A. Borde, A. Guth and A. Vilenkin, "Inflationary spacetimes are not past-complete," *Phys. Rev. Lett.* **90** (2003) 151301.
[149] M. Sami, P. Singh and S. Tsujikawa, "Avoidance of future singularities in loop quantum cosmology," *Phys. Rev. D* **74** (2006) 043514.
[150] P. Singh and F. Vidotto, "Exotic singularities and spatially curved loop quantum cosmology," *Phys. Rev. D* **83** (2011) 064027.
[151] D. Samart and B. Gumjudpai, "Phantom field dynamics in loop quantum cosmology," *Phys. Rev. D* **76** (2007) 043514.
[152] T. Naskar and J. Ward, "Type I singularities and the phantom menace," *Phys. Rev. D* **76** (2007) 063514.
[153] G. F. R. Ellis and B. G. Schmidt, "Singular spacetimes," *Gen. Rel. Grav.* **8** (1977) 915.

[154] F. J. Tipler, "Singularities in conformally flat spacetimes," *Phys. Lett. A* **64** (1977) 8.
[155] C. J. S. Clarke and A. Królak, "Conditions for the occurence of strong curvature singularities," *J. Geom. Phys.* **2** (1985) 127.
[156] T. Cailleteau, A. Cardoso, K. Vandersloot and D. Wands, "Singularities in loop quantum cosmology," *Phys. Rev. Lett.* **101** (2008) 251302.
[157] B. Gupt and P. Singh, "Contrasting features of anisotropic loop quantum cosmologies: the role of spatial curvature," *Phys. Rev. D* **85** (2012) 044011.
[158] A. Joe and P. Singh, *Class. Quant. Grav.* **32** (2015) 015009.
[159] A. Corichi and A. Karami, "Loop quantum cosmology of Bianchi IX: Inclusion of inverse triad corrections," arXiv:1605.01383 [gr-qc].
[160] P. Tarrio, M. F. Mendez and G. A. Mena Marugan, *Phys. Rev. D* **88** (2013) 084050.
[161] P. Singh, K. Vandersloot and G. V. Vereshchagin, "Non-singular bouncing universes in loop quantum cosmology," *Phys. Rev. D* **74** (2006) 043510.
[162] L. Linsefors and A. Barrau, "Duration of inflation and conditions at the bounce as a prediction of effective isotropic loop quantum cosmology," *Phys. Rev. D* **87** (2013) 123509.
[163] E. Ranken and P. Singh, "Non-singular power-law and assisted inflation in loop quantum cosmology," *Phys. Rev. D* **85** (2012) 104002.
[164] H. H. Xiong and J. Y. Zhu, "Tachyon field in loop quantum cosmology: Inflation and evolution picture," *Phys. Rev. D* **75** (2007) 084023.
[165] M. Artymowski, A. Dapor and T. Pawlowski, "Inflation from non-minimally coupled scalar field in loop quantum cosmology," *JCAP* **1306** (2013) 010.
[166] B. Gupt and P. Singh, "A quantum gravitational inflationary scenario in Bianchi-I spacetime," *Class. Quant. Grav.* **30** (2013) 145013.
[167] L. Kofman, A. Linde and V. Mukhanov, "Inflationary theory and alternative cosmology," *JHEP* **0210** (2002) 057.
[168] G. W. Gibbons and N. Turok, "The measure problem in cosmology," *Phys. Rev. D* **77** (2008) 063516.
[169] D. W. Chiou and K. Vandersloot, "Behavior of non-linear anisopropies in bouncing Bianchi I models of loop quantum cosmology," *Phys. Rev. D* **76** (2007) 084015.
[170] T. Cailleteau, P. Singh and K. Vandersloot, "Non-singular ekpyrotic/cyclic model in loop quantum cosmology," *Phys. Rev. D* **80** (2009) 124013.
[171] A. Corichi and E. Montoya, "Effective dynamics in Bianchi type II loop quantum cosmology," *Phys. Rev. D* **85** (2012) 104052.
[172] A. Corichi, A. Karami and E. Montoya, "Loop quantum cosmology: Anisotropy and singularity resolution," `1210.7248`.
[173] A. Corichi and E. Montoya, "Loop quantum cosmology of Bianchi IX: Effective dynamics," arXiv:1502.02342 [gr-qc].
[174] R. Maartens and K. Vandersloot, "Magnetic Bianchi I universe in loop quantum cosmology," arXiv:0812.1889.
[175] A. Henderson and T. Pawlowski (In preparation).
[176] B. Gupt and P. Singh, "Quantum gravitational Kasner transitions in Bianchi-I spacetime," *Phys. Rev. D* **86** (2012) 024034, `arXiv:1205.6763`.
[177] A. Ashtekar, A. Henderson and D. Sloan, "Hamiltonian formulation of general relativity and the Belinksii, Khalatnikov, Lifshitz conjecture," *Class. Quant. Grav.* **26** (2009) 052001; "A Hamiltonian formulation of the BKL conjecture," *Phys. Rev. D* **83** (2011) 084024.
[178] L. Linsefors and A. Barrau, "Modified Friedmann equation and survey of solutions in effective Bianchi-I loop quantum cosmology," *Class. Quant. Grav.* **31** (2014) 015018.

[179] T. Thiemann, "QSD V: Quantum gravity as the natural regulator of matter quantum field theories," *Class. Quant. Grav.* **15** (1998) 1281–1314.
[180] A. R. Liddle and D. H. Lyth, *Cosmological Inflation and Large-scale Structure* (Cambridge University press, Cambridge (2000)).
[181] S. Dodelson, *Modern Cosmology* (Academic Press, Amsterdam (2003)).
[182] V. Mukhanov, *Physical Foundations of Cosmology* (Cambridge University Press, Cambridge (2005)).
[183] S. Weinberg, *Cosmology* (Oxford University Press, Oxford (2008)).
[184] R. M. Wald, *Quantum Field Theory in Curved Spacetime and Black Hole Thermodynamics* (University of Chicago Press (1994)).
[185] N. D. Birrell and P. C. W. Davies, *Quantum Fields in Curved Space* (Cambridge University Press (1982)).
[186] L. Parker and D. Toms, *Quantum Field Theory in Curved Spacetime* (Cambridge UP, Cambridge (2009)).
[187] I. Agullo, B. Gupt and E. Mato, in preparation.
[188] Y. Cai, R. H. Brandenberger and X. Zhang, "The matter bounce curvaton scenario," *JCAP* **1103** (2011) 003.
[189] J. Khoury, B. A. Ovrut, P. J. Steinhardt and N. Turok, "Density perturbations in the ekpyrotic scenario," *Phys. Rev. D* **66** (2002) 046005.
[190] A. Ashtekar, W. Kaminski and J. Lewandowski, "Quantum field theory on a cosmological, quantum spacetime," *Phys. Rev. D* **79** (2009) 064030, arXiv:0901.0933.
[191] J. Puchta, "Quantum fluctuations in quantum spacetime," MSc Thesis under supervision of Jerzy Lewandowski, University of Warsaw 2009.
[192] A. Dapor, J. Lewandowski and J. Puchta, "QFT on quantum spacetime: A compatible classical framework," *Phys. Rev. D* **87** (2013) 104038.
[193] A. Dapor, J. Lewandowski and J. Puchta, "QFT on quantum spacetime: A compatible classical framework," *Phys. Rev. D* **87** (2013) 104038.
[194] A. Dapor, J. Lewandowski and Y. Tavakoli, "Lorentz symmetry in QFT on quantum Bianchi I spacetime," *Phys. Rev. D* **86** (2012) 064013.
[195] R. Gambini and J. Pullin, "Hawking radiation from a spherical loop quantum gravity black hole," *Class. Quant. Grav.* **31** (2014) 115003; "A scenario for black hole evaporation on a quantum geometry," arXiv:1408.3050.
[196] V. F. Mukhanov, H. A. Feldman and R. H. Brandenberger, "Theory of cosmological perturbations," *Phys. Rep.* **215** (1992) 203.
[197] I. Agullo, A. Ashtekar and B. Gupt, "Phenomenology with fluctuating quantum geometries in loop quantum cosmology," `arXiv 1611.09810`.
[198] E. Wilson-Ewing, "Separate universes in loop quantum cosmology: Framework and applications," *Int. J. Mod. Phys. D* **25** (2016) 1642002.
[199] L. Castelló Gomar, M. Martín-Benito and G. A. Mena Marugán, "Gauge-invariant perturbations in hybrid quantum cosmology," *JCAP* **1506** (2015) 045.
[200] F. Benítez Martínez and J. Olmedo, "Primordial tensor modes of the early Universe," *Phys. Rev. D* **93** (2016) 124008.
[201] D. Martín de Blas and J. Olmedo, "Primordial power spectra for scalar perturbations in loop quantum cosmology," *JCAP* **1606** (2016) 029.
[202] M. Bojowald and G. M. Paily, *Phys. Rev. D* **87** (2013) 044044.
[203] S. Schander, A. Barrau, B. Bolliet, L. Linsefors, J. Mielczarek and J. Grain, "Primordial scalar power spectrum from the Euclidean big bounce," *Phys. Rev. D* **93** (2016) 023531.
[204] B. Bolliet, A. Barrau, J. Grain and S. Schander, "Observational exclusion of a consistent loop quantum cosmology scenario," *Phys. Rev. D* **93** (2016) 124011.

[205] A. Barrau and J. Grain, "Cosmology without time: What to do with a possible signature change from quantum gravitational origin?" arXiv:1607.07589.
[206] Y. Cai and E. Wilson-Ewing, "Non-singular bounce scenarios in loop quantum cosmology and th effective description," *JCAP* **1403** (2014) 026.
[207] Y. Cai, J. Quintin, E. N. Saridakis and E. Wilson-Ewing, "Nonsingular bouncing cosmologies in the light of BICEP2," *JCAP* **1407** (2014) 033.
[208] Y. Cai and E. Wilson-Ewing, "A ΛCDM bounce scenario," *JCAP* **1503** (2015) 006.
[209] I. Agullo and L. Parker, "Non-gaussianities and the stimulated creation of quanta in the inflationary universe," *Phys. Rev. D* **83** (2011) 063526; "Stimulated creation of quanta during inflation and the observable universe," *Gen. Rel. Grav.* **43** (2011) 2541-2545.
[210] I. Agullo, J. Navarro-Salas and L. Parker, "Enhanced local-type inflationary trispectrum from a non-vacuum initial state," *JCAP* **1205** (2012) 019.
[211] J. D. Barrow, "The premature recollapse problem in closed inflationary universes," *Nucl. Phys. B* **296** (1988) 697-709.
[212] J. D. Barrow and S. Cotsakis, "Inflation and the conformal structure of higher order gravity theories," *Phys. Lett. B* **214** (1988) 515.
[213] B. Gupt and B. Bonga, "Inflation with the Starobinsky potential in loop quantum cosmology," *Gen. Rel. Grav.* **48** (2016) 71.
[214] B. Gupt and B. Bonga, "Phenomenological investigations of a quantum gravity extension of inflation with the Starobinsky potential," *Phys. Rev. D* **93** (2016) 063513.
[215] I. Agullo and N. Morris, "Detailed analysis of the predictions of loop quantum cosmology for the primordial power spectrum," *Phys. Rev. D* **92** (2015) 124040.
[216] I. Agullo, W. Nelson and A. Ashtekar, "Preferred instantaneous vacuum for linear scalar fields in cosmological spacetimes," *Phys. Rev. D* **91** (2015) 064051.
[217] A. Ashtekar and B. Gupt, "Quantum gravity in the sky: Interplay between fundamental theory and observations," *Class. Quant. Grav.* **34** (2017) 014002.
[218] A. Ashtekar and B. Gupt, "Initial conditions for cosmological perturbations," *Class. Quant. Grav.* **34** (2017) 035004.
[219] Planck Collaboration, Planck 2015 results XVI: Isotropy and Statistics of the CMB, *Astron. and Astroph.* **594** (2016) A16.
[220] I. Agullo, "Loop quantum cosmology, non-Gaussianity, and CMB power asymmetry," *Phys. Rev. D* **92** (2015) 064038.
[221] F. Schmidt and L. Hui, "CMB power asymmetry from Gaussian modulation," arXiv:1210.2965.
[222] I. Agullo, B. Bolliet and S. Vijayakumar, in preparation.
[223] D. Oriti, L. Sindoni and E. Wilson-Ewing, Emergent Friedmann dynamics with a quantum bounce from quantum gravity condensates," *Class. Quant. Grav.* **33** (2016) 224001.
[224] J. Engle, "Relating loop quantum cosmology to loop quantum gravity: Symmetric sectors and embeddings," *Class. Quant. Grav.* **24** (2007) 5777; "Piecewise linear loop quantum gravity," *Class. Quant. Grav.* **27** (2010) 035003.
[225] J. Brunnemann and C. Fleischhack, "On the configuration spaces of homogeneous loop quantum cosmology and loop quantum gravity," arXiv:0709.1621 [math-ph].
[226] T. A. Koslowski, "A cosmological sector in loop quantum gravity," arXiv:0711.1098 [gr-qc].
[227] E. Alesci and F. Cianfrani, "Quantum-reduced loop gravity: Cosmology," *Phys. Rev. D* **87** (2013) 083521.
[228] E. Alesci and F. Cianfrani, "Improved regularization from quantum reduced loop gravity," arXiv:1604.02375 [gr-qc].

[229] D. Oriti, L. Sindoni and E. Wilson-Ewing, "Emergent Friedmann dynamics with a quantum bounce from quantum gravity condensates," *Class. Quant. Grav.* **33** (2016) 224001.

[230] P. Diener, A. Joe, M. Megevand and P. Singh, "Numerical simulations of loop quantum Bianchi-I spacetime," (To appear).

Chapter 7

Quantum Geometry and Black Holes

J. Fernando Barbero G.

Instituto de Estructura de la Materia, CSIC, Serrano 123, 28006 Madrid, Spain

Alejandro Perez

†*Centre de Physique Théorique, Campus de Luminy, 13288 Marseille, France*

1. Discussion of the Conceptual Issues

Black holes are remarkable solutions of classical general relativity describing important aspects of the physics of gravitational collapse. Their existence in our nearby universe is supported by a great amount of observational evidence [103]. When isolated, these systems are expected to be simple for late and distant observers. Once the initial very dynamical phase of collapse has passed, the system should settle down to a stationary situation completely described by the Kerr-Newman solution[a] labelled by three macroscopic parameters: the mass M, the angular momentum J, and the electromagnetic charge Q.

The fact that the final state of gravitational collapse is described by only three macroscopic parameters, independently of the details of the initial conditions leading to the collapse, could be taken as a first indication of the thermodynamical nature of black holes (which as we will see below is really of quantum origin). In fact the statement in the first paragraph contains the usual coarse graining perspective of thermodynamical physics in the assertion that for sufficiently long times after collapse *the system should settle down to a stationary situation... described by three parameters*. The details about how this settling down takes place depend indeed on the initial conditions leading to the collapse (the microstates of the system). The coarse graining consists of neglecting these details in favor of the idealization of stationarity.

---

*Grupo de Teorias de Campos y Fisica Estadistica, Instituto Universitario Gregorio Millán Barbany, Universidad Carlos III de Madrid, Unidad Asociada al IEM-CSIC.
†Unité Mixte de Recherche (UMR 6207) du CNRS et des Universités Aix-Marseille I, Aix-Marseille II, et du Sud Toulon-Var; laboratoire afilié à la FRUMAM (FR 2291).
[a]Such scenario is based on physical grounds, some concrete indications from perturbation theory, and the validity of the so-called no-hair theorem.

Another classical indication is Hawking area theorem [84] stating that for mild energy conditions (satisfied by classical matter fields) the area of a black hole horizon can only increase in any physical process. Namely, the so-called second law of black hole mechanics holds:

$$\delta A \geq 0. \tag{1}$$

This brings in the irreversibility characteristic of thermodynamical systems to the context of black hole physics and motivated Bekenstein [33] to associate with BHs a notion of entropy proportional to their area. Classically, one can also prove the so-called first law of BH mechanics [31] relating different nearby stationary BH spacetimes of Einstein-Maxwell theory

$$\delta M = \frac{\kappa}{8\pi}\delta A + \Omega \delta J + \Phi \delta Q, \tag{2}$$

where Ω is the angular velocity of the horizon, Φ is the horizon electric potential, and κ is the surface gravity.

The realization that black holes can indeed be considered (in the semiclassical regime) as thermodynamical systems came with the discovery of black hole radiation [85]. In the mid 70's Hawking considered the scattering of a quantum test field on a space time background geometry representing gravitational collapse of a compact source. Assuming that very early observers far away from the source prepare the field in the vacuum state he showed that, after the very dynamical phase of collapse is replaced by a stationary quasi equilibrium situation, late observers in the future measure an afterglow of particles of the test field coming from the horizon with a temperature

$$T_H = \frac{\kappa}{2\pi}. \tag{3}$$

As black holes radiate, the immediate conclusion is that they must evaporate through the (quantum phenomenon of) emission of Hawking radiation. The calculation of Hawking neglects such back reaction but provides a good approximation for the description of black holes that are sufficiently large, for which the radiated power is small relative to the scale defined by the mass of the black hole. These black holes are referred to as *semiclassical* in this chapter.

This result, together with the validity of the first and second laws, suggest that semiclassical black holes should have an associated entropy (here referred to as the Bekenstein–Hawking entropy) given by

$$S_H = \frac{A}{4\ell_{Pl}^2} + S_0 \tag{4}$$

where S_0 is an integration constant that cannot be fixed by the sole use of the first law. In fact, as in any thermodynamical system, entropy cannot be determined only by the use of the first law. Entropy can either be measured in an experimental setup (this was the initial way in which the concept was introduced) or calculated from the basic degrees of freedom by using statistical mechanical methods once a model for the fundamental building blocks of the system is available.

More precisely, even though the thermodynamical nature of semiclassical black holes is a robust prediction of the combination of general relativity and quantum field theory as a first approximation to quantum gravity, the precise expression for the entropy of black holes is a question that can only be answered within the framework of quantum gravity in its semiclassical regime. This is a central question for any proposal of quantum gravity theory.

This chapter will mainly deal with the issue of computing black hole entropy for semiclassical black holes which, as we will argue here, already presents an important challenge to quantum gravity but seems realistically within reach at the present stage of development of the approach. The formalism applies to physical black holes of the kind that can be formed in the early primordial universe or other astrophysical situations (no assumption of extremality or supersymmetry is needed).

Questions related to the information loss paradox, or the fate of unitarity are all issues that necessitate full control of the quantum dynamics in regimes far away from the semiclassical one. For that reason we designate this set of questions as the *hard problem*. These involve in particular the understanding of the dynamics near and across (what one would classically identify with) the interior singularity. There are studies of the quantum dynamics through models near the (classically apparent) singularities of general relativity indicating that not only the quantum geometry is well defined at the classically pathological regions, but also the quantum dynamics is perfectly determined across them. For the variety of results concerning cosmological singularities we refer the reader to *Chapter 6*. Similar results have been found in the context of black holes [12]. These works indicate that singularities are generically avoided due to quantum effects at the deep Planckian regime. Based on these results new paradigms have been put forward concerning the *hard problem* [11]. The key point is that the possibility of having physical dynamics beyond the apparent classical singularities allows for information to be lost into causally disconnected worlds (classical singularities as sinks of information) or to be recovered in subtle ways during and after evaporation as suggested by results in 2d black hole systems [22–24]. All these scenarios would be compatible with a local notion of unitarity [127]. The information paradox could also be solved [106] if quantum correlations with the (discrete) UV Planckian degrees of freedom remain hidden to low energy (semiclassical) observers. This possibility is appealing in an approach such as LQG where continuum space-time is obtained by coarse graining [18, 21, 123]. Space limitations prevent us from discussing the *hard problem* further in this chapter.

To date, investigations within the LQG framework, can be divided into the following categories: isolated horizons and their quantum geometry (Sections 2 and 3); rigorous counting of micro-states (Section 4); semiclassical quasi-local formulation (Section 5); spin foam dynamical accounts and low energy dynamical counterparts (Section 6.3); and the Hawking effect phenomenology and insights from symmetry reduced models (Section 6.5). The different sections are largely self-contained so they can be read independently.

2. Isolated Horizons

The model employed to describe black holes in loop quantum gravity is based on the use of *isolated horizons* (IH), a concept introduced around the year 2000 by Ashtekar and collaborators [8–10, 17] and developed by a number of other researchers [99, 101].[b] The main goal of this line of work was to find a *quasilocal* notion of horizon that could be used in contexts where the teleological nature of event horizons (i.e. the need to know the whole spacetime in order to determine if they are present) is problematic.

The most important features of isolated horizons are: their quasilocal nature, the availability of a Hamiltonian formulation for the sector of general relativity containing IH's, the possibility of having physically reasonable versions of some of the laws of black hole thermodynamics and the existence of quasilocal definitions for the energy and angular momentum. It is important to remark, already at this point, the striking interplay between the second and the third issues.

The quasilocality of isolated horizons reflects itself in the fact that they can be described by introducing an inner spacetime boundary and imposing boundary conditions on the gravitational field defined on it (either in a metric or a connection formulation). As we want to describe black holes in equilibrium, it is natural to look for particular boundary conditions compatible with a static horizon but allowing the geometry outside to be dynamical (admitting, for example, gravitational radiation). This will lead us to consider a sector of general relativity significantly larger than the one consisting of standard black hole solutions.

The sector of the gravitational phase space that we will be dealing with admits a Hamiltonian, hence, it is conceivable to quantize it to gain an understanding of quantum black holes. This is one of the advantages of working with isolated horizons and a very non-trivial fact because such a Hamiltonian formalism is not always available for interesting sectors of general relativity. The approach that we will follow is somehow reminiscent of the study of symmetry reductions of general relativity (mini and midisuperspaces). As the sector of the phase space of the reduced system is large enough — actually infinite dimensional — it seems reasonable to expect that the quantum model that we consider will provide a good physical approximation for the equilibrium phenomena that we want to discuss,[c] in particular the microscopic description of black hole entropy and the Bekenstein-Hawking area law.

We review next the construction of isolated horizons justifying, along the way, the conditions that have to be incorporated during the process. The main results regarding the geometry of isolated horizons can be found in [20]. We will be defining different types of null hypersurfaces until we arrive at the concept of isolated horizon. In the process we will introduce the notation that will be used in the following.

---

[b]The mathematical foundations of the subject were developed by Kupeli in Ref. [22].
[c]The quasilocal description of *dynamical* black hole behaviors can be achieved by using the so-called dynamical horizons [19].

Null hypersurfaces: Let $\mathcal{M}$ be a 4-dim manifold and $g_{\mu\nu}$ a Lorentzian metric on $\mathcal{M}$. A 3-dimensional embedded submanifold $\Delta \subset \mathcal{M}$ will be called a null hypersurface if the pull-back g_{ab}^Δ of $g_{\mu\nu}$ onto Δ is degenerate. This condition implies the existence of a null normal ℓ^a *tangent* to Δ. Notice that there is not a *unique* projection of tangent vectors X^μ sitting on $p \in \Delta$ onto the tangent space $T_p \Delta$ and, hence, it is impossible to define an induced connection on Δ.

Non-expanding null hypersurfaces: The degeneracy of the metric g_{ab}^Δ implies that there is not a *unique* inverse metric, however it is always possible to find g_Δ^{ab} such that $g_{ab}^\Delta = g_{aa'}^\Delta g_\Delta^{a'b'} g_{b'b}^\Delta$. If ℓ^a is a field tangent to Δ consisting of null normals, we define its *expansion* θ_ℓ associated with a particular choice of g_Δ^{ab} as $\theta_\ell := g_\Delta^{ab} \mathcal{L}_\ell g_{ab}^\Delta$. The invariance under rescalings implies that this expansion cannot be associated in an intrinsic way to the null hypersurface Δ unless it is zero. Null hypersurfaces with zero expansion in the previous sense will be referred to as *non-expanding*.

Non-expanding horizons (NEH): As we want to model black holes in four dimensions — for which the horizons have a simple geometry — we will require that: (i) Δ is diffeomorphic to $\mathbb{S}^2 \times (0,1)$ where $\mathbb{S}^2$ is a 2-sphere. (ii) For each $x \in \mathbb{S}^2$ this diffeomorphism maps $\{x\} \times (0,1)$ to null geodesics on Δ. (iii) For each $t \in (0,1)$, $\mathbb{S}^2 \times \{t\}$ is mapped onto a spacelike 2-surface in Δ. We impose now a key physical condition by requiring that the metric $g_{\mu\nu}$ be a solution to the Einstein field equations and demanding that the pull back of the stress-energy-momentum tensor $T_{\mu\nu}$ on Δ satisfies the condition $T_{ab}^\Delta \ell^a \ell^b \geq 0$. This is equivalent to the condition $R_{ab}^\Delta \ell^a \ell^b \geq 0$ on the pull-back of the Ricci tensor to Δ. The preceding conditions on non-expanding null surfaces define *non-expanding horizons*. An important feature of them is that, as a consequence of the non-expansion condition, cross-sections are marginally trapped surfaces and have constant area. Also, the Raychaudhuri equation together with the non-expansion condition implies that $R_{ab}^\Delta \ell^a \ell^b = T_{ab}^\Delta \ell^a \ell^b = 0$ and $\mathcal{L}_\ell g_{ab}^\Delta = 0$. This can be interpreted as the fact that non-expanding horizons *are in equilibrium*.

Weakly isolated horizons (WIH): In order to incorporate the laws of black hole mechanics to the present quasilocal framework we need to add additional structure to the preceding constructions. For example, the notion of temperature for ordinary black holes relies on the concept of surface gravity κ (see, for example, [124]). We can introduce now a rather similar concept by imposing additional requirements to NEH's. Given a non-expanding horizon, it can be shown [see [10, 17, 20, 98]] that the spacetime connection ∇ induces a unique connection $\mathcal{D}$ compatible with the induced metric g_{ab}^Δ. We also declare as equivalent all the normal null fields related by constant rescalings (we denote these equivalence classes as $[\ell]$). A weakly isolated horizon is now a pair $(\Delta, [\ell])$ consisting of a non-expanding horizon Δ and class of null normals $[\ell]$ such that

$$(\mathcal{L}_\ell \mathcal{D}_a - \mathcal{D}_a \mathcal{L}_\ell)\ell^b = 0. \tag{5}$$

Geometrically this requirement is equivalent to the condition that *some* components of the connection defined by $\mathcal{D}$ are left invariant by the diffeomorphisms defined by ℓ

on Δ (or roughly speaking are "time independent"). This means that $\nabla_\ell \ell = \kappa \ell$ with κ *constant* for each $\ell \in [\ell]$. It is important to mention here that different choices of $[\ell]$ lead to inequivalent weakly isolated structures on the same non-expanding horizon.

Isolated horizons (IH): Isolated horizons are weakly isolated horizons $(\Delta, [\ell])$ for which

$$(\mathcal{L}_\ell \mathcal{D}_a - \mathcal{D}_a \mathcal{L}_\ell)\tau^b = 0, \qquad (6)$$

for every tangent field τ^a on Δ. This condition can be read as $[\mathcal{L}, \mathcal{D}]|_\Delta = 0$. An important difference between WIH's and IH's is that, whereas a given non-expanding horizon admits infinitely many weakly isolated horizon structures, for an isolated horizon the only freedom in the choice of null normals consists of constant rescalings [20].

Non-trivial examples of all these types of horizons can be found in the extensive literature available on the subject (see [20] and references therein); in any case it is important to keep in mind that any Killing horizon diffeomorphic to $\mathbb{S}^2 \times \mathbb{R}$ is an isolated horizon so the concept is a genuine — and useful — generalization that encompasses all the globally stationary black holes.

Multipole moments can be used to define spherically symmetric isolated horizons in an intrinsic way [15, 16]. They are useful because, for stationary spacetimes in vacuum, they determine the near horizon geometry. Concrete expressions for these objects can be written in terms of the Ψ_2 Newmann-Penrose component of the Weyl tensor. In the spherically symmetric case $\mathrm{Im}\Psi_2 = 0$ and $\mathrm{Re}\Psi_2$ is constant which implies that the only non-zero multipole moment is M_0. This condition provides the intrinsic characterization mentioned above.

The zeroth and first laws of black hole mechanics have interesting generalizations for weakly isolated and isolated horizons. In the case of the zero law the geometric features of weakly isolated horizons guarantee that a suitable concept of surface gravity can be introduced. This is done as follows [20]. For a non-expanding horizon Δ the null normal ℓ^a has vanishing expansion, shear and twist. It is then straightforward to show that there must exist a 1-form ω_a on Δ such that $\nabla_a \ell^b = \omega_a \ell^b$ and $(\mathcal{L}_\ell \omega)_a = 0$ (the last condition as a consequence of the definition of weakly isolated horizon). Defining now the surface gravity associated with the null normal ℓ^a as $\kappa_\ell := \ell^a \omega_a$ we have $d\kappa_\ell = d(\omega_a \ell^a) = (\mathcal{L}_\ell \omega)_a = 0$ and hence κ_ℓ is constant on the horizon. This is analogous to the behavior of the surface gravity for Killing horizons and provides us with the sought for generalized zeroth law.

The generalization of the first law of black hole dynamics requires the definition of a suitable energy associated with the isolated horizon. A way to proceed is to look for a Hamiltonian description for the sector of general relativity containing IHs. The availability of such a formulation is a very non-trivial and remarkable fact, and it is a necessary first step towards quantization. In general covariant theories the Hamiltonian generating time translations is given by a surface integral (once the constraints are taken into account). In the present case there will be, hence, an energy associated with the isolated horizon (and an extra ADM term corresponding to the boundary at infinity).[d] In practice, associating a Hamiltonian to the boundary Δ requires the choice of an appropriate concept of time evolution defined by vector field t^a with appropriate values t^a_Δ at the horizon. In simple examples (for instance, non-rotating isolated horizons) it is natural to take t^a_Δ proportional to the null normal ℓ^a, however, there is some freedom left in the choice of ℓ^a by the IH boundary conditions. By choosing t^a_Δ in such a way that the surface gravity is a specific function of the area (and other charges) one finds that the first law is a necessary and sufficient condition for the evolution generated by t^a_Δ to be Hamiltonian [20]. In this way, there is a family of mathematically consistent first laws parametrized by these choices.

The textbook approach to obtain the Hamiltonian would consist in starting from a suitable action principle for general relativity in a spacetime manifold with an inner boundary where the isolated horizon boundary conditions are enforced. This action can be written in principle both in terms of connection or metric variables. The standard Dirac approach to deal with constrained systems (or more sophisticated formalisms such as the one given in [81]) can then be used to get the phase space of the model, the symplectic structure, the constraints and the Hamiltonian [43]. Notice that owing to the presence of boundaries one should expect, in principle, a non-zero Hamiltonian consisting both in horizon contributions (defining the horizon energy E_Δ in terms of which the first law is spelled) and the standard ADM energy associated with the boundary at infinity. A different approach that has some computational advantages relies on the covariant methods proposed and developed in [52, 53]. Their essence is to directly work in the space of solutions to the Einstein field equations with fields subject to the appropriate boundary conditions (in particular the isolated horizon ones). Despite the fact that the solutions to the field equations in most field theories are not known it is possible to obtain useful information about the space of solutions and, in particular, the symplectic form defined in it.

As this is a crucial ingredient to understand the quantization of the model and the quantum geometry of isolated horizons we sketch now the derivation of the symplectic structure based on covariant phase space methods. Let us suppose that we have a local coframe e^I_μ, $I = 1, \ldots, 4$ in the spacetime[e] $(\mathcal{M}, g_{\mu\nu})$ and the frame connection Γ^I_J defined by $de^I + \Gamma^I_J \wedge e^J_\nu = 0$ with $\Gamma_{IJ} + \Gamma_{JI} = 0$. If we denote tangent vectors (at a certain solution e^I) as δe^I it is straightforward to show that the 3-form defined on $\mathcal{M}$ by

$$\omega(\delta_1, \delta_2) := \frac{1}{2}\varepsilon_{IJKL}\delta_{[1}(e^I \wedge e^J) \wedge \delta_{2]}\Gamma^{KL} - \frac{1}{\gamma}\delta_{[1}(e^I \wedge e^J) \wedge \delta_{2]}\Gamma_{IJ} \qquad (7)$$

is closed if e^I is a solution to the Einstein field equations (in the previous expression γ is the Barbero-Immirzi parameter). This means that if we have two 3-surfaces Σ_1

---
[d]Similar arguments apply to the angular momentum.
[e]η_{IJ} denotes the Minkowski metric. In the following we will suppress spacetime indices when working with differential forms.

and Σ_2 defining the boundary of a 4-dim submanifold of $\mathcal{M}$ then

$$\Omega(\delta_1,\delta_2) = \int_\Sigma \omega(\delta_1,\delta_2) \tag{8}$$

is independent of Σ. If an inner boundary, such as an isolated horizon, is present then a similar argument leads to the obtention of the symplectic form. Indeed, let us take a region of $\mathcal{M}$ with an inner boundary Δ (a causal 3-surface) and a family of spatial 3-surfaces Σ such that every pair Σ_1 and Σ_2, defines a 4-dim spacetime region bounded by Σ_1, Σ_2 and the segment of the surface Δ contained between the 2-surfaces $\Sigma_1 \cap \Delta$ and $\Sigma_2 \cap \Delta$. Let us suppose also that, for every pair of tangent vectors δ_1, δ_2, there is a 2-form $\alpha(\delta_1,\delta_2)$ on Δ such that the pullback of ω onto Δ is exact $[\omega^\Delta(\delta_1,\delta_2) = d\alpha(\delta_1,\delta_2)]$. When these conditions are satisfied it is possible to generalize (8) in such a way that, in addition to the bulk term obtained above, it also has a surface term and the resulting expression is still independent of the choice of Σ

$$\Omega(\delta_1,\delta_2) = \int_\Sigma \omega(\delta_1,\delta_2) + \int_{\Sigma \cap \Delta} \alpha(\delta_1,\delta_2)\,. \tag{9}$$

These types of surface terms are defined both for weakly isolated horizons or spherical isolated horizons as inner boundaries. In the case of weakly isolated horizons of fixed area A it is possible to perform a gauge fixing such that the only symmetry left is a $U(1)$ symmetry. In such a situation it is possible to see that the surface contribution to (9) has the form

$$\frac{A}{\pi\gamma} \int_S \delta_{[1} V \wedge \delta_{2]} V \tag{10}$$

where V is a $U(1)$ connection on the spheres S that foliate the horizon. It is important to notice that this is a $U(1)$ Chern–Simons symplectic form. It is convenient now to rewrite the bulk term by using Ashtekar variables as

$$2 \int_\Sigma \delta_{[1} E^a_i \wedge \delta_{2]} A^i_a\,. \tag{11}$$

It is necessary to mention at this point [13] that the values of the $U(1)$ connection and the pullbacks of the connection/triad variables are not independent but are connected through a horizon constraint of the form[f]

$$(dV)_{ab} + \frac{2\pi\gamma}{A} \epsilon_{abc}(E^c_i r^i)\bigg|_\Delta = 0\,. \tag{12}$$

The quantum version of this condition plays a central role in the quantization of this model.

For spherical isolated horizons it is possible to define the Hamiltonian framework without gauge fixing on the horizon [58–60]. In such formulation the symplectic form in the field space has an $SU(2)$ Chern–Simons surface term of the form

$$\frac{A}{8\pi^2(1-\gamma^2)\gamma} \int_S \delta_1 A_i \wedge \delta_2 A_i\,, \tag{13}$$

---

[f] Here r^i denotes a fixed internal vector and we have used units such that $8\pi G = 1$.

where A_i denotes the pullback of the $SU(2)$ connection to the horizon. Now the horizon constraint is not a single condition but the three conditions

$$\frac{1}{2}\epsilon_{abc}E^{ci} + \frac{A}{8\pi^2(1-\gamma^2)\gamma}F^i_{ab}\bigg|_\Delta = 0, \qquad (14)$$

written in terms of the curvature F^i_{ab}. The difference between the $U(1)$ and the $SU(2)$ approaches stems, mainly, from this fact but it is important to mention that the physical assumptions used to define both models are slightly different.

A remark is in order when comparing equations (12) and (14). At first sight there seems to be a mismatch in the number of conditions. In fact, as a consequence of the gauge fixing that reduces the $SU(2)$ triad rotation in the bulk to $U(1)$ on the boundary, one has two extra conditions on the fluxes corresponding to

$$E^c_i y^i = 0 = E^c_i x^i \qquad (15)$$

where x^i and y^i are internal directions orthogonal to each other and to r^i. We see that the three conditions in (14) are recovered. Due to the non-commutativity of the E^c_i the previous conditions cannot be satisfied in the quantum theory: only (12) is imposed in the $U(1)$ framework. As a result the $U(1)$ framework slightly over counts states, a fact which (under qualifications that are discussed at the end of Section 3) is reflected in the form of logarithmic corrections to the micro canonical entropy (see table in Section 4).

3. Quantum Geometry of Weakly Isolated Horizons

The formulation put forward in the preceding section can be used to identify the degrees of freedom that account for the black hole entropy and understand their quantum origin. It is precisely the quantum geometry associated with weakly isolated horizons that will let us understand the origin of black hole entropy in the LQG framework. As we will discuss in this section a special role will be played by the quantum horizon boundary conditions. For simplicity of exposition we will restrict ourselves to the setting provided by Type I WIH's and suppose that we do not have matter nor extra charges. The starting point of the following construction is a WIH of fixed area[g] a. As we mentioned in the preceding section the sector of general relativity consisting of solutions to the Einstein field equations on regions bounded by weakly isolated horizons admits a Hamiltonian formulation so that its quantization can be considered in principle. It is important to point out, however, that the following construction must be based on the use of connection-triad variables of the Ashtekar type (*see Chapter 1*). To our knowledge such a construction is not available in the geometrodynamical framework [h]. One of the reasons for this is the central role that Chern-Simons theories play in the following arguments.

---

[g]And fixed charges, in general.

[h]As mentioned in Section 2 there is a boundary contribution to the gravity symplectic form that can be written as an $SU(2)$ Chern-Simons symplectic form in connection variables. In triad variables e^i_a this is [59] $\gamma^{-1}\kappa^{-1}\int \delta_1 e_i \wedge \delta_2 e^i$. If we define smeared fluxes in the usual way $E(S,\alpha) = \int_S \epsilon_{ink}\alpha^i e^j \wedge e^k$ then it follows that $\{E(S,\alpha), E(S',\beta)\} = \gamma\kappa E(S \cap S', [\alpha,\beta])$. The previous non-commutativity of fluxes is characteristic of the bulk holonomy flux algebra [14].

We have chosen to describe with some degree of detail the $U(1)$ gauge fixed formulation of quantum IHs. The $SU(2)$ invariant framework [58, 59] can be constructed along similar lines. In addition to the point discussed at the end of the previous section, the main advantage of the latter is that both boundary and bulk fields possess the same gauge symmetry. This allows the IH quantum constraints to be interpreted as first class constraints generating the common symmetry [59]. Results of the $SU(2)$ invariant formulation will be presented without details at the end of the section.

A very striking feature of the construction that we have discussed at the end of the preceding section is the presence of a surface term in the symplectic structure. Such surface terms are usually absent for field theories with boundaries (at least in simple models, see [26]). They are a very distinctive feature of the present approach. From the point of view of the quantization of the model this surface term strongly suggests the necessity to introduce a Hilbert space associated with the boundary. The fact that it corresponds to a Chern-Simons model directly leads to the consideration of a Chern-Simons quantization.

In statistical mechanics, the classical and quantum degrees of freedom that account for the entropy of a thermodynamical system are usually the same. For example the atoms in a gas, interpreted as point particles in a box, are in one to one correspondence with the quantum degrees of freedom used to model the gas as an ensemble of particles in an infinite potential well. In the present case the logical interpretation of the results about the specification of spacetimes with isolated horizons [99] implies that there are no classical degrees of freedom associated with them. What is then the origin of the entropy of black holes in this setting? The answer lies in the nature of equations (12) and (14). More precisely, the intersections of the edges of the spin network (excitations of the field E_i^a) used to represent a suitable quantum bulk state are treated as point particle defects at the horizon — effectively excising them. The degrees of freedom of the horizon Chern-Simons theory created in this way are responsible for the entropy.

The construction of the LQG Hilbert spaces has been summarized in *Chapter 1*. In the present context we will import results from these constructions — for the bulk degrees of freedom — and also from the quantization of Chern-Simons theories to deal with the horizon [7]. As mentioned before it is natural to introduce a Hilbert space $\mathcal{H} = \mathcal{H}_S \otimes \mathcal{H}_V$ where the Hilbert spaces $\mathcal{H}_{Hor}$ and $\mathcal{H}_{Bulk}$ are associated with the horizon and the bulk spacetime respectively.

The volume or bulk Hilbert space $\mathcal{H}_{Bulk}$ is a subspace of the usual LQG Hilbert space $L^2(\bar{\mathcal{A}}, \mu_{AL})$ defined in a suitable space of generalized connections with the help of the uniquely defined Ashtekar-Lewandowski measure (see *Chapter 1*). A useful orthonormal basis for this type of Hilbert space is provided by *spin networks* with edges that (may) transversally pierce the inner spacetime boundary that models the black hole. These points will be referred to as *punctures*; they are endowed with the quantum numbers that label the edges defining them. By using these punctures it is possible to represent the bulk Hilbert space as an orthogonal sum [55]

$$\mathcal{H}_{Bulk} = \bigoplus_{(P,j,m)} \mathcal{H}_{Bulk}^{P,j,m} \qquad (16)$$

extended to all the possible finite sets $P = \{P_1, \ldots, P_n\}$ consisting of points at the spherical sections of the horizon. The (j, m) labels correspond to edges piercing the horizon transversally and the empty set corresponds to spin networks that do not pierce the horizon.

In order to construct the surface Hilbert space it is necessary to excise the punctures from the sphere S at the horizon and study the quantization of a Chern-Simons model in the resulting punctured surface. From a classical point of view this modification of the horizon topology has the effect of introducing topological degrees of freedom in the model (that can be thought of as the holonomies around closed loops surrounding the punctures of the otherwise flat connection), however one has to keep in mind that these punctures are induced by spin network states defined in the bulk, hence, they have a quantum origin.

The Chern-Simons quantization requires us to impose a prequantization condition on the classical horizon area. In the present situation, it reads [7] $A_\kappa = 4\pi\gamma\ell_{Pl}^2\kappa$ with $\kappa \in \mathbb{N}$. In analogy with the bulk Hilbert space $\mathcal{H}_V$ the surface Hilbert space can be conveniently written as an orthogonal sum in the form

$$\mathcal{H}_{Hor} = \bigoplus_{(\vec{P},b)} \mathcal{H}_{Hor}^{\vec{P},b} \qquad (17)$$

where now $\vec{P}$ stands for an ordered n-tuple of points on the "horizon" S labeled by integers mod κ ($b_i \in \mathbb{Z}_\kappa$, $i = 1, \ldots, n$) satisfying the condition $b_1 + \cdots + b_n = 0$. Here $\mathcal{H}^\emptyset = \{\vec{0}\}$.

At this stage in the process both spaces are completely independent. The key element that establishes a relationship between them is the quantized version of the horizon boundary conditions that we have discussed at the end of the preceding Sections 12, 14. It is very important to highlight here the fact that the operators that appear in these quantized boundary conditions are defined in completely unrelated Hilbert spaces; hence, the fact that there exist solutions to these quantum boundary conditions is highly non-trivial. Of course, one has also to take into account the quantized constraints in the bulk Hilbert space by using the standard LQG methods (Dirac quantization, group averaging, etc., see *Chapters 1 and 2*). The implementation of the quantum boundary conditions leads to a subspace consisting of orthogonal sums of elements of the form

$$\mathcal{H}_{Bulk}^{P,j,m} \otimes \mathcal{H}^{\vec{P},b} \qquad (18)$$

such that the points in the set P coincide with those in the vector $\vec{P}$ and the b_i labels associated with the punctures satisfy the condition $b_i = -2m_i (mod\,\kappa)$ for $i = 1, \ldots, n$.

Up to this point the construction has given us some kind of kinematical Hilbert space adapted to the present situation where we have inner spacetime boundaries.[i] We still have to take into account the rest of the constraints in the model. This is done by following the standard procedure (*see Chapters 1 and 2*) and making some assumptions — presumed mild — regarding solutions to the Hamiltonian constraint [7].

One of the key insights in the development of the present framework was the introduction by Krasnov [95–97] of the *area ensemble*. In the absence of a suitable notion of energy such definition seemed natural: the area is an extensive quantity with a well understood discrete spectrum. This state of affairs has evolved due to results [65] that provide an interpretation of the horizon area as a quasilocal notion of energy. This will be discussed in the last two sections of this chapter. It is important to highlight, at this point, that area and angular momentum play a fundamental role already at the classical level in the IH framework whereas mass is a derived physical magnitude.

The customary way to define the entropy starts by considering the prequantized value of the area A_κ and introducing an area interval $[A_\kappa - \delta, A_\kappa + \delta]$ of width δ of the order of the Planck length.[j] Once this is done the entropy can be computed by tracing out the bulk degrees of freedom to define a density matrix describing a maximal mixture of states on the horizon surface S with area eigenvalues in the previous interval. In order to count the number of states in $[A_\kappa - \delta, A_\kappa + \delta]$ we have to find out how many lists of non-zero elements of $\mathbb{Z}_\kappa$ satisfy the condition $\sum_{i=1}^{n} b_\kappa = 0$ with $b_i = -2m_i (mod \, \kappa)$ for a permissible list of labels $m_1, \ldots$ By permissible we mean that there must exist a list of non-vanishing spin labels j_i such that each m_i is a spin component of j_i ($m_i \in \{-j_i, -j_i + 1 \ldots, j_i\}$) and the following inequality holds

$$A_\kappa - \delta \leq 8\pi\gamma\ell_{Pl}^2 \sum_{i=1}^{n} \sqrt{j_i(j_i + 1)} \leq A_\kappa + \delta. \qquad (19)$$

In principle the preceding discussion gives a concrete prescription that defines the counting (combinatorial) problem that has to be solved in order to compute the entropy for a given value of the prequantized area A_κ. This is generalized to arbitrary values of the area by allowing A_κ to be replaced by any arbitrary value A.

The preceding combinatorial problem can be considered as it is (and, in fact, when the flux operator is used it can be solved in a relatively straightforward way). However, there is a neat way to simplify it known as the Domagala-Lewandowski (DL) approach [55]. By carefully considering the details of the problem it is possible

---

[i] Notice, however, that the quantum boundary conditions, arising from consistency requirements for the Hamiltonian formulation of the sector of general relativity that we are considering here, can also be thought of as constraints and, from this perspective, what we have really done is to implement them *à la Dirac*.

[j] A different construction is possible if one uses the so-called flux operator [25] to define the entropy. In this case there is no need to introduce an area interval to solve the quantum matching conditions though, on physical grounds, it is useful to introduce it afterwards.

to pose it in such a way that only one type of labels appear (instead of the three labels in the original formulation, viz. j_i, m_i, b_i). In the new rephrasing the entropy is computed as $\log n(A)$ where $n(A)$ is 1 plus the number of finite sequences of non-zero integers or half-integers satisfying the following two conditions

$$\sum_{i=1}^{n} \sqrt{|m_i|(|m_i|+1)} \leq \frac{A}{8\pi\gamma\ell_{Pl}^2}, \qquad (20)$$

and the so-called projection constraint

$$\sum_{i=1}^{n} m_i = 0. \qquad (21)$$

A different approach corresponds to the models described by Ghosh and Mitra (GM) [71, 73] leading to the definition of the entropy as $\log n(A)$ where $n(A)$ is 1 plus the number of all finite, arbitrarily long sequences $((j_1, m_1), \ldots, (j_N, m_N))$ of ordered pairs of non-zero, positive half integers j_i and spin components $m_i \in \{-j_i, -j_i+1, \ldots, j_i\}$ satisfying

$$\sum_{i=1}^{n} \sqrt{j_i(j_i+1)} \leq \frac{A}{8\pi\gamma\ell_{Pl}^2}, \qquad \sum_{i=1}^{n} m_i = 0. \qquad (22)$$

The difference between the DL and the GM definitions of the counting problem resides in the following technical point. As two punctures with different spins $j \neq j'$ but with the same magnetic number m are, from the boundary $U(1)$ Chern-Simons theory, indistinguishable, they are considered as physically equivalent in the DL prescription. In the GM prescription the previous two configurations are considered as different and counted individually. This apparent ambiguity of prescriptions disappears in the $SU(2)$ invariant formulation where, roughly speaking, the states of the Chern-Simons boundary connection depend both on j and m. To leading order the counting in the $SU(2)$ invariant formulation agrees with the GM prescription (see table in Section 4).

Up to this point we have described the $U(1)$ framework, which among other things, involves the quantization of condition (12) and its variants. Let us now briefly present the $SU(2)$ framework following from the quantization of the system containing (14). The first models using $SU(2)$ Chern-Simons theory were proposed by Kaul and Majumdar [93]. The complete $SU(2)$ framework, including the classical description of the theory, was proposed by Engle, Noui and Perez in [58–60]. The entropy in this case is computed as $\log n(A)$ where $n(A)$ is 1 plus the number of all finite, arbitrarily long sequences $(j_1, \ldots, j_N)$ of non-zero, positive half-integers j_i satisfying the inequality

$$\sum_{i=1}^{n} \sqrt{j_i(j_i+1)} \leq \frac{A}{8\pi\gamma\ell_{Pl}^2}, \qquad (23)$$

and counted with multiplicity given by the dimension of the invariant subspace $\text{Inv} \otimes_i [j_i]$.

The next section will be devoted to introducing efficient methods to solve the different types of combinatorial problems involved in the computation of the entropy in the different proposals. These methods are based in number-theoretic ideas and provide a powerful setup to deal with the broad class of problems arising in the study of black holes in LQG.

4. Counting and Number Theory

The different models for semiclassical black holes described in the preceding section provide concrete examples of the kind of counting problems that have to be solved in order to compute the black hole entropy as a function of the area (and other physical features such as angular momentum). They are remarkable for several reasons. First, they are relatively easy to state and, in fact, reduce to the counting of specific types of finite sequences of integers or half integers subject to simple conditions. Furthermore their resolution can be tackled by using methods that combine known types of Diophantine equations, the use of generating functions and Laplace transforms.

As we will show in the following all the black hole models that have been discussed so far in the LQG framework lead to the Bekenstein-Hawking law. Some of them, in particular the older ones [6, 74, 93], require the fine tuning of the Immirzi parameter to get the correct proportionality factor between area and entropy; others give results consistent with the first law and equation (4) without any fine tuning [75, 76] (see Section 5). It is important to point out at this point that the fact that the entropy grows linearly in the asymptotic limit of large areas is not a *generic* behavior. At first sight the situation seems to be quite similar to that of a sufficiently regular real function $f(A)$ satisfying $f(0) = 0$ and $f'(0) \neq 0$ for which Taylor's theorem implies that in the $A \to 0$ asymptotic limit $f(A) \propto A$. However the limit that we are considering is $A \to \infty$ and the function of interest (the entropy) is not analytic but, actually, has a staircase form (though it can be written in terms of non-trivial integral expressions). In such circumstances the linear asymptotic behavior for large areas is certainly significant and becomes a genuine nontrivial prediction of the model.

We want to make some additional comments regarding the counting entropy before describing in some detail the mathematical methods necessary to efficiently solve the combinatorial problems involved in its computation. The first has to do with its behavior for *small areas* that was considered in detail by Corichi, Diaz Polo and Fernandez Borja [50, 51]. Quite unexpectedly one finds a regular step structure that persists for a reasonably wide interval—microscopic in any case—of areas (a detailed account of the mathematical reasons for this phenomenon can be found in [27]). This is mildly reminiscent of the predictions by Bekenstein, Mukhanov and others [34, 35] regarding a "quantized" area spectrum. In the face of it this does not seem to be utterly unexpected because the area operator, with its discrete spectrum, plays a central role in the formalism. However, the eigenvalues of the area are not equally spaced and their density (as a function of the area) grows very fast whereas the width of the steps seen in the entropy is both exact and persistent (although they eventually disappear).

A second relevant comment has to do with the rigorous notion of thermodynamic limit [82]. This has important implications for the mathematical properties of the entropy as a function of its natural variables (the energy in the case of statistical mechanics). In this limit (that can be computed by working with the counting entropy that we are considering here) the entropy is smooth almost everywhere — which implies that standard thermodynamical formulae can be used — and is *concave* (downwards). A consequence of this last fact is that the step structure for small areas should not be directly observable (although it can possibly have some kind of impact on its properties). Another important consequence of this is the change in the predictions for the subdominant corrections to the entropy for large areas (that actually disappear for some models [58–60]).

The general structure of the combinatorial problems that have to be solved is the following. In all the cases one must count the number of finite, arbitrarily long, sequences of non-zero half-integers satisfying an inequality condition involving the horizon area. These numbers are associated with spin network edges that pierce the horizon and quantum numbers coming from the Chern-Simons sector at the horizon. In the case of the original $U(1)$ proposal of Ashtekar, Baez, Corichi and Krasnov [6] the associated combinatorial problem was rephrased in a convenient simplified way [55, 102] that did not involve directly the spin labels j_i associated with the punctures at the horizon but, rather, the magnetic quantum numbers m_i (satisfying the condition $-j_i < m_i \leq j_i$). For a given value of the horizon area these numbers have to satisfy the inequality

$$\sum_{i=1}^{N} \sqrt{|m_i|(|m_i|+1)} \leq \frac{A}{8\pi\gamma\ell_{Pl}^2}, \qquad (24)$$

and the *projection constraint*

$$\sum_{i=1}^{N} m_i = 0, \qquad (25)$$

In other proposals, such as the GM prescription [74], the combinatorial problem is expressed in terms of both the spin labels j_i of the edges that pierce the horizon and the m_i labels. There is an inequality (similar to (24)) and a projection constraint with the same form as before

$$\sum_{i=1}^{N} \sqrt{j_i(j_i+1)} \leq \frac{A_k}{8\pi\gamma\ell_{Pl}^2}, \quad \sum_{i=1}^{N} m_i = 0. \qquad (26)$$

Notice how these two counting problems are different: in the first one both conditions involve the m_i labels whereas in the second the j_i and m_i labels are quite independent (though they must satisfy the restriction $-j_i \leq m_i \leq j_i$). In the $SU(2)$

models [58–60] the projection constraint is replaced by a condition involving the dimension of the invariant subspace. Lack of space precludes us from delving into the details of all the different cases and proposals so we will describe only the DL approach in some detail and refer the reader to the literature for the rest. In the rest of this section we will use units such that $4\pi\gamma\ell_{Pl}^2 = 1$.

In order to count the number of sequences as required by the previous prescription it is convenient to adopt a stepwise approach. This has been explained in detail elsewhere [2, 3] so we give here a summary of the procedure. The main steps are:

1. For each fixed value of the area a obtain all the possible choices for the positive half-integers $|m_i|$ *compatible* with it in the sense that they satisfy

$$\sum_{i=1}^{N} \sqrt{|m_i|(|m_i|+1)} = \frac{A}{2}. \quad (27)$$

At this stage the numbers $|m_i|$ can repeat themselves and are not ordered. In other words, in this first step we just want to find out how many times each spin component appears (how many 1/2's, how many 1's, and so on).

2. Count the different ways in which the multiset just described can be reordered.
3. Count all the different ways of introducing signs in the sequences of the previous step in such a way that the condition $\sum_{i=1}^{N} m_i = 0$ is satisfied.
4. Repeat this procedure for all the eigenvalues of the area operator smaller than A and add up the number of sequences thus obtained.

The first step is a characterization of the part of the spectrum of the area operator relevant to the computation of black hole entropy, in particular the degeneracy of the area eigenvalues. The condition (27) can be rewritten as

$$\sum_{k=1}^{k_{\max}} N_k \sqrt{(k+1)^2 - 1} = A \quad (28)$$

where we have introduced integer labels $k_i := 2|m_i|$. The non-negative integers N_k (that will be allowed to be zero) in the last sum tell us the number of times that the label $k/2 \in \mathbb{N}/2$ appears in the sequence. We also denote as $k_{\max} = k_{\max}(A)$ the maximum value of the positive integer k compatible with the given area A. The problem that we need to solve at this step can be rephrased as that of finding all the sets of pairs $\{(k, N_k) : k \in \mathbb{N}, N_k \in \mathbb{N} \cup \{0\}\}$ satisfying (28). It is important to notice now that (28) implies that the area eigenvalue a must be an integer linear combination of square roots of squarefree numbers of the form $A = \sum_{i=1}^{i_{\max}} q_i \sqrt{p_i}$, $q_i \in \mathbb{N} \cup \{0\}$, so that we have the condition

$$\sum_{k=1}^{k_{\max}} N_k \sqrt{(k+1)^2 - 1} = \sum_{i=1}^{i_{\max}} q_i \sqrt{p_i}. \quad (29)$$

where the right-hand side is fixed from the initial choice of area eigenvalue A. The resolution of the previous equation is quite direct although the procedure, that involves the solution of the quadratic Diophantine equation known as the Pell equation and an auxiliary set of linear Diophantine equations, is somewhat lengthy. The interested reader is referred to [2, 3] for details. The final result of the analysis sketched at this step is a characterization of the number of times that each spin label corresponding to a puncture can appear for a given area eigenvalue.

The second step simply requires us to count the number of ordered sequences containing the number of each label obtained in the previous step and is completely straightforward. Once we have found all the possible sequences of positive half-integers $|m_i|$ satisfying condition (27), the third step asks for the computation of the number of ways to introduce *signs* in each m_i in such a way that the condition $\sum_{i=1}^{N} m_i = 0$ is satisfied. There are several ways to solve this problem as described in [3]. The simplest one makes use of generating functions and is actually the preferred one as generating functions play a fundamental role in this framework (as first explored and explained by Sahlmann [112, 113]). The other methods are interesting because they suggest deep connections between the ideas presented here with other physical problems, in particular those involving conformal field theories [4].

The last step requires us to add up the number of configurations corresponding to all the area eigenvalues smaller than or equal to A. The best way to do this makes again use of generating functions [2, 112, 113] and Laplace transforms (see references [2, 102]). Generating functions are a very powerful tool in combinatorics because they can encode a lot of useful information about a particular problem and can be manipulated with very simple analytical tools. In the present case the step by step procedure described above leads to concrete forms for the generating functions for all the problems described before and others considered in the literature [58–60]. In the specific example of the DL counting the generating function is [28]

$$G^{\mathrm{DL}}(z, x_1, x_2, \ldots) = \left(1 - \sum_{i=1}^{\infty} \sum_{\alpha=1}^{\infty} \left(z^{k_\alpha^i} + z^{-k_\alpha^i}\right) x_i^{y_\alpha^i}\right)^{-1}. \tag{30}$$

Here the pairs (k_α^i, y_α^i) are solutions to the Pell equation defined by the ith square free integer. The coefficients $[z^0][x_1^{q_1} x_2^{q_2} \cdots] G^{\mathrm{DL}}(z, x_1, x_2, \ldots)$ contain the information on the number of configurations compatible with a certain value of the area $\sum q_i \sqrt{p_i}$. Once the generating function is at hand it is possible to use it to get a very useful integral representation [29, 102] that takes the form of an double inverse Laplace-Fourier transform.

The usefulness of Laplace transforms to deal with counting problems in this setting was pointed out by Meissner in reference [102]. In addition to providing a way to effectively deal with step 4 in our scheme it is important also from the point of view of statistical mechanics and has been used to gain some understanding about the thermodynamic limit for black holes [30]. The underlying reason is the fact that the passage from the microcanonical to the canonical ensembles can be understood *precisely* in terms of Laplace transforms. This way we get the following expression for the entropy

$$\exp S(A) = \qquad (31)$$
$$\frac{1}{(2\pi)^2 i} \int_0^{2\pi} \int_{x_0-i\infty}^{x_0+i\infty} s^{-1}\left(1 - 2\sum_{k=1}^{\infty} e^{-s\sqrt{k(k+2)}}\cos\omega k\right)^{-1} e^{As}\, ds\, d\omega\,,$$

where x_0 is a real number larger than the real part of all the singularities of the integrand.[k] The treatment of other models such as the ones proposed by Ghosh, Mitra (GM), and Engle, Noui, Perez (ENP) basically differ only on the treatment of the projection constraint. Relevant details can be found in reference [3].

In addition to the cases mentioned above it is sometimes useful to consider the simplified model in which the projection constraint is ignored. Physically this corresponds to a situation in which the entropy satisfies the Bekenstein-Hawking law with no logarithmic corrections. In this simplified example the generating function is just

$$G_{DL(0)}(x_1, x_2, \dots) = \left(1 - 2\sum_{i=1}^{\infty}\sum_{\alpha=1}^{\infty} x_i^{y_\alpha^i}\right)^{-1}. \qquad (32)$$

leading to the following expression for the entropy

$$\exp S(A) = \frac{1}{2\pi i}\int_{x_0-i\infty}^{x_0+i\infty} s^{-1}\left(1 - 2\sum_{k=1}^{\infty} e^{-s\sqrt{k(k+2)}}\right)^{-1} e^{As}\, ds\,. \qquad (33)$$

Let us now briefly explain how the asymptotic behavior of the entropy is obtained. To this end one should remember that whenever a function is represented as an inverse Laplace transform (a so-called *Bromwich integral* such as (33)) its asymptotic behavior as a function of the independent variable (the area A in this case) is determined by the analytic structure of the integrand, specifically the position of the singularity s_0 with the largest real part. In the present case, after reintroducing units for the sake of the argument, we get

$$S(A) = S_0 + \frac{\mathrm{Re}(s_0)}{\pi\gamma}\frac{A}{4\ell_{Pl}^2}\,. \qquad (34)$$

where S_0 is a constant independent of the area A. The preceding expression tells us that we exactly recover the Bekenstein-Hawking law by choosing γ such that

$$\gamma = \frac{\mathrm{Re}(s_0)}{\pi}\,. \qquad (35)$$

The analytic structure of the integrand in (33) has some very interesting features such as the accumulation of the real parts of its singularities [29] that reflect themselves in the behavior of the entropy. The expression (33) is very useful to explore

---

[k]This expression is actually valid only for those values of the area $a \geq 0$ *that do not belong to the spectrum of area operator* whereas for a_n in the spectrum of the area operator it gives the arithmetic mean of the left and right limits when $a \to a_n^{\pm}$. In practice the integral representation contains all the information about the entropy in a useful form.

these issues as only the complex variable s is relevant (the discussion when the projection constraint is also taken into account is slightly more involved).

In the present example the entropy $S(A)$ displays a simple linear growth for large values of the area (without any logarithmic corrections) with a slope that depends on the Immirzi parameter γ. An interesting fact is that the value of the parameter γ is the same for a number of different types of black holes although different proposals such as [6, 58–60, 73] lead to different values for it. The larger ones correspond to those cases where the number of microstates is larger as can be easily deduced from (34). At variance with this behavior it is interesting to mention that the subdominant logarithmic corrections are independent of γ.

The values of the Immirzi parameter leading to the Bekenstein-Hawking law and the logarithmic corrections for the different models and proposals are the following:

Approach T	γ T	Log correction T	Log corr. therm. limit
DL(0)	$\gamma_{DL} = 0.237 \cdots$	0	$\log(A/\ell_{Pl}^2)$
DL	$\gamma_{DL} = 0.237 \cdots$	$-\frac{1}{2}\log(A/\ell_{Pl}^2)$	$\frac{1}{2}\log(A/\ell_{Pl}^2)$
GM	$\gamma_{GM} = 0.274 \cdots$	$-\frac{1}{2}\log(A/\ell_{Pl}^2)$	exercise
ENP	$\gamma_{ENP} = \gamma_{GM}$	$-\frac{3}{2}\log(A/\ell_{Pl}^2)$	0

The first column refers to the four different models considered (the one provided by the DL prescription without and with the projection constraint, the GM approach and the ENP model). The difference between the results for the DL prescription with and without the projection constraint is the presence of a negative logarithmic correction for the latter. This is to be expected as the incorporation of the projection constraints eliminates some microstates that are taken into account in the DL(0) model. A similar argument applies to the GM and ENP cases; the $-3/2$ coefficient for the logarithmic correction in the latter case means that the number of microstates allowed is smaller than for the GM proposal. See the discussion at the end of Section 3.

5. Semiclassical Advances

The indeterminacy, mentioned in Section 2, of the quantities appearing in the first law for IHs disappears if one changes the point of view and assumes that the near horizon geometry corresponds to that of a stationary black hole solution and shifts the perspective to that of a suitable family of stationary nearby local observers. As explained in Section 2 the whole idea behind isolated horizons is to describe a sector of the phase space of gravity containing a boundary with the geometric properties of a BH horizon in equilibrium and infinitely many bulk degrees of freedom. In such context no condition in the definition requires the near-horizon geometry to be that of a stationary black hole. A key point is that the systems that behave thermodynamically are those solutions in the phase space of IH whose *near horizon geometry* (NHG) is that of a stationary black hole solution [64, 65, 75, 76]. In the quantum theory this would amount to selecting a bulk quantum state that is semiclassical and peaked on the stationary black hole configuration near the isolated horizon.

At present there is not enough control on the nature of the physical Hilbert space to be able to describe such states in detail (current status is reported in *Chapter 2*). Nevertheless, one can assume that such states exist and bring in their semiclassical properties into the analysis. This *semiclassical input* has led to interesting new insights into the black hole entropy calculation that we will briefly review here. This perspective opens a variety of new questions and tensions waiting to be resolved. We shall discuss them in the following section.

Assume that the NHG to be isometric to that of a Kerr-Newman BHs.[1] A family of stationary observers $\mathscr{O}$ located right outside the horizon at a small proper distance $\ell \ll \sqrt{A}$ is defined by those following the integral curves of the Killing vector field

$$\chi = \xi + \Omega \psi = \partial_t + \Omega \partial_\phi, \tag{36}$$

where ξ and ψ are the Killing fields associated with the stationarity and axisymmetry of Kerr-Newman spacetime respectively, while Ω is the horizon angular velocity. The four-velocity of $\mathscr{O}$ is given by

$$u^a = \frac{\chi^a}{\|\chi\|}. \tag{37}$$

It follows from this that $\mathscr{O}$ are uniformly accelerated with an acceleration $a = \ell^{-1} + o(\ell)$ in the normal direction. These observers are the unique stationary ones that coincide with the *locally non-rotating observers* [124] or ZAMOs [121] as $\ell \to 0$. As a result, their angular momentum is not exactly zero, but $o(\ell)$. Thus $\mathscr{O}$ are at rest with respect to the horizon which makes them the preferred observers for studying thermodynamical issues from a local perspective.

It is possible to show that the usual first law (2) translates into a much simpler relation among quasilocal physical quantities associated with $\mathscr{O}$ [65]. As long as the spacetime geometry is well approximated by the Kerr Newman BH geometry in the local outer region between the BH horizon and the world-sheet of local observers at proper distance ℓ, and, in the leading order approximation for $\ell/\sqrt{A} \ll 1$, the following local first law holds

$$\delta E = \frac{\bar{\kappa}}{8\pi} \delta A, \tag{38}$$

where $\delta E = \int_W T_{\mu\nu} u^\mu dW^\nu = \|\chi\|^{-1} \int_W T_{\mu\nu} \chi^\mu dW^\nu$ represents the flow of energy across the world-sheet W defined by the local observers, and $\bar{\kappa} \equiv \kappa/(\|\chi\|)$. The above result follows from the conservation law $\nabla^a(T_{ab}\chi^b) = 0$ that allows one to write δE as the flux of $T_{ab}\chi^b$ across the horizon. This, in turn, can be related to changes in its area using the optical Raychaudhuri equations [65].

---
[1]Such assumption is physically reasonable due to the implications of the no-hair theorem.

Two important remarks are in order: First, there is no need to normalize the Killing generator χ in any particular way. The calculation leading to (38) is invariant under the rescaling $\chi \to \alpha\chi$ for α a non-vanishing constant. This means that the argument is truly local and should be valid for more general black holes with a Killing horizon that are not necessarily asymptotically flat. This rescaling invariance of the Killing generator corresponds precisely to the similar arbitrariness of the generators of IHs as described in Section 2. The fact that equation (38) does not depend on this ambiguity implies that the local first law makes sense in the context of the IH phase space as long as one applies it to those solutions that are isometric to stationary black hole solutions in the thin layer of width ℓ outside the horizon. The semiclassical input is fully compatible with the notion of IHs.

Second, the local surface gravity $\bar{\kappa}$ is universal $\bar{\kappa} = \ell^{-1}$ in its leading order behavior for $\ell/\sqrt{A} \ll 1$. This is not surprising and simply reflects the fact that in the limit $\sqrt{A} \to \infty$ with ℓ held fixed the NHG in the thin layer outside the horizon becomes isometric to the corresponding thin slab of Minkowski spacetime outside a Rindler horizon: the quantity $\bar{\kappa}$ is the acceleration of the stationary observers in this regime. Therefore, the local surface gravity loses all memory of the macroscopic parameters that define the stationary black hole (see Section 5.2.1 for further discussion). This implies that, up to a constant which one sets to zero, equation (38) can be integrated, thus providing an effective notion of horizon energy

$$E = \frac{A}{8\pi G_N \ell}, \tag{39}$$

where G_N is Newton's constant. Such energy notion is precisely the one to be used in statistical mechanical considerations by local observers. Similar energy formulae have been obtained in the Hamiltonian formulation of general relativity with boundary conditions imposing the presence of a stationary bifurcate horizon [45]. The area as the macroscopic variable defining the ensemble has been always used in the context of BH models in loop quantum gravity. The new aspect revealed by the previous equation is its physical interpretation as energy for the local observers.

The thermodynamical properties of quantum IHs satisfying the NHG condition can be described using standard statistical mechanical methods with the effective Hamiltonian that follows from equation (39) and the LQG area spectrum (*see Chapter 1*), namely

$$\widehat{H}|j_1, j_2 \cdots\rangle = \left(\gamma \frac{\ell_{Pl}^2}{2G_N \ell} \sum_p \sqrt{j_p(j_p+1)}\right) |j_1, j_2 \cdots\rangle \tag{40}$$

where j_p are positive half-integer spins of the p-th puncture and $\ell_{Pl} = \sqrt{G\hbar}$ is the fundamental Planck length associated with the gravitational coupling G in the deep Planckian regime. The analysis that follows can be performed in both the microcanonical ensemble or in the canonical ensemble; ensemble equivalence is granted in this case because the system is simply given by a set of non-interacting units with discrete energy levels.

5.1. Pure Quantum Geometry Calculation

In this section we compute black hole entropy first in the microcanonical ensemble following a simplified (physicist) version [72] of the rigorous detailed counting of the previous section. As the canonical ensemble becomes available with the notion of Hamiltonian (40), we will also derive the results in the canonical ensemble framework. The treatment in terms of the grand canonical ensemble as well as the equivalence of the three ensembles has been shown [76].

Denote by s_j the number of punctures of the horizon labelled by the spin j. Ignoring the closure constraint, and in the $SU(2)$ Chern-Simons formulation of quantum IHs, the number of states associated with a distribution of distinguishable punctures $\{s_j\}_{j=\frac{1}{2}}^{\infty}$ is

$$n(\{s_j\}) = \prod_{j=\frac{1}{2}}^{\infty} \frac{N!}{s_j!} (2j+1)^{s_j}, \tag{41}$$

where $N \equiv \sum_j s_j$ is the total number of punctures. The leading term of the microcanonical entropy can be associated with $S = \log(n(\{\bar{s}_j\}))$, where $\bar{s}_j$ are the solutions of the variational condition

$$\delta \log(n(\{\bar{s}_j\})) + 2\pi\gamma_0 \delta C_1(\{\bar{s}_j\}) + \sigma C_2(\{\bar{s}_j\}) = 0 \tag{42}$$

where $2\pi\gamma_0$ (the 2π factor is introduced for later convenience) and σ are Lagrange multipliers for the constraints

$$C_1(\{\bar{s}_j\}) = \sum_j \sqrt{j(j+1)} s_j - \frac{A}{8\pi\gamma\ell_{Pl}^2} = 0,$$

$$C_2(\{\bar{s}_j\}) = \sum_j s_j - N = 0. \tag{43}$$

In words, $\bar{s}_j$ is the configuration maximizing $\log(n(\{s_j\}))$ for fixed macroscopic area A and number of punctures N. Notice that C_1 was not imposed in the treatment of Section 4.[m] Ignoring C_1 amounts to setting the punctures chemical potential $\bar{\mu} = 0$. However, as we will show here, allowing for non-vanishing chemical potential provides a whole new look at the question of the dependence of entropy on the Barbero-Immirzi parameter.

A simple calculation shows that the solution to the variational problem (42) is

$$\frac{\bar{s}_j}{N} = (2j+1) \exp(-2\pi\gamma_0 \sqrt{j(j+1)} - \sigma), \tag{44}$$

from which it follows, by summing over j, that the Lagrange multipliers are not independent

$$\exp\sigma(\gamma_0) = \sum_j (2j+1) \exp(-2\pi\gamma_0 \sqrt{j(j+1)}). \tag{45}$$

---
[m]The physicist method of this section can be made precise using the counting techniques of Section 4. The counting with fixed N is proposed as an exercise to the reader who is referred to [61] for relevant equations.

It also follows from (44), and the evaluation of $S = \log(n(\{\bar{s}_j\}))$, that

$$S = \frac{\gamma_0}{\gamma} \frac{A}{4\ell_{Pl}^2} + \sigma(\gamma_0) N. \qquad (46)$$

What is the value of the Lagrange multiplier γ_0? As in standard thermal systems the value of γ_0 is related to the temperature of the system. Its value is fixed by the requirement that

$$\left.\frac{\partial S}{\partial E}\right|_N^{-1} = T = \frac{\hbar}{2\pi\ell}, \qquad (47)$$

where E is the energy measured by quasilocal observers (39) and the last equality on the right is the condition that the temperature be the Unruh temperature (as measured by the same semiclassical observers). The previous condition allows one to express the Lagrange multiplier γ_0 in terms of the (otherwise arbitrary) Barbero-Immirzi parameter γ, G, and G_N, namely

$$\gamma_0 = \gamma \frac{G}{G_N}, \qquad (48)$$

and thus

$$S = \frac{A}{4\hbar G_N} + \sigma(\gamma) N. \qquad (49)$$

where

$$\sigma(\gamma) = \log\left[\sum_j (2j+1) e^{-2\pi\gamma \frac{G}{G_N}}\right]$$

Notice that the first term in the entropy formula is given by the Bekenstein-Hawking area law with the low energy value of Newton constant G_N; in other words it does not depend explicitly on the fundamental Planck length ℓ_{Pl} appearing in the area spectrum. Even though this is to be expected as the Bekenstein-Hawking term is a semiclassical quantity, the above result sheds new light on a long standing discussion in the community as to which is the value of Newton's constant that should go into the area spectrum. Due to quantum effects Newton's constant is expected to flow from the IR regime to the deep Planckian one. The Planckian value of the gravitational coupling should be defined in terms of the fundamental quantum of area predicted by LQG yet the low energy value should appear in the entropy formula. The semiclassical input that enters the derivation of the entropy through the assumption of (39) is the ingredient that bridges the two regimes.

Finally, punctures are associated with a chemical potential which is given by

$$\bar{\mu} = -T \left.\frac{\partial S}{\partial N}\right|_E = -\frac{\ell_{Pl}^2}{2\pi\ell} \sigma(\gamma) \qquad (50)$$

which depends on the fiducial length scale ℓ and the Barbero-Immirzi parameter, and where one is again evaluating the equation at the Unruh temperature $T = \hbar/(2\pi\ell)$.

The above derivation can be done in the framework of the canonical and grand canonical ensembles. From the technical perspective it would have been simpler to do it using one of those ensembles. In particular basic formulae allow for the calculation of the energy fluctuations which at the Unruh temperature are such that $(\Delta E)^2/\langle E \rangle^2 = \mathcal{O}(1/N)$. The specific heat at T_U is $C = N\gamma_0^2 d^2\sigma/d\gamma^2$ which is positive. This implies that as a thermodynamic system the IH is locally stable. The specific heat tends to zero in the large γ limit for fixed N and diverges as $\hbar \to 0$. The three ensembles give equivalent results [76].

5.1.1. The thermodynamical vs. the geometric first law

By simply computing the total differential of the entropy (49) one finds the thermodynamical first law

$$\delta E = \frac{\bar{\kappa}\hbar}{2\pi}\delta S + \bar{\mu}\delta N \tag{51}$$

In order to find a relationship with the geometric first law (2), one needs to assume that the spacetime geometry corresponds to that of a stationary black hole (for which (2) applies). If one does so then one can show (by simply reverting the argument that took one from (2) to (38) [65]) that (51) is equivalent to

$$\delta M = \frac{\kappa\hbar}{2\pi}\delta S + \Omega\delta J + \Phi\delta Q + \mu\delta N, \tag{52}$$

where $\mu = -\ell_{Pl}^2 \kappa \sigma(\gamma)/(2\pi)$ (the redshifted version of $\bar{\mu}$). At first sight the previous equation does not look like (2). However, it is immediate to check that the exotic chemical potential term in (52) cancels the term proportional to the number of punctures in the entropy formula (49). For (51) this is due to the equation of state (50); for (52) this is due to the form of μ. Therefore, the above balance equation is just exactly the same as (2). The different versions of the first law are presented in Table 1. Notice that only those on the left column are to be interpreted thermodynamically. Assuming the validity of semiclassical consistency discussed here for general accelerated observers in arbitrary local neighborhoods [90], the emergence of general relativity directly from the statistical mechanics of the polymer like structures of LQG has been argued [116].

5.1.2. Recovering the results of Section 3

As we mentioned above the key difference with the counting of Section 3 is the imposition of the constraint C_2 in (43). One can therefore recover the results by simply setting the Lagrange multiplier $\sigma = 0$ from the onset of the calculation in Section 5.1. What happens then is that equation (45) completely fixes γ_0 to the numerical value: in the present case $\gamma_0 = 0.274...$. Equation (48) — which continues to hold — introduces a strong constraint between fundamental constants; namely

$$\gamma\frac{G}{G_N} = \gamma_0 = 0.274..., \tag{53}$$

Table 1. Different versions of balance equations. On the left column one has the results coming from quantum geometry involving a chemical potential term. The semiclassical input of the area effective Hamiltonian in the quantum geometry statistical mechanics calculation leads to results that are consistent with the geometry first laws shown on the right column.

	Quantum Statistical Mechanics		Classical Einstein gravity
Local	$\delta E = \frac{\bar{\kappa}\hbar}{2\pi}\delta S + \bar{\mu}\delta N$	$\Longleftrightarrow$	$\delta E = \frac{\bar{\kappa}}{8\pi}\delta A$
	$\Updownarrow$		$\Updownarrow$
Global	$\delta M = \frac{\kappa\hbar}{2\pi}\delta S + \Omega\delta J + \Phi\delta Q + \mu\delta N$	$\Longleftrightarrow$	$\delta M = \frac{\kappa\hbar}{2\pi}\delta A + \Omega\delta J + \Phi\delta Q$

[a]Moving along horizontally in this table is a trivial identity; moving vertically requires the background geometry to be a stationary black hole solution.

which corresponds to equation (35) with the identification $\gamma_0 = \text{Re}(s_0)/\pi$. The previous equation implies that $S = A/(4G_N\hbar)$. Therefore, by declaring that the chemical potential of punctures vanishes $\bar{\mu} = 0$ (equivalently $\sigma = 0$) the semiclassical consistency, equation (47), is satisfied at the price of restricting the fundamental constants as above. It has been proposed that the previous equation, relating low energy G_N with the fundamental couplings G and γ, could be interpreted in the context of the renormalization group flow [91]. However, due to the completely combinatorial way in which γ_0 arises (which does not make reference to any dynamical notion) it is so far unclear how such scenario could be realized. The contribution of matter degrees of freedom ('vacuum fluctuations') to the degeneracy of the area spectrum has been neglected in the derivation leading to (53).

5.2. Matter and Holography

In the framework based on quantum geometry, the imposition of the diffeomorphism constraint on the IH is subtle because it is represented as a punctured 2-sphere. A careful analysis [7] shows that the punctures have to be treated as distinguishable. This is the paradigm used in treatments mentioned so far. Nonetheless it is of interest to see what indistinguishability would change. Instead of the microcanonical ensemble, we use now the grand canonical ensemble as this will considerably shorten the derivations (keep in mind that all ensembles are equivalent). Thus we start from the canonical partition function which for a system of non-interactive punctures is $Q(\beta, N) = q(\beta)^N/N!$ where the $N!$ in the denominator is the Gibbs factor that effectively enforces indistinguishability, and the one-puncture partition function $q(\beta)$ is given by

$$q(\beta) = \sum_{j=\frac{1}{2}}^{\infty} d_j \exp(-\frac{\hbar\beta\gamma_0}{\ell}),$$

where d_j is the degeneracy of the spin j state (for instance $d_j = (2j+1)$ as in the $SU(2)$ Chern-Simons treatment). The grand canonical partition function is

$$\mathscr{Z}(\beta, z) = \sum_{N=1}^{\infty} \frac{z^N q(\beta)^N}{N!} = \exp(zq(\beta)). \tag{54}$$

From the equations of state $E = -\partial_\beta \log(\mathscr{Z})$, and $N = z\partial_z \log(\mathscr{Z})$ one gets

$$\frac{A}{8\pi G_N \ell} = -z\partial_\beta q(\beta)$$
$$N = zq(\beta) = \log(\mathscr{Z}). \tag{55}$$

In thermal equilibrium at the Unruh temperature one has $\beta = 2\pi\ell\hbar^{-1}$ and the ℓ dependence disappears from the previous equations. However, for d_j that grow at most polynomially in j, the BH area predicted by the equation is just Planckian and the number of punctures N of order one. Therefore, indistinguishability with degeneracies d_j of the kind we find in the pure geometry models is ruled out because it cannot predict semiclassical BH's.

An interesting perspective [75] arises in the framework of quasilocal observers. If one only restricts to quantum geometry degrees of freedom then $d_j = 2j+1$ in the $SU(2)$ ENP treatment or $d_j = 1$ in the GM and DL models. Now, from the local observers perspective, the quantum state of the system close to the horizon appears as a highly excited state at inverse temperature $\beta = 2\pi\ell/\ell_{Pl}^2$. Of course this state looks like the vacuum state for freely falling observers (at scales smaller than the size of the BH). These two dual versions of the same physics tell us that the quantum state describing the near-horizon physics contains more than just pure quantum geometric excitations. Very general results from quantum field theory on curved spacetimes imply that the quasilocal observers close to the horizon would find that the number of degrees of freedom grows exponentially with the horizon area according to (see for instance [118])

$$D \propto \exp(\lambda A/(\hbar G_N)), \tag{56}$$

where λ is an unspecified dimensionless constant that cannot be determined due to two related issues: On the one hand UV divergences of standard QFT introduce regularization ambiguities affecting the value of λ; on the other hand, the value of λ depends on the number of species of fields considered. For that reason, here we only assume the qualitative exponential growth and will prove below that the ambiguity in λ is completely removed by non-perturbative quantum gravity considerations.

From (56) $D[\{s_j\}] = \prod_j d_j$ with $d_j = \exp(\lambda 8\pi\gamma_0)$. For simplicity let us take $\approx j + 1/2$. We also introduce two dimensionless variables δ_β and δ_h and write $\beta = \beta_U(1 + \delta_\beta)$ — where $\beta_U = 2\pi\ell/\hbar$ — and $\lambda = (1 - \delta_h)/4$. A direct calculation of the geometric series that follows from (54) yields

$$q(\beta) = \frac{\exp(-\pi\gamma_0 \delta(\beta))}{\exp(\pi\gamma_0 \delta(\beta)) - 1}, \tag{57}$$

where $\delta(\beta) = \delta_h + \delta_\beta$. The equations of state now predict large semiclassical BHs: for large $A/(\hbar G_N)$ equation (55) can be used to determine δ as a function of A and z. The result is $\delta = 2\sqrt{G_N \hbar z/(\pi \gamma_0 A)} \ll 1$. For semiclassical BHs $\delta_\beta \ll 1$ since the temperature measured by quasilocal observers must be close to the Unruh temperature, this, together with the previous equation for δ, implies $\delta_h \ll 1$. In other words semiclassical consistency implies that the additional degrees of freedom producing the degeneracy (56) must saturate the *holographic bound* [75], i.e. $\lambda = 1/4$ up to quantum corrections.

The entropy is given by the formula $S = \beta E - \log(z) N + \log(\mathscr{Z})$ which upon evaluation yields

$$S = \frac{A}{4 G_N \hbar} - \frac{1}{2}(\log(z) - 1)\left(\frac{zA}{\pi \gamma \ell_{Pl}^2}\right)^{\frac{1}{2}} \tag{58}$$

This gives the Bekenstein-Hawking entropy to leading order plus quantum corrections. If one sets the chemical potential of the punctures to zero (as for photons or gravitons) then these corrections remain. One can get rid of the corrections by setting the chemical potential $\mu = T_U$. Such possibility is intriguing, yet the physical meaning of such a choice is not clear at this stage. The thermal state of the system is dominated by large spins as the mean spin $\langle j \rangle = A/(N \ell_{Pl}^2)$ grows like $\sqrt{A/\ell_{Pl}^2}$. The conclusions of this subsection hold for arbitrary puncture statistics. This is to be expected because the system behaves as if it were at a very high effective temperature (the Unruh temperature is the precise analog of the Hagedorn temperature of particle physics) [75]. Because it will be important for further discussion we write the partition function corresponding to the choice of Bosonic statistics of punctures explicitly and for $z = 1$, namely

$$\mathscr{Z}(\beta) = \prod_{j=\frac{1}{2}}^{\infty} \sum_{s_j} \exp(2\pi\ell - \beta)\frac{a_j}{8\pi\ell G_N}, \tag{59}$$

where $a_j = 8\pi\gamma\ell_{Pl}^2$ are the area eigenvalues, and we have assumed for simplicity $\lambda = 1/4$ in (56), namely $d_j = \exp(a_j/(4 G_N \hbar))$. Interestingly, such exact holographic behavior of the degeneracy of the area spectrum can be obtained from an analytic continuation of the dimension of the boundary Chern-Simons theory by sending the spins $j_i \to is - 1/2$ with $s \in \mathbb{R}^+$ [62, 83]. The new continuous labels correspond to $SU(1,1)$ unitary representations that solve the $SL(2,C)$ self(antiself)-duality constraints $L^i \pm K^i = 0$ (*see Chapter 3*), which in addition comply with the necessary reality condition $E \cdot E \geq 0$ for the fields E_i^a (*see Chapter 1*). All this suggests that the holographic behavior postulated in (56) with $\lambda = 1/4$ would naturally follow from the definition of LQG in terms of self(antiself)-dual variables, i.e. $\gamma = \pm i$. The same holographic behavior of the number of degrees of freedom available at the horizon surface is found from a conformal field theoretical perspective for $\gamma = \pm i$ [78]. A relationship between the thermal nature of BH horizons and self dual variables seems also valid according to similar analytic continuation arguments [108]. The analytic continuation technique has also been applied in the context of lower dimensional BHs [63].

5.2.1. *What is the ensemble in the quasilocal treatment*

The quasi local perspective provides a description complementary to the isolated horizon definition of the horizon Hilbert space. It allows one to perform manipulations in the canonical ensemble language. At the basic level the ensemble is still defined by the details of the isolated horizon boundary conditions which tell us whether we are dealing with a spherical, distorted, rotating or static BH horizon. Even when charge and angular momentum do not appear in the expression of the quasi local first law these parameters (and all multipole moments in the case of distorted isolated horizons [15, 16]) are encoded implicitly in the form of the boundary condition used to define the quantum theory of the horizon. Notice also that the usual canonical ensemble is ill defined [86] because the number of states grows too fast as a function of the ADM energy. This problem disappears in the quasi local treatment where the area ensemble plays the central role.

6. Synergy as well as Tension between the Microscopic and Semi-classical Descriptions

6.1. *Spinfoams*

In the covariant path integral representation of loop quantum gravity the state of a puncture (open spin network link) $|j, m\rangle$ is embedded in the unitary representations of $SL(2, C)$ (whose basis vectors can be written as $|p, k; j, m\rangle$ for $p \in \mathbb{R}^+$ and $k \in \mathcal{N}$) according to $|j, m\rangle \to |\gamma(j+1), j; j, m\rangle$. The maximum weight states $m = j$ define a puncture state which is in turn a coherent state peaked along the z-axis which is assumed (through an implicit gauge fixing) to correspond to the normal to the horizon. We denote such states as follows

$$|\mathbf{j}\rangle \equiv |\gamma(j+1), j; j, j\rangle. \tag{60}$$

These states satisfy the simplicity constraints $L^i = \gamma K^i$ in a weak sense. (*See Chapter 3. Note however that our L^i, K^i are denoted there by B^i, E^i respectively.*) One postulates [40] that quantum horizon states (in the infinite area limit, i.e. Rindler states) evolve in the *time* of stationary observers (37) — uniformly accelerated with $a = \ell^{-1}$ — according to

$$|\mathbf{j_t}\rangle = \exp(iHt)|\mathbf{j}\rangle, \tag{61}$$

with $H = aK_z = \ell^{-1}K_z$ the Rindler Hamiltonian. This time evolution is consistent with the semiclassical condition (39). More precisely from the simplicity constraints one has that $\langle \mathbf{j}|H|\mathbf{j}\rangle = \hbar\gamma j\ell^{-1}$ which coincides with the eigenvalue of $E = A/(8\pi G_N\ell)$ for a single plaquette in the large j limit. By coupling the system with an idealized detector modeled by a two-level system [122] with energy separation $T_U = \hbar/(2\pi\ell) \ll \Delta\epsilon$ it is shown that the population of the excited state in the stationary state is [40]

$$p_1 \approx \exp(-\frac{2\pi\ell}{\hbar}\Delta\epsilon), \tag{62}$$

which is the Wien distribution at temperature $T_U = \hbar/(2\pi\ell)$. A key property [89] leading to this result is the fact that

$$|\langle\lambda|\mathbf{j}\rangle|^2 \approx \lambda^{2j}\exp(-\pi\lambda) \tag{63}$$

where $|\lambda\rangle = |\gamma(j+1), j; \lambda, j\rangle$ is an eigenstate of K_z and L_z with eigenvalues λ and j respectively. In relation to this it has been postulated [47] that the one puncture reduced density matrix measuring the *inside-outside* correlations in spin foams is given by

$$\rho_p = \frac{\exp(-2\pi K_z)}{Z}, \tag{64}$$

where $Z_p = \text{Tr}[\exp 2\pi K_z]$. The single puncture entropy $S_p = -\text{Tr}(\rho_p \log(\rho_p)) = a_p/(4G\hbar) + \log(Z_p)$; by adding this result for all punctures one gets a result of the form (49) with Z_p playing the role of σ (notice that both are the single puncture partition function). In this way the results of the covariant and canonical approach are consistent. Notice that the fundamental input in the derivation of the temperature is that quantum horizon physical states are of the form (61).

6.2. Entanglement Entropy Perturbations and Black Hole Entropy

Starting from a pure state $|0\rangle\langle 0|$ ("vacuum state") one can define a reduced density matrix $\rho = \text{Tr}_{in}(|0\rangle\langle 0|)$ by taking the trace over the degrees of freedom *inside* the BH horizon. The entanglement entropy is defined as $S_{ent}[\rho] = -\text{Tr}(\rho\log(\rho))$. In four dimensions [117] the leading order term of entanglement entropy in standard QFT goes like

$$S_{ent} = \lambda\frac{A}{\epsilon^2} + \text{corrections}, \tag{65}$$

where ϵ is an UV cut-off, and λ is left undetermined in the standard QFT calculation due to UV divergences and associated ambiguities. An important one is that λ is proportional to the number of fields considered; this is known as the *species problem*. These ambiguities disappear if one studies perturbations of (65) when gravitational effects are taken into account [37, 39]. The analysis is done in the context of perturbations of the vacuum state in Minkowski spacetime as seen by accelerated Rindler observers. Entanglement entropy is defined by tracing out degrees of freedom outside the Rindler wedge. Such system reflects some of the physics of stationary black holes in the infinite area limit. A key property [126] is that, formally,

$$\rho = \frac{\exp(-2\pi\int_\Sigma \hat{T}_{\mu\nu}\chi^\mu d\Sigma^\nu)}{\text{Tr}[\exp(-2\pi\int_\Sigma \hat{T}_{\mu\nu}\chi^\mu d\Sigma^\nu)]}, \tag{66}$$

where Σ is any Cauchy surface of the Rindler wedge. If one considers a perturbation of the vacuum state $\delta\rho$ then the first interesting fact is that the (relative entropy)

$\delta S_{ent} = S_{ent}[\rho+\delta\rho] - S_{ent}[\rho]$ is UV finite and hence free of regularization ambiguities [46]. The second fact is that due to (66) one has

$$\delta S_{ent} = 2\pi \text{Tr}(\int_\Sigma \delta\langle T_{\mu\nu}\rangle \chi^\mu d\Sigma^\nu). \tag{67}$$

Now from semiclassical Einstein's equations $\nabla^\mu \delta\langle T_{\mu\nu}\rangle = 0$, this (together with the global properties of the Rindler wedge) implies that one can replace the Cauchy surface Σ by the Rindler horizon H in the previous equation. As in the calculation leading to (38) one can use the Raychaudhuri equation (i.e. semiclassical Einstein's equations) to relate the flux of $\delta\langle T_{\mu\nu}\rangle$ across the Rindler horizon to changes in its area. The result is that $\delta S_{ent} = \frac{\delta A}{4G_N\hbar}$ independently of the number of species. The argument can be generalized to static black holes [106] where a preferred vacuum state exists (the Hartle-Hawking state). However, due to the fact that the BH horizon is no longer a good initial value surface the resulting balance equation is

$$\delta S_{ent} = \frac{\delta A}{4G_N\hbar} + \delta S_\infty, \tag{68}$$

where $\delta S_\infty = \delta E/T_H$, and δE is the energy flow at $\mathscr{I}^+ \cup i^+$. Changes of entanglement entropy match changes of Hawking entropy plus an entropy flow to infinity. These results shed light on the way the species problem could be resolved in quantum gravity. However, as the concept of relative entropy used here is insensitive to the UV degrees of freedom, the key question [106] is whether the present idea can be extrapolated to the Planck scale. The results described in Section 5.2 go in this direction.[n]

6.3. Entanglement Entropy vs. Statistical Mechanical Entropy

One can argue that the perspective that BH entropy should be accounted for in terms of entanglement entropy [117] and the statistical mechanical derivation presented in this chapter are indeed equivalent in a suitable sense [106]. The basic reason for such equivalence resides in the microscopic structure predicted by LQG [36, 38, 47]. In our context, the appearance of the UV divergence in (65) tells us that the leading contribution to S_{ent} must come from the UV structure of LQG close to the boundary separating the two regions. Consider a basis of the subspace of the horizon Hilbert space characterized by condition (19), and assume the discrete index a labels the elements of its basis. Consider the state

$$|\Psi\rangle = \sum_a \alpha_a |\psi_{int}^a\rangle |\psi_{ext}^a\rangle, \tag{69}$$

where $|\psi_{int}^a\rangle$ and $|\psi_{ext}^a\rangle$ denote physical states compatible with the IH boundary data a, and describing the interior and the exterior states of matter and geometry of the

---

[n]Since this article was completed, using coherent states Madhavan Varadarajan has pointed out in *Gen. Rel. Grav.* **48**, (2016) 35 that there is a conceptual difficulty in the identification of variation of the entanglement entropy with that of the horizon area, made in Refs. [88,89]. Further work is needed to settle this issue.

BH respectively. The assumption that such states exist is a basic input of Section 3. In the form of the equation above we are assuming that correlations between the outside and the inside at Planckian scales are mediated by the spin-network links puncturing the separating boundary. This encodes the idea that vacuum correlations are ultra-local at the Planck scale. This assumption is implicit in the recent treatments [36] based on the analysis of a single quantum of area correlation and it is related to the (Planckian) Hadamard condition as defined in [47]. We also assume states to be normalized as follows: $\langle \psi^a_{ext} | \psi^a_{ext} \rangle = 1$, $\langle \psi^a_{int} | \psi^a_{int} \rangle = 1$, and $\langle \Psi | \Psi \rangle = 1$. The reduced density matrix obtained from the pure state by tracing over the interior observables is

$$\rho_{ext} = \sum_a p_a |\psi^a_{ext}\rangle\langle\psi^a_{ext}|, \qquad (70)$$

with $p_a = |\alpha_a|^2$. It follows from this that the entropy $S_{ext} \equiv -\text{Tr}[\rho_{ext}\log(\rho_{ext})]$ is bounded by micro-canonical entropy of the ensemble (19) as discussed in Section 3. If instead one starts from a mixed state encoding an homogeneous statistical mixture of quantum states compatible with (19), then the reduced density matrix leads to an entropy that matches the microcanonical one [106].

6.4. *Euclidean Path Integral (the Quasi Local Treatment) and Logarithmic Corrections*

Here we review some basic features of the Euclidean path integral approach to the computation of BH entropy. Although the method is formal, as far as the contribution of geometric degrees of freedom is concerned, it allows one to study the contributions of matter degrees of freedom in the vicinity of the horizon. The formalism is relevant for discussing two important points. On the one hand it allows one to compare the partition function obtained in Section 5.2 with the field theoretical formal expression (providing in this way another test for semiclassical consistency). On the other hand it provides one with the tools that are necessary for comparison and discussion of the issue of logarithmic corrections in LQG and in other approaches.

There is a well known relationship between the statistical mechanical partition function and the Euclidean path integral on a flat background. One has:

$$Z_{sc}(\beta) = \int D\phi \exp\{-S[\phi]\}, \qquad (71)$$

where field configurations are taken to be periodic in Euclidean time with period β. Such expression can be formally extended to the gravitational context at least in the treatment of stationary black holes. One starts from the formal analog of the previous expression and immediately uses the stationary phase approximation to make sense of it on the background of a stationary black hole. Namely

$$Z_{sc}(\beta) = \int DgD\phi \exp\{-S[g,\phi]\}$$

$$\approx \exp\{-S[g_{cl},0]\} \int D\eta \exp\left[-\int dxdy\, \eta(x)\left(\frac{\delta^2 \mathscr{L}}{\delta\eta(x)\delta\eta(y)}\right)\eta(y)\right] \qquad (72)$$

where the first term depends entirely on the classical BH solution g_{cl} while the second term represents the path integral over fluctuation fields, both of the metric as well as the matter, that we here schematically denote by η. For local field theories $\delta_{\eta(x)}\delta_{\eta(y)}\mathscr{L} = \delta(x,y)\Box_{gc}$ where $\Box_{gc}$ is a Laplace-like operator. (Possible gauge symmetries, in particular diffeomorphisms, must be gauge fixed to make sense of such a formula).

Let us first concentrate on the evaluation of the classical action. In the quasi local treatment, the Euclidean space time region, where the fields η are supported, is given by a $D \times S^2$ where D is a disk in a plane orthogonal to and centered at the horizon radius and having a proper radius ℓ (recall that in the Euclidean case the BH horizon shrinks to a point, represented here by the center of D). Using the Gibbons-Hawking prescription for the boundary term [79], the action $S[g_{cl}, 0]$ is

$$S[g_{cl}, 0] = \frac{1}{8\pi G_N}\left[\int_{D \times S^2} \sqrt{g}\, R + \int_{\partial D \times S^2} (K - K_0)\, d\Sigma\right] \tag{73}$$

On shell the bulk term in the previous integral vanishes. However, unless $\beta_H = 2\pi\kappa^{-1}$, the geometry has a conical singularity at the centre of the disk and the first term will contribute. The boundary term is the usual one with K the extrinsic curvature of the boundary, $d\Sigma$ its volume form, and $K_0 = -1/\ell$ is the value of the extrinsic curvature at the boundary in the $A \to \infty$ limit (Rindler space-time). The subtraction of the counter term K_0 has the same effect as replacing the inner conical singularity by an inner boundary with a boundary term of the form $\beta_H A/(8\pi)$ [45]. A direct calculation gives the semi-classical free energy

$$-S[g_{cl}, 0] = \log(Z_{cl}) = (2\pi\ell - \beta)\frac{A}{8\pi G_N \ell}, \tag{74}$$

where $\beta = \beta_H \|\chi\|$ is the local energy. The equation of state $E = -\partial_\beta \log(Z_{cl})$ reproduces the quasilocal energy (39) — this is a consequence of the subtraction of K_0 [128]. The entropy is $S = \beta E + \log(Z) = A/(4\ell_{Pl}^2)$ when evaluated at the inverse Unruh temperature $\beta_U = 2\pi\ell$. Notice that in the quasi-local framework used here, entropy grows linearly with energy (instead of quadratically as in the usual Hawking treatment). This means that the usual ill behavior of the canonical ensemble of the standard global formulation [86] is cured by the quasilocal treatment.

Notice that equation (74) matches in form the partition function (59). In other words, the inclusion of the holographic degeneracy (56) plus the assumption of Bosonic statistics for punctures makes the results of Section 5.2 compatible with the continuous formal treatment of the Euclidean path integral. In essence (59) is a regularization of (72).

Quantum corrections to the entropy come from the fluctuation factor which can formally be expressed in terms of the determinant of a second order local (elliptic) differential operator $\Box_{g_{cl}}$

$$F = \int D\eta \exp\left[-\int dx\, \eta(x)\Box_{g_{cl}}\eta(x)\right] = [\det(\Box_{g_{cl}})]^{-\frac{1}{2}}. \tag{75}$$

The determinant can be computed from the identity (the heat kernel expansion)

$$\log\left[\det(\Box_{g_{cl}})\right] = \int_{\epsilon^2}^{\infty} \frac{ds}{s} \operatorname{Tr}\left[\exp(-s\,\Box_{g_{cl}})\right], \qquad (76)$$

where ϵ is a UV cut-off needed to regularize the integral. We will assume here that it is proportional to ℓ_{Pl}. In the last equality we have used the heat kernel expansion in d dimensions

$$\operatorname{Tr}\left[\exp(-s\,\Box_{g_{cl}})\right] = (4\pi s)^{-\frac{d}{2}} \sum_{n=0}^{\infty} a_n s^{\frac{n}{2}}, \qquad (77)$$

where the coefficients a_n are given by integrals in $D \times S^2$ of local quantities.

At first sight the terms a_n with $n \leq 2$ produce potential important corrections to BH entropy. All of these suffer from regularization ambiguities with the exception of the term a_2 which leads to logarithmic corrections. Moreover, contributions coming from a_0 and a_1 can be shown to contribute to the renormalization of various couplings in the underlying Lagrangian [115]; for instance a_0 contributes to the cosmological constant renormalization. True loop corrections are then encoded in the logarithmic term a_2 and for that reason it has received great attention in the literature (see Ref. [96] and references therein). Another reason is that its form is regularization independent. According to Ref. [66] there are no logarithmic corrections in the $SU(2)$ pure geometric model once the appropriate smoothing is used (canonical ensemble). From this we conclude that the only possible source of logarithmic corrections in the $SU(2)$ case must come from the non-geometric degrees of freedom that produce the so-called matter degeneracy that plays a central role in Section 5.2. A possible way to compute these corrections is to compute the heat kernel coefficient a_2 for a given matter model. This is the approach taken in Ref. [96] One can argue [75] that logarithmic corrections in the one-loop effective action are directly reflected as logarithmic corrections in the LQG BH entropy. The preceding considerations partially dissipate the perceived tensions between the LQG approach and others. This is an important question that deserves further attention.

6.5. Hawking Radiation

Detailed derivation of Hawking radiation from first principles in LQG remains an open problem, this is partly due to the difficulty associated with the definition of semiclassical states approximating space-time backgrounds. Without a detailed account of the emission process it is still possible to obtain information from a spectroscopical approach that uses as an input the details of the area spectrum in addition to some semiclassical assumptions [32]. The status of the question has improved with the definition and quantization of spherical symmetric models [66–68, 77]. The approach uses techniques from 'hybrid quantization' and quantum field theory on quantum space-times used in loop quantum cosmology (*see Chapter 6*). More precisely, the quantum spherical background space-time is defined using LQG techniques, whereas perturbations, accounting for Hawking radiation, are described by a quantum test field (defined by means of a Fock Hilbert space) which now propagate on the quantum, rather than classical, space-time.

References

[1] I. Agullo, A. Ashtekar and W. Nelson, "Extension of the quantum theory of cosmological perturbations to the Planck era," *Phys. Rev. D* **87** (2013) 043507.

[2] I. Agulló, J. Fernando Barbero G., Jacobo Diaz-Polo, Enrique F. Borja, and E. J. S. Villaseñor, "Black hole state counting in loop quantum gravity: A number-theoretical approach," *Phys. Rev. Lett.* **100** (2008) 211301-1–4.

[3] I. Agulló, J. Fernando Barbero G., Jacobo Diaz-Polo, Enrique F. Borja, and E. J. S. Villaseñor, "Detailed black hole state counting in loop quantum gravity," *Phys. Rev. D* **82** (2010) 084029-1–31.

[4] I. Agullo, E. F. Borja and J. Diaz-Polo, "Computing black hole entropy in loop quantum gravity from a conformal field theory perspective," *JCAP* **0907** (2009) 016.

[5] I. Agullo, J. Navarro-Salas, G. J. Olmo and L. Parker, "Insensitivity of Hawking radiation to an invariant Planck-scale cutoff," *Phys. Rev. D* **80** (2009) 047503.

[6] A. Ashtekar, J. Baez, A. Corichi and K. Krasnov, "Quantum geometry and black hole entropy," *Phys. Rev. Lett.* **80** (1998) 904–907.

[7] A. Ashtekar, J. Baez and K. Krasnov, "Quantum geometry of isolated horizons and black hole entropy," *Adv. Theor. Math. Phys.* **4** (2000) 1–94.

[8] A. Ashtekar, C. Beetle and S. Fairhurst, "Isolated horizons: A generalization of black hole mechanics," *Class. Quant. Grav.* **16** (1999) L1–L7.

[9] A. Ashtekar, C. Beetle and S. Fairhurst, "Mechanics of isolated horizons," *Class. Quant, Grav.* **17** (2000) 253–298.

[10] A. Ashtekar, C. Beetle and J. Lewandowski, "Mechanics of rotating isolated horizons," *Phys. Rev. D* **64** (2001) 044016-1–17.

[11] A. Ashtekar and M. Bojowald, "Black hole evaporation: A paradigm," *Class. Quant. Grav.* **22** (2005) 3349–3362.

[12] A. Ashtekar and M. Bojowald, "Quantum geometry and the Schwarzschild singularity," *Class. Quant. Grav.* **23** (2006) 391–411.

[13] A. Ashtekar, A. Corichi and K. Krasnov, "Isolated horizons: The classical phase space," *Adv. Theor. Math. Phys.* **3** (1999) 419–478.

[14] A. Ashtekar, A. Corichi and J. A. Zapata, "Quantum theory of geometry III: Noncommutativity of Riemannian structures," *Class. Quant. Grav.* **15** (1998) 2955–2972.

[15] A. Ashtekar, J. Engle, T. Pawlowski and C. Van Den Broeck, "Multipole moments of isolated horizons," *Class. Quant. Grav.* **21** (2004) 2549–2570.

[16] A. Ashtekar, J. Engle and C. Van Den Broeck, "Quantum horizons and black hole entropy: Inclusion of distortion and rotation," *Class. Quant. Grav.* **22** (2005) L27–L34.

[17] A. Ashtekar, S. Fairhurst and B. Krishnan, "Isolated horizons: Hamiltonian evolution and the first law," *Phys. Rev. D* **62** (2000) 104025-1–29.

[18] A. Ashtekar, S. Fairhurst and J. L. Willis, "Quantum gravity, shadow states, and quantum mechanics," *Class. Quant. Grav.* **20** (2003) 1031–1062.

[19] A. Ashtekar and B. Krishnan, "Dynamical horizons and their properties," *Phys. Rev. D* **68** (2003) 104030-1–25.

[20] A. Ashtekar and B. Krishnan, "Isolated and dynamical horizons and their applications," *Living Rev. Relat.* **7** (2004) 1–91.

[21] A. Ashtekar and J. Lewandowski, "Relation between polymer and Fock excitations," *Class. Quant. Grav.* **18** (2001) L117–L128.

[22] A. Ashtekar, F. Pretorius and F. M. Ramazanoglu, "Evaporation of 2-dimensional black holes," *Phys. Rev. D* **83** (2011) 044040.

[23] A. Ashtekar, F. Pretorius and F. M. Ramazanoglu, "Surprises in the evaporation of 2-dimensional black holes," *Phys. Rev. Lett.* **106** (2011) 161303.

[24] A. Ashtekar, V. Taveras and M. Varadarajan, "Information is not lost in the evaporation of 2-dimensional black holes," *Phys. Rev. Lett.* **100** (2008) 211302.

[25] J. Fernando Barbero G., J. Lewandowski and E. J. S. Villaseñor, "Flux-area operator and black hole entropy," *Phys. Rev. D* **80** (2009) 044016-1–15.

[26] J. Fernando Barbero G., J. Prieto and E. J. S. Villaseñor, "Hamiltonian treatment of linear field theories in the presence of boundaries: A geometric approach," *Class. Quant. Grav.* **31** (2014) 045021.

[27] J. Fernando Barbero G. and E. J. S. Villaseñor, "Statistical description of the black hole degeneracy spectrum," *Phys. Rev. D* **83** (2011) 104013-1–21.

[28] J. Fernando Barbero G. and E. J. S. Villaseñor, "Generating functions for black hole entropy in loop quantum gravity," *Phys. Rev. D* **77** (2008) 121502(R)-1–5.

[29] J. Fernando Barbero G. and E. J. S. Villaseñor, "On the computation of black hole entropy in loop quantum gravity," *Class. Quant. Grav.* **26** (2009) 035017-1–22.

[30] J. Fernando Barbero G. and E. J. S. Villaseñor, "The thermodynamic limit and black hole entropy in the area ensemble," *Class. Quant. Grav.* **28** (2011) 215014-1–15.

[31] J. M. Bardeen, B. Carter and S. W. Hawking, "The four laws of black hole mechanics," *Commun. Math. Phys.* **31** (1973) 161–170.

[32] A. Barrau, T. Cailleteau, X. Cao, J. Diaz-Polo and J. Grain, "Probing loop quantum gravity with evaporating black holes," *Phys. Rev. Lett.* **107** (2011) 251301.

[33] J. D. Bekenstein, "Black holes and entropy," *Phys. Rev. D* **7** (1973) 2333–2346.

[34] J. Bekenstein, "Black hole entropy quantization," *Lett. Nuovo Cim.* **11** (1974) 467–470.

[35] J. Bekenstein and V. F. Mukhanov, "Spectroscopy of the quantum black hole," *Phys. Lett. B* **360** (1995) 7–12.

[36] E. Bianchi, *Entropy of Non-Extremal Black Holes from Loop Gravity* (2012).

[37] E. Bianchi, *Horizon Entanglement Entropy and Universality of the Graviton Coupling* (2012).

[38] E. Bianchi and R. C. Myers, *On the Architecture of Spacetime Geometry* (2012).

[39] E. Bianchi and A. Satz, "Mechanical laws of the Rindler horizon," *Phys. Rev. D* **87** (2013) 124031.

[40] E. Bianchi and W. Wieland, *Horizon Energy as the Boost Boundary Term in General Relativity and Loop Gravity* (2012).

[41] N. Bodendorfer, *A Note on Entanglement Entropy and Quantum Geometry* (2014).

[42] L. Bombelli, R. K. Koul, J. Lee and R. D. Sorkin, "A quantum source of entropy for black holes," *Phys. Rev. D* **34** (1986) 373–383.

[43] I. S. Booth, "Metric-based hamiltonians, null boundaries and isolated horizons," *Class. Quantum Grav.* **18** (2001) 4239–4264.

[44] R. Bousso, H. Casini, Z. Fisher and J. Maldacena, *Proof of a Quantum Bousso Bound* (2014).

[45] S. Carlip and C. Teitelboim, "The Off-shell black hole," *Class. Quant. Grav.* **12** (1995) 1699–1704.

[46] H. Casini, "Relative entropy and the Bekenstein bound," *Class. Quant. Grav.* **25** (2008) 205021.

[47] G. Chirco, H. M. Haggard, A. Riello and C. Rovelli, *Spacetime Thermodynamics without Hidden Degrees of Freedom* (2014).

[48] J. Collins, A. Perez, D. Sudarsky, L. Urrutia and H. Vucetich, "Lorentz invariance and quantum gravity: an additional fine-tuning problem?" *Phys. Rev. Lett.* **93** (2004) 191301.

[49] A. Connes and C. Rovelli, "Von Neumann algebra automorphisms and time thermodynamics relation in general covariant quantum theories," *Class. Quant. Grav.* **11** (1994) 2899–2918.

[50] A. Corichi, J. Diaz-Polo and E. F.-Borja, "Black hole entropy quantization," *Phys. Rev. Lett.* **98** (2007) 181301-1–4.

[51] A. Corichi, J. Diaz-Polo and E. F.-Borja, "Quantum geometry and microscopic black hole entropy," *Class. Quant. Grav.* **24** (2007) 243–251.

[52] C. Crnković, "Symplectic geometry of the convariant phase space," *Class. Quant. Grav.* **5** (1988) 1557–1575.

[53] C. Crnković and E. Witten, "Covariant description of canonical formalism in geometrical theories (in 300 years of gravitation)," eds. S. W. Hawking and W. Israel (1987).

[54] G. Date, R. K. Kaul and S. Sengupta, "Topological interpretation of Barbero-Immirzi parameter," *Phys. Rev. D* **79** (2009) 044008.

[55] M. Domagala and J. Lewandowski, "Black-hole entropy from quantum geometry," *Class. Quant. Grav.* **21** (2004) 5233–5243.

[56] W. Donnelly, "Entanglement entropy in loop quantum gravity," *Phys. Rev. D* **77** (2008) 104006.

[57] J. Eisert, M. Cramer and M. B. Plenio, "Area laws for the entanglement entropy - a review," *Rev. Mod. Phys.* **82** (2010) 277–306.

[58] J. Engle, K. Noui and A. Perez, "Black hole entropy and SU(2) Chern-Simons theory," *Phys. Rev. Lett.* **105** (2010) 031302-1–4.

[59] J. Engle, K. Noui and A. Perez, "Black hole entropy from the SU(2)-invariant formulation of type I isolated horizons," *Phys. Rev. D* **82** (2010) 044050-1–23.

[60] J. Engle, K. Noui and A. Perez, "The SU(2) black hole entropy revisited," *JHEP* **05** (2011) 1–26.

[61] J. Fernando Barbero G. and E. J. S. Villaseñor, "Statistical description of the black hole degeneracy spectrum," *Phys. Rev. D* **83** 104013 (2011).

[62] E. Frodden, M. Geiller, K. Noui and A. Perez, *Black Hole Entropy from Complex Ashtekar Variables* (2012).

[63] E. Frodden, M. Geiller, K. Noui and A. Perez, "Statistical entropy of a BTZ black hole from loop quantum gravity," *JHEP* **1305** (2013) 139.

[64] E. Frodden, A. Ghosh and A. Perez, "Black hole entropy in LQG: Recent developments," *AIP Conf. Proc.* **1458** (2011) 100–115.

[65] E. Frodden, A. Ghosh and A. Perez, "Quasilocal first law for black hole thermodynamics," *Phys. Rev. D* **87** (2013) 121503.

[66] R. Gambini, J. Olmedo and J. Pullin, "Quantum black holes in loop quantum gravity," *Class. Quant. Grav.* **31** (2014) 095009.

[67] R. Gambini and J. Pullin, *An Introduction to Spherically Symmetric Loop Quantum Gravity Black Holes* (2013).

[68] R. Gambini and J. Pullin, *Hawking Radiation from a Spherical Loop Quantum Gravity Black Hole* (2013).

[69] R. Gambini and J. Pullin, "Loop quantization of the Schwarzschild black hole," *Phys. Rev. Lett.* **110** (2013) 211301.

[70] M. Geiller and K. Noui, "Near-horizon radiation and self-dual loop quantum gravity," *Europhys. Lett.* **105** (2014) 60001.

[71] A. Ghosh and P. Mitra, "An Improved lower bound on black hole entropy in the quantum geometry approach," *Phys. Lett. B* **616** (2005) 114–117.

[72] A. Ghosh and P. Mitra, "Fine-grained state counting for black holes in loop quantum gravity," *Phys. Rev. Lett.* **102** (2009) 141302.

[73] A. Ghosh and P. Mitra, "A Bound on the log correction to the black hole area law," *Phys. Rev. D* **71** (2005) 027502.

[74] A. Ghosh and P. Mitra, "Counting black hole microscopic states in loop quantum gravity," *Phys. Rev. D* **74** (2006) 064026–1–5.

[75] A. Ghosh, K. Noui and A. Perez, "Statistics, holography, and black hole entropy in loop quantum gravity," *Phys. Rev. D* **89** (2013) 084069.

[76] A. Ghosh and A. Perez, "Black hole entropy and isolated horizons thermodynamics," *Phys. Rev. Lett.* **107** (2011) 241301.

[77] A. Ghosh and A. Perez, *The Scaling of Black Hole Entropy in Loop Quantum Gravity* (2012).

[78] A. Ghosh and D. Pranzetti, "CFT/Gravity correspondence on the isolated horizon," *Nucl. Phys. B* **889** (2014) 1–24.

[79] G. W. Gibbons and S. W. Hawking, "Action integrals and partition functions in quantum gravity," *Phys. Rev. D* **15** (1977) 2752–2756.

[80] G. W. Gibbons, S. W. Hawking and M. J. Perry, "Path integrals and the indefiniteness of the gravitational action," *Nucl. Phys. B* **138** (1978) 141.

[81] M. J. Gotay, J. N. Nester and G. Hinds, "Presymplectic manifolds and the Dirac Bergmann theory of constraints," *J. Math. Phys.* **19** (1978) 2388–2399.

[82] R. B. Griffiths, "Microcanonical ensemble in quantum statistical mechanics," *J. Math. Phys.* **6** (1965) 1447–1461.

[83] M. Han, *Black Hole Entropy in Loop Quantum Gravity, Analytic Continuation, and Dual Holography* (2014).

[84] S. W. Hawking, "Gravitational radiation from colliding black holes," *Phys. Rev. Lett.* **26** (1971) 1344–1346.

[85] S. W. Hawking, "Particle creation by black holes," *Commun. Math. Phys.* **43** (1975) 199–220.

[86] S. W. Hawking, "Black holes and thermodynamics," *Phys. Rev. D* **13** (1976) 191–197.

[87] Z. Huang, A. Perez and S. Speziale, "Hawking radiaiton: An exactly solvable model," in progress.

[88] V. Husain, "Apparent horizons, black hole entropy and loop quantum gravity," *Phys. Rev. D* **59** (1999) 084019.

[89] M. Huszar, "Angular momentum and unitary spinor bases of the lorentz group," *Acta Phys. Hung.* **30** (1971) 241–251.

[90] T. Jacobson, "Thermodynamics of space-time: The Einstein equation of state," *Phys. Rev. Lett.* **75** (1995) 1260–1263.

[91] T. Jacobson, "Renormalization and black hole entropy in loop quantum gravity," *Class. Quant. Grav.* **24** (2007) 4875–4879.

[92] T. Jacobson, "Boundary unitarity and the black hole information paradox," *Int. J. Mod. Phys. D* **22** (2013) 1342002.

[93] R. K. Kaul and P. Majumdar," Quantum black hole entropy," *Phys. Lett. B* **3** (1998) 267–270.

[94] B. S. Kay and R. M. Wald, "Theorems on the uniqueness and thermal properties of stationary, nonsingular, quasifree states on space-times with a bifurcate Killing horizon," *Phys. Rept.* **207** (1991) 49–136.

[95] K. V. Krasnov, "Counting surface states in the loop quantum gravity," *Phys. Rev. D* **55** (1997) 3505–3513.

[96] K. V. Krasnov, "On Quantum statistical mechanics of Schwarzschild black hole," *Gen. Rel. Grav.* **30** (1998) 53–68.

[97] K. V. Krasnov, "Quantum geometry and thermal radiation from black holes," *Class. Quant. Grav.* **16** (1999) 563–578.
[98] D. N. Kupeli, "On null submanifolds in spacetimes." *Geometriae Dedicata* **23** (1987) 33–51.
[99] J. Lewandowski, "Spacetimes admitting isolated horizons," *Class. Quant. Grav.* **17** (2000) L-53–L-59.
[100] J. Lewandowski, A. Okolow, H. Sahlmann and T. Thiemann, "Uniqueness of diffeomorphism invariant states on holonomy-flux algebras," *Commun. Math. Phys.* **267** (2006) 703–733.
[101] J. Lewandowski and T. Pawlowski, "Extremal isolated horizons: A local uniqueness theorem," *Class. Quant. Grav.* **20** (2003) 587–606.
[102] K. A. Meissner, "Black-hole entropy in loop quantum gravity," *Class. Quant. Grav.* **21** (2004) 5245–5251.
[103] R. Narayan and J. E. McClintock, *Observational Evidence for Black Holes* (2013).
[104] A. Perez, *Introduction to Loop Quantum Gravity and Spin Foams* (2004).
[105] A. Perez, *Entanglement vs. Statistical Entropy in Quantum Gravity* (2014).
[106] A. Perez, *Statistical and Entanglement Entropy for Black Holes in Quantum Geometry* (2014).
[107] J. Polchinski, "Comment on [arXiv:1106.1417] 'Small Lorentz violations in quantum gravity: Do they lead to unacceptably large effects?'" *Class. Quant. Grav.* **29** (2012) 088001.
[108] D. Pranzetti, "Black hole entropy from KMS-states of quantum isolated horizons," *Phys. Rev. D* **89** (2014) 104046.
[109] D. Jimenez Rezende and A. Perez, "4d Lorentzian Holst action with topological terms," *Phys. Rev. D* **79** (2009) 064026.
[110] C. Rovelli, "Black hole entropy from loop quantum gravity," *Phys. Rev. Lett.* **77** (1996) 3288–3291.
[111] C. Rovelli, *Quantum Gravity* (Cambridge University Press (2004)).
[112] H. Sahlmann, "Toward explaining black hole entropy quantization in loop quantum gravity," *Phys. Rev. D* **76** (2007) 104050-1–7.
[113] H. Sahlmann, "Entropy calculation for a toy black hole," *Class. Quant. Grav.* **25** (2008) 055004-1–14.
[114] H. Sahlmann, "Black hole horizons from within loop quantum gravity," *Phys. Rev. D* **84** (2011) 044049.
[115] A. Sen, "Logarithmic corrections to Schwarzschild and other non-extremal black hole entropy in different dimensions," *JHEP* **1304** (2013) 156.
[116] L. Smolin, *General Relativity as the Equation of State of Spin Foam* (2012).
[117] S. N. Solodukhin, "Entanglement entropy of black holes," *Living Rev. Rel.* **14** (2011) 8.
[118] G. 't Hooft, "On the quantum structure of a black hole," *Nucl. Phys. B* **256** (1985) 727.
[119] C. Teitelboim, "Statistical thermodynamics of a black hole in terms of surface fields," *Nucl. Phys. Proc. Suppl.* **57** (1997) 125–130.
[120] T. Thiemann, *Modern Canonical Quantum General Relativity* (Cambridge University Press (2001)).
[121] K. S. Thorne, R. H. Price and D. A. Macdonald, *Black Holes: The Membrane Paradigm* (Yale University Press (1986)).
[122] W. G. Unruh, "Notes on black hole evaporation," *Phys. Rev. D* **14** (1976) 870.
[123] M. Varadarajan, "Fock representations from U(1) holonomy algebras," *Phys. Rev. D* **61** (2000) 104001.

[124] R. M. Wald, *General Relativity* (Chicago University Press (1984)).
[125] R. M. Wald, "Black hole entropy is the Noether charge," *Phys. Rev. D* **48** (1993) 3427–3431.
[126] R. M. Wald, *Quantum Field Theory in Curved Space-time and Black Hole Thermodynamics* (1995).
[127] R. M. Wald, "The thermodynamics of black holes," *Living Rev. Rel.* **4** (2001) 6.
[128] J. W. York Jr., "Black hole thermodynamics and the Euclidean Einstein action," *Phys. Rev. D* **33** (1986) 2092–2099.

Chapter 8

Loop Quantum Gravity and Observations

Aurélien Barrau

Laboratoire de Physique Subatomique et de Cosmologie,
Université Grenoble-Alpes, CNRS-IN2P3,
53 avenue des Martyrs, 38026 Grenoble Cedex, France

Julien Grain

Institut d'Astrophysique Spatiale, Université Paris-Sud 11, CNRS
Bâtiments 120-121, 91405 Orsay Cedex, France

1. Introduction

Building a quantum theory of space-time might be the most outstanding problem of contemporary fundamental theoretical physics. Probably this is not mainly because unification is necessary and unavoidable. Unification is unquestionably a useful guide that has indeed helped a lot in the past but that might very well not be the final word on what physics should look like. After all, it could be that different subfields of physics are described by different theories. The key issue has more to do with consistency. In some physical circumstances quantum mechanics and strong gravity are both important. In addition, the quantum world has interactions with the gravitational field itself, which automatically requires gravity to be understood in a quantum language, as can be demonstrated by appropriate thought experiments. Furthermore, because of the nonlinear nature of gravity, as soon as a strong gravitational field is involved, the coupling to gravitons also becomes strong, whence one cannot ignore quantum effects of gravity. The very existence of singularities in general relativity (GR) also requires a quantum extension. Finally, although the signal measured by the BICEP2 experiment [1] was not from cosmological origin, there is a reasonable hope that primordial gravitational waves will be soon seen through B-modes in the cosmological microwave background: this would be the first direct observation of a quantum gravity phenomenon, albeit at a linear level, in the history of science.

Several non-perturbative and background-independent approaches have been developed in the last decades. Among them, loop quantum gravity (LQG) may be the most advanced one (see [2] for introductions). One of the main achievements of LQG is that it has led an interplay between theory and experiments. At this stage, none of these ideas has been tested and some of them are still controversial. There are even tensions between different approximation schemes within LGQ. Still, it is a remarkable achievement that a quantum theory of gravity is now able to produce a set of predictions that might be tested in a foreseeable future.

In this brief overview, we will first focus on cosmology, considering different probes, both direct and indirect. We will then consider possible consequences of a possible Lorentz invariance violation. Evaporating black holes will be reviewed next and, finally, we will mention new ideas about 'Planck stars'.

2. Cosmology: Indirect Probes

When assumed to be isotropic and homogeneous, the Universe is sufficiently symmetric to be a quite easy system to quantize. As explained in *Chapter 6*, and as reviewed in [3], LQG ideas have been successfully applied to this specific situation: this is what loop quantum cosmology (LQC) describes. Although a rigorous derivation of LQC from LQG is still missing, it is now fairly believed to capture effectively most quantum effects from the mother theory. Recent progress was reported, e.g. in [4]. The most important result is probably the singularity resolution: the Big Bang is replaced by a Big Bounce and the LQC dynamics is different from the Wheeler-DeWitt one.

It is difficult but possible to make predictions for perturbations in LQC. Two main paths are followed at this stage. On the one hand a 'dressed metric approach' [5] was developed. It tries to deal deeply with quantum fields on a quantum background geometry. On the other hand, an 'effective approach' [6] was investigated. It tries to avoid fixing or assuming any background structure but instead derives it from effective equations. Both deserve to be seriously considered.[a] In this section we therefore first focus on more 'reliable' predictions related to the background evolution. Holonomy corrections appear in the theory because there is an operator that can only be associated with the holonomy of the Ashtekar connection, rather than the connection itself. Although the way those corrections are implemented, leading to the bounce, can of course be questioned, the main picture is now consistent and well established.

2.1. *Isotropic Case*

2.1.1. *Initial conditions at the bounce*

A first approach, developed in [7], assumes that the bounce is the appropriate time to set initial conditions. This is reasonable as the bounce is the only 'preferred'

---

[a]Since this overview was written, new results on the observational consequences of the deformed algebra approach were obtained in B. Boillet, A. Barrau, J. Grain and S. Schander, *Phys. Rev. D* **93** (2016) 124011.

point in time. The early Universe is also assumed to be filled with a massive scalar field, as usually done in inflation.

The idea is to solve, thanks to the bounce, the ambiguity that usually appears in the construction of a measure on the space of initial data. The space of solutions is isomorphic to a gauge fixed surface, i.e., a 2-surface $\hat{\Gamma}$ which is intersected by each dynamical trajectory only once. Since b, the conjugate momentum to the volume of the fixed fiducial cell used in the quantization, is monotonic in each solution, the strategy is to choose for $\hat{\Gamma}$ an appropriate 2-surface $b = b_o$. Symplectic geometry considerations unambiguously equip $\hat{\Gamma}$ with an induced Liouville measure $d\hat{\mu}_{\mathcal{L}}$. A natural choice is to set $b_o = \pi/2\lambda$, the value that characterizes the bounce, so that $\hat{\Gamma}$ is naturally coordinatized by $(\bar{\varphi}_B, v_B)$, the scalar field and the volume at the bounce. The induced measure is given by $d\hat{\mu}_{\mathcal{L}} = \frac{\sqrt{3\pi}}{\lambda}\left[1 - x_B^2\right]^{\frac{1}{2}} d\bar{\varphi}_B \, dv_B$, where x_B^2 is the value of x^2 at the bounce (with $x^2 = m^2\bar{\varphi}^2/(2\rho_c)$), that is the fraction of total energy density in form of potential energy at the bounce. After factoring out the gauge orbits the fractional volumes of physically relevant sub-regions of $\hat{\Gamma}$ can be calculated. The main results of the study performed in [7], depending on 3 different possible regimes, are:

- for $x_B^2 < 10^{-4}$, the number of e-folds during slow roll is given approximately by $N \approx 2\pi\left(1 - \frac{\bar{\varphi}_o^2}{\bar{\varphi}_{\max}^2}\right)\bar{\varphi}_o^2 \ln \bar{\varphi}_o$, where $\bar{\varphi}_o$ is the value of the scalar field at the onset of inflation and $\bar{\varphi}_{\max} = 1.5 \times 10^6$. For $\bar{\varphi}_B = 0.99$, one has $\bar{\varphi}_o = 3.24$ and $N = 68$. Thus, there is a slow roll inflation with over 68 e-foldings for all $\bar{\varphi}_B > 1$, i.e., if $x_B^2 > 4.4 \times 10^{-13}$.
- for $10^{-4} < x_B^2 < 0.5$, the LQC departures from GR are now significant. The Hubble parameter is essentially frozen at a very high value. Throughout this range of x_B^2 there are more than 68 e-foldings.
- for $0.5 < x_B^2 < 1$, the LQC effects strongly dominate. Again, because $\dot{\bar{\varphi}} > 0$, the inflaton climbs up the potential but the turn around ($\dot{\bar{\varphi}} = 0$) occurs during super-inflation. The Hubble parameter freezes at the onset of inflation and the slow roll conditions are easily met as $\dot{H}/H^2$ is less than 1×10^{-11} when $\ddot{\bar{\varphi}} = 0$. There are many more than 68 e-foldings already in the super-inflation phase. The friction term is large and the inflation enters a long (more than 68 e-folds) slow roll inflationary phase.

Basically all LQC dynamical trajectories are funneled to conditions which virtually guarantee slow-roll inflation with more than 68 e-foldings, without any input from the pre-big bang regime. This work was developed further, using analytical and numerical methods, to calculate the *a priori* probability of realizing a slow-roll phase compatible with CMB. It was found that the probability is greater than 0.999997 in LQC. This can be considered as a good indirect — although not definitive — test of LQC.

2.1.2. *Initial conditions in the remote past*

In [7], the probability distribution is assumed to be flat and defined at the bounce (the first attempts in this direction were performed in [8]). It is however possible to make a very different assumption: the phase of the oscillations of the field in the remote past can also be considered as a very natural random variable [9]. The choice of what is a natural measure depends heavily on when one decides to set initial conditions [10]. It is important to consider seriously the meaning of an 'initial' condition in a Universe that has a contracting branch before the bounce. In this approach one does not focus on the initial data at the bounce as in [7], but rather derives a probability distribution for them as a prediction of the model.

The approach consists in calculating the probability distribution for x_B, the square root of the fraction of potential energy at the bounce, and N, the number of e-folds of slow-roll inflation. The most natural and consistent assumption is to set the initial probability distribution in the pre-bounce oscillatory phase where the Universe is in addition classical and therefore well under control. The evolution in this phase is described by: $\rho = \rho_0 \left(1 - \frac{1}{2}\sqrt{3\kappa\rho_0}\left(t + \frac{1}{2m}\sin(2mt + 2\delta)\right)\right)^{-2}$, with $x = \sqrt{\frac{\rho}{\rho_c}}\sin(mt+\delta)$, $y = \sqrt{\frac{\rho}{\rho_c}}\cos(mt+\delta)$. In fact, due to hidden symmetries, δ can be shown to be the only parameter.

In addition of being the obviously expected distribution for any oscillatory process of this kind, a flat probability for δ will be preserved over time during the pre-bounce oscillations, making it a very natural choice. This is not a trivial point as any other probability distribution would be distorted over time, meaning that the final result in the full numerical analysis would depend on the choice of ρ_0. Starting with a flat probability distribution for δ, the probability for different values of x_B can be calculated numerically. In [7], x_B is considered as unknown whereas, in this second approach [9], it is shown to be sharply peaked around 3.55×10^{-6} (this value scales with m as $m\log\left(\frac{1}{m}\right)$, where we assumed that $m \ll 1$ in Plank units). The most likely solutions are exactly those that have no slow-roll deflation. The probability density for N can also be computed and is given in Figure 1, showing that the model leads to a slow-roll inflation of about 140 e-folds. This becomes, as shown in [9], a *prediction* of effective LQC: inflation and its duration are not arbitrary anymore.

2.2. Anisotropic Case

In bouncing cosmologies, either from the loop approach or any other, the question of anisotropies is very important for a clear reason: the shear term varies as $1/a^6$ where a is the 'mean' scale factor of the Universe. When the Universe is contracting, the shear term becomes more and more important and eventually drives the dynamics. The reason for which the shear can be neglected in standard cosmology is precisely the reason why it becomes important in bouncing models. The question of predicting the duration of inflation in LQC was studied in the Bianchi-I case.

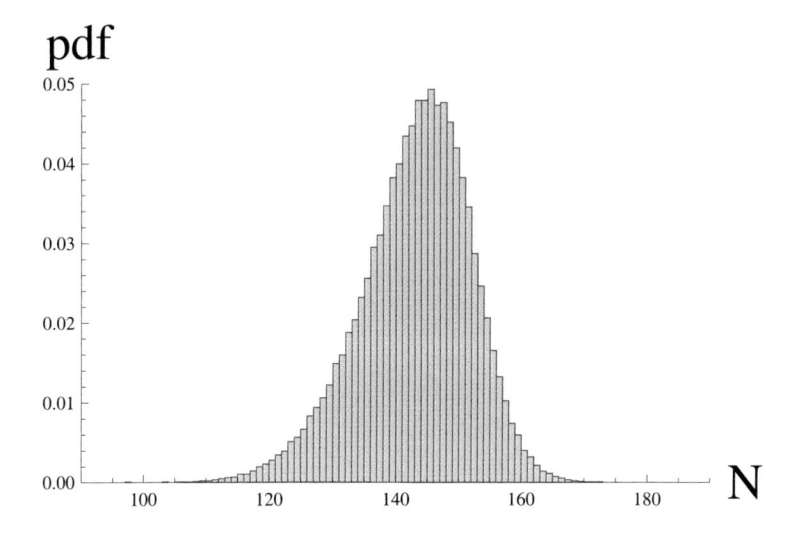

Fig. 1. Probability distribution function of the number of e-folds of slow-roll inflation (from [9]).

The metric is given by $ds^2 := -N^2 d\tau^2 + a_1^2 dx^2 + a_2^2 dy^2 + a_3^2 dz^2$, where a_i denotes the directional scale factors. The classical gravitational Hamiltonian is $\mathcal{H}_G = \frac{N}{\kappa \gamma^2} \left(\sqrt{\frac{p_1 p_2}{p_3}} c_1 c_2 + \sqrt{\frac{p_2 p_3}{p_1}} c_2 c_2 + \sqrt{\frac{p_3 p_1}{p_2}} c_2 c_3 \right)$, with Poisson brackets $\{c_i, p_j\} = \kappa \gamma \delta_{ij}$. The classical directional scale factors can be written as $a_1 = \sqrt{\frac{p_2 p_3}{p_1}}$ and cyclic expressions. The holonomy correction is implemented to account for specific LQG effects with the usual prescription (the framework was introduced in [11]) $c_i \to \frac{\sin(\bar{\mu}_i c_i)}{\bar{\mu}_i}$. The $\bar{\mu}_i$ are given by $\bar{\mu}_1 = \lambda \sqrt{\frac{p_1}{p_2 p_3}}$ and cyclic expressions, where λ is the square root of the minimum area eigenvalue of the LQG area operator ($\lambda = \sqrt{\Delta}$). The quantum corrected gravitational Hamiltonian is:

$$\mathcal{H}_G = -\frac{N \sqrt{p_1 p_2 p_3}}{\kappa \, \gamma^2 \lambda^2} \Big[\sin(\bar{\mu}_1 c_1) \sin(\bar{\mu}_2 c_2) + \sin(\bar{\mu}_2 c_2) \sin(\bar{\mu}_3 c_3) + \sin(\bar{\mu}_3 c_3) \sin(\bar{\mu}_1 c_1) \Big]. \quad (1)$$

In the gravitational sector, all the information is contained in the h_i: $h_1 = \bar{\mu}_1 c_1 = \lambda \sqrt{\frac{p_1}{p_2 p_3}} c_1$ and cyclic expressions. By defining the quantum shear by

$$\sigma_Q^2 := \frac{1}{3\gamma^2 \lambda^2} \left(1 - \frac{1}{3} \Big[\cos(h_1 - h_2) + \cos(h_2 - h_3) + \cos(h_3 - h_1) \Big] \right), \quad (2)$$

one can show [12] that LQC-modified generalized Friedman equation is: $H^2 = \sigma_Q^2 + \frac{\kappa}{3}\rho - \lambda^2 \gamma^2 \left(\frac{3}{2} \sigma_Q^2 + \frac{\kappa}{3} \rho \right)^2$.

In [13], exhaustive numerical simulations to investigate the duration of inflation as a function of the different variables entering the dynamics in Bianchi-I LQC were carried out. As the shear is initially small compared to everything else, the initial conditions for the matter content are chosen [9] as

$\rho(0) = \rho_0 \left(1 - \frac{1}{2}\sqrt{3\kappa\rho_0}\frac{1}{2m}\sin(2\delta)\right)^{-2}$, $m\phi(0) = \sqrt{2\rho(0)}\sin(\delta)$, and $\dot{\phi}(0) = \sqrt{2\rho(0)}\cos(\delta)$, where ρ_0 is the initial energy density up to a small correction, and δ is the phase of the oscillations between the kinetic and the potential energies. The phase and shear are the initial variables to set.

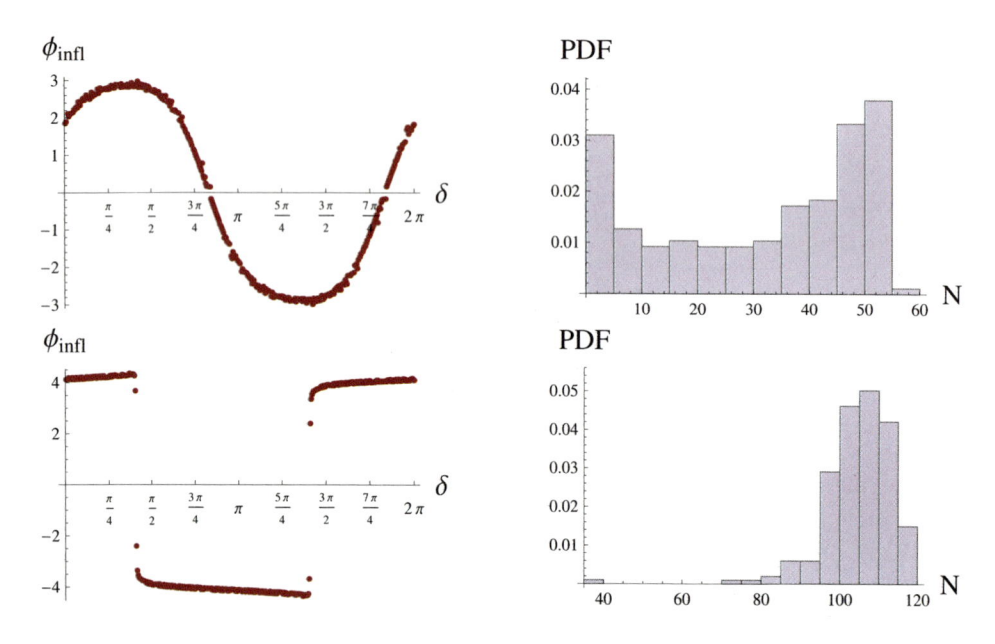

Fig. 2. Results of the simulations carried out in [13]. From top to bottom : $\sigma_Q(0) = 10^{-2}\frac{\kappa}{3}\rho_0$ and $\sigma_Q(0) = 10^{-6}\frac{\kappa}{9}\rho_0$. The first column is ψ at the start of slow-roll inflation and the second column corresponds to the numerically calculated probability distribution function of the number of e-folds of inflation.

Some results of the simulations are showed in Figure 2. The main conclusion is that, in general, the number of e-folds decreases when the shear increases. But a greater shear will also lead to a larger spread in the number of e-folds, depending on the initial angle δ. The number of e-folds of slow-roll inflation depends strongly on $\rho_{\max}$ which is fixed only when the shear vanishes. At the bounce, the dynamics is completely driven by the kinetic energy and the shear. The kinetic energy grows a lot in a very short time, which gives the scalar field a boost, and lifts it up to create the initial conditions for slow-roll inflation. If the shear is important, the bounce will happen at a lower value of the kinetic energy, and the scalar field potential will not 'climb' as high as in the isotropic case.

Anisotropies lead to fewer e-folds of slow-roll inflation. It is however interesting that for a wide range of parameters, the probability distribution for the number of e-folds is peaked at values compatible with data, between 70 and 130 e-folds. It is worth noticing that whereas any value between 0 and $N_{\max} = 2\pi\sqrt{2\rho_c}m^{-2} = 3.9 \times 10^{12}$ is *a priori* possible for N, the favored value is very close to the minimum required value. This makes the bounce/inflation scenario particularly appealing for phenomenology: all the quantum information from the bounce might not have been washed out by inflation. Having N close to 70 is exactly what is required to lead to measurable effects in the CMB spectrum. An important issue however remains: what would be a 'natural' initial value for the shear?

3. Cosmology: Direct Probes

Directly probing LQC modeling of the universe from astronomical observations follows the standard procedure used in classical cosmology to probe e.g. the physics of inflation. Any observer is confined within the Universe and one relies on cosmic inhomogeneities (whose evolution across cosmic times depends on the dynamics of the Universe) as internal tracers. They are revealed by the observed galaxies and large scale structures, and by the anisotropies of the cosmic microwave background (CMB). This however tells us that our Universe is *statistically* homogeneous and isotropic, being filled with inhomogeneities, and can be modeled by a *perturbed* FLRW metric, for those inhomogeneities are small in the primordial Universe.

In classical cosmology, inhomogeneities are produced during inflation from the gravitational amplification of the fluctuations of the quantum vacuum. In the context of single field inflation, the perturbations are of two types: scalar modes corresponding to perturbations of the scalar 3-dimensional curvature, denoted $\mathcal{R}$, and, tensor modes h_a^i corresponding to primordial gravitational waves. They are commonly described by the Mukhanov–Sasaki, gauge-invariant variables, $v_S = z\mathcal{R}$ and $v_T = ah$ with $z = (a\bar{\varphi}')/\mathcal{H}$ and $\mathcal{H} = a'/a$ is the Hubble parameter in conformal time, i.e. $ds^2 = a^2(\eta)\left(d\eta^2 - dx_a dx^a\right)$. The quantum fluctuations of these two fields are dynamically amplified during the accelerated expanding inflationary phase. Since they originate from the quantum vacuum which is Gaussian (and assuming linear evolution for simplicity), the perturbations at the end of inflation, η_e, are fully described by their 2-points correlation function or, in Fourier space, by their primordial power spectrum:

$$\mathcal{P}_S = \frac{k^3}{2\pi} \langle \mathcal{R}(k)\, \mathcal{R}^\star(k) \rangle_{\eta_e} \quad \text{and} \quad \mathcal{P}_T = (16Gk^3) \sum_{s=1}^{2} \left\langle h_{a,(s)}^i(k)\, h_{i,(s)}^a(k) \right\rangle_{\eta_e}, \quad (3)$$

where the average is a quantum expectation value over the vacuum state. For tensor modes, the sum is over the two helicity degrees of freedom. At the end of inflation, our Universe is then filled with inhomogeneities of quantum origin: scalar perturbations serve as the primordial seeds for structures formation, and, both scalar *and* tensor perturbations leave their footprint in the CMB in the form of anisotropies of temperature and linear polarization. The latter is decomposed into two modes dubbed E and B modes. The statistics of these anisotropies follows the statistics of the cosmological perturbations and is Gaussian (primordial non-Gaussianities are observationally constrained to be extremely small). The observed information contained in the CMB is then compressed into six angular power spectra measuring the power of the T, E and B auto- and cross-correlations. These are estimated from the CMB observations and are theoretically related to the primordial power spectra via the line-of-sight solution of the Boltzmann equations [14]:

$$C_\ell^{XY} = \int_0^\infty dk \int_{\eta_e}^{\eta_0} d\eta \left[\Delta_\ell^{X,\mathrm{S}}(k,\eta) \Delta_\ell^{Y,\mathrm{S}}(k,\eta) \mathcal{P}_\mathrm{S}(k) + \Delta_\ell^{X,\mathrm{T}}(k,\eta) \Delta_\ell^{Y,\mathrm{T}}(k,\eta) \mathcal{P}_\mathrm{T}(k) \right], \tag{4}$$

with X, Y running over T, E and B. The time integration is performed from the end of inflation to today, η_0. The functions $\Delta_\ell^{X,\mathrm{S(T)}}$ are the transfer functions encoding the evolution of scalar(tensor) perturbations and the primordial power spectra are source terms. Fitting the predicted angular power spectra on the estimated ones allows for setting constraints on both cosmological parameters driving the dynamics of the homogeneous Universe via the transfer functions and cosmological parameters driving the shape of the primordial power spectra. Since the later are classically derived from the inflationary dynamics, any constraints on $\mathcal{P}_\mathrm{S(T)}$ from the CMB measurements can be translated into constraints on inflationary models.

In the context of LQC, the cosmological perturbations evolve through the contracting phase and the bounce prior to inflation. Because of that, one can expect some distortions in the predicted $\mathcal{P}_\mathrm{S(T)}$ as compared to the standard prediction of pure inflation. The shape of primordial power spectra now contains information about the contracting phase and the quantum bounce in addition to information about inflation, and this will inevitably translate into distortions of the angular power spectra of the CMB anisotropies, leading to possible direct probes of this quantum gravity modeling of the Universe. The main prediction is therefore the primordial power spectra from which CMB angular power spectra are derived. Preliminary results were obtained by solely considering the change in the background evolution, the Universe passing through a contraction phase and bounce prior to inflation [15]. The distortions on the polarized CMB anisotropies could be observed from a clear inspection of those anisotropies and used to constrain e.g. the fraction of potential energy in the scalar field at the time of the bounce [16]. However, the very fact that cosmological perturbations are to be constructed from a quantum theory of gravity was not properly taken into account, though the change of the Universe history was. Indeed, cosmological perturbations are perturbations of the gravitational field itself (as well as perturbations of the matter content). This means that the classical theory of cosmological perturbations (consisting in linearizing the Einstein's field equations around the FLRW solution) should be amended first for accounting that perturbations live in a *quantum* background.

3.1. *Cosmological Perturbations in LQC*

Different approaches to treat cosmological perturbations in a LQC-derived cosmological background have been developed recently. The dressed metric approach, discussed in *Chapter 6*, adopts a strategy in which the minisuperspace homogeneous and isotropic degrees of freedom *and* the infinitely many inhomogeneous degrees of freedom (considered as perturbations) are quantized [5]. The former is obtained by the loop quantization and the latter is obtained from a Fock quantization on a *quantum* background space-time. The physical inhomogeneous degrees of freedom are given by the Mukhanov-Sasaki variables derived from the linearized classical constraints. The second order Hamiltonian (restricted to the square of the first order perturbations) is promoted to be an operator and the quantization is performed using techniques suitable for the quantization of a test field evolving in a quantum background [17]. The Hilbert space is a tensor product $\Psi(\nu, v_{S(T)}, \varphi) = \Psi_{\text{FLRW}}(\nu, \bar{\varphi}) \otimes \Psi_{\text{pert}}(v_S, v_T, \bar{\varphi})$ with ν the homogeneous and isotropic degrees of freedom and $v_{S(T)}$ the degrees of freedom for perturbations. In the interaction picture, so long as the backreaction of the perturbations on Ψ_{FLRW} remains negligible, the Schrödinger equation for the perturbations is shown to be identical to the Schrödinger equation for the quantized perturbations evolving in a classical background but using a *dressed* metric encoding the quantum nature of the background (for tensor modes):

$$i\hbar \partial_{\bar{\varphi}} \Psi_{\text{pert}} = \frac{1}{2} \int \frac{d^3k}{(2\pi)^3} \left\{ \frac{32\pi G}{\tilde{p}_\varphi} \left| \hat{\pi}_{T,\vec{k}} \right|^2 \Psi_{\text{pert}} + \frac{k^2}{32\pi G} \frac{\tilde{a}^4(\bar{\varphi})}{\tilde{p}_\varphi} \left| \hat{v}_{T,\vec{k}} \right|^2 \Psi_{\text{pert}} \right\}, \quad (5)$$

with

$$(\tilde{p}_\varphi)^{-1} = \left\langle \hat{H}_{\text{FLRW}}^{-1} \right\rangle \text{ and } \tilde{a}^4 = \frac{\left\langle \hat{H}_{\text{FLRW}}^{-1/2} \hat{a}^4(\bar{\varphi}) \hat{H}_{\text{FLRW}}^{-1/2} \right\rangle}{\left\langle \hat{H}_{\text{FLRW}}^{-1} \right\rangle}. \quad (6)$$

In the above, $(\hat{v}_{T,\vec{k}}, \hat{\pi}_{T,\vec{k}})$ are the configuration and momentum operators of the perturbations while $\hat{H}_{\text{FLRW}}$ is the Hamiltonian operator of the isotropic and homogeneous background. The dressed metric is in principle *neither* equal to the classical metric *nor* equal to the metric traced by the peak of the sharply peaked background state. This is finally translated into a Fock quantization for which the mode functions (providing the evolution of scalar and tensor perturbations in a quantum background, here expressed in the spatial Fourier space) are solutions of

$$Q_k'' + 2\left(\frac{\tilde{a}'}{\tilde{a}}\right) Q_k' + \left(k^2 + \tilde{U}\right) Q_k = 0, \quad (7)$$

$$h_k'' + 2\left(\frac{\tilde{a}'}{\tilde{a}}\right) h_k' + k^2 h_k = 0. \quad (8)$$

The gauge-invariant variable Q_k is related to the Mukhanov-Sasaki variables for scalar modes via $Q_k = v_{S,k}/a$, and, $\tilde{U}$ is a dressed potential-like term given by

$$\tilde{U}(\bar{\varphi}) = \frac{\left\langle \hat{H}_{\text{FLRW}}^{-1/2} \hat{a}^2(\bar{\varphi}) \hat{U}(\bar{\varphi}) \hat{a}^2(\bar{\varphi}) \hat{H}_{\text{FLRW}}^{-1/2} \right\rangle}{\left\langle \hat{H}_{\text{FLRW}}^{-1/2} \hat{a}^4(\bar{\varphi}) \hat{H}_{\text{FLRW}}^{-1/2} \right\rangle}, \quad (9)$$

the quantum counterpart of

$$U(\bar{\varphi}) = a^2 \left(fV(\bar{\varphi}) - 2\sqrt{f}\partial_{\bar{\varphi}}V + \partial_{\bar{\varphi}}^2 V \right), \tag{10}$$

with $f = 24\pi G(\dot{\bar{\varphi}}^2/\rho)$, the fraction of kinetic energy in the scalar field.

A second approach developed in Ref. [18] consists in perturbing the semi-classical, effective space-time whose dynamics is given by the modified Friedmann equations. The idea is to start from the classical perturbed Hamiltonian and to introduce corrections taking into account at the effective level the quantum nature of the background. For the zeroth-order Hamiltonian, providing the dynamics of the background, such a modification is easily obtained from the fact that the quantization being based on holonomies, the connection $\bar{k}$ is replaced by $\left(\sin(\gamma\bar{\mu}\bar{k})/\gamma\bar{\mu}\right)$, yielding the modified Friedmann equations. Similar effective modifications are introduced to the first and second order perturbation Hamiltonians. Though there is a priori much more freedom for those modifications, there expressions are univocally derived by requiring that first, the classical Hamiltonian is recovered in the limit of large volumes (i.e. $\bar{\mu} \to 0$), and, second, that the algebra of the truncated scalar, diffeomorphism and Gauss constraints is still closed, as is the case for truncated constraints in the classical theory of cosmological perturbations. This second requirement fixes all the ambiguities of the introduced quantum corrections (at least for the case of holonomy corrections). Moreover, the set of effective constraints is first class and can be used to generate the gauge transformations to derive the effective gauge-invariant variables for the cosmological perturbations. The dynamics is generated by the second-order, effective Hamiltonian. Those perturbations are finally quantized à la Fock using the techniques developed for quantum fields in curved spaces. In that process, it appears that the anomaly-free algebra of effective constrained is deformed compared to the classical algebra of constraints by [19]:

$$\{D[M^a], D[N^a]\} = D[M^b\partial_b N^a - N^b\partial_b M^a], \tag{11}$$

$$\{D[M^a], S^Q[N]\} = S^Q[M^a\partial_a N - N\partial_a M^a], \tag{12}$$

$$\{S^Q[M], S^Q[N]\} = D\left[\Omega q^{ab}(M\partial_b N - N\partial_b M)\right], \tag{13}$$

with D the diffeomorphism constraint and S^Q the scalar constraint. The deformation is encoded in Ω which depends on the background phase-space variables, $\Omega = \cos(2\gamma\bar{\mu}\bar{k}) = 1 - 2\rho/\rho_c$. In this deformed algebra approach, the mode functions describing the dynamics of the scalar and tensor modes (in terms of *effective* Mukhanov-Sasaki variables) are solutions of

$$v''_{S(T),k} + \left[\Omega k^2 - \frac{z''_{S(T)}}{z_{S(T)}}\right] v_{S(T),k} = 0, \tag{14}$$

with $z_S = (a\bar{\varphi}')/\mathcal{H}$ and $z_T = a/\sqrt{\Omega}$. Those functions encode the impact of the effective background on the perturbations.

3.2. Primordial Power Spectrum in Loop Quantum Cosmology

The primordial power spectra are the sources of the CMB anisotropies and are the key quantities to compute. Assuming some initial conditions for the mode functions, thus fixing the choice of the initial quantum states for perturbations, the primordial power spectra are determined by the knowledge of the mode functions at the end of inflation. A first choice of initial conditions for perturbations is a fourth order WKB vacuum at the time of the bounce. Such a choice is however only possible in the dressed metric approach. For the deformed algebra, Ω is negatively valued at the time of the bounce which prevents the existence of standard oscillatory solutions for the mode functions. An example of the resulting primordial power spectra for scalar and tensor perturbations in the dressed metric approach and setting the initial conditions for perturbations at the time of the bounce is displayed in Figure 3. This shows that the bounce leaves a characteristic length scale $(k_\star)^{-1}$ as a typical footprint. For shorter length scales, $k > k_\star$, the predicted primordial power spectrum coincides with the prediction of standard inflationary cosmology since the slightly red-tilted power law is recovered. However for larger length scales, LQC predicts a different power spectrum (which can be viewed as a running of the spectral index in the language of inflation). This typical scale can be intuitively understood for tensor modes by a clear inspection of $(\tilde{a}''/\tilde{a})$, tracing the effective 'curvature' of the background. For sharply peaked states, the dressed scale factor $\tilde{a}$ is very well approximated by the scale factor traced by the peak of the background quantum states, a, which is the solution of the modified Friedmann equations. At the time of the bounce, $a''/a = 8\pi G \rho_c$ and rapidly decreases in the beginning of the expansion. Then, this quantity rapidly increases once the Universe enters its inflationary phase. The shape of the primordial power spectrum is driven by $(k^2 - a''/a)$: if $k^2 > a''/a$, the modes are oscillatory whereas in the opposite case, the mode functions are a linear combination of growing and decreasing modes. As a consequence, modes at very short scales, $k \gg k_\star$ with $k_\star = \sqrt{8\pi G \rho_c}$, are affected by the background 'curvature' during inflation only, explaining why the standard power law is recovered for the primordial power spectrum at these scales. However, the dynamics of modes such that $k \sim k_\star$ is also affected by the background 'curvature' at the time of the bounce and one should expect for those modes a discrepancy as compared to the standard prediction of inflation.

Such a length scale translates into a characteristic *angular* scale in the CMB angular power spectra. By denoting $k_H(t_0) = 2.3 \times 10^{-4} \text{Mpc}^{-1}$ the wavenumber corresponding to the Hubble distance *today*, the characteristic angular scales is given, in terms of multipole $\ell \sim 1/\theta$, by $\ell_\star \approx k_\star(t_0)/k_H(t_0)$. This angular scale lies in the range of scales observed in the CMB anisotropies if $k_\star(t_0) > k_H(t_0)$. The characteristic length scale $k_\star$ is set at the time of the bounce and is inevitably stretched by the following cosmic expansion leading to $k_\star(t_0) = \sqrt{8\pi G \rho_c} \times e^{-N}$ with N the number of e-folds from the bounce to today. From the fact that $k_\star$ is of the order of the inverse of the Planck length at the time of the bounce and from the

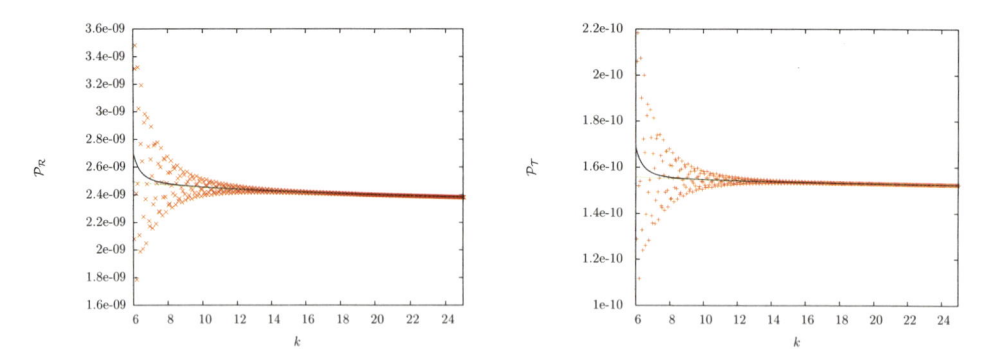

Fig. 3. Primordial power spectra for scalar (left) and tensor (right) modes in the dressed-metric approach. Initial conditions are set at the time of the bounce (from [5]).

knowledge of the number of e-folds from the *end* of inflation to today, this scale set by the bounce enters in the observable range if the number of e-folds during inflation is smaller than ~ 90. If such a characteristic length scale is indeed in the range observed with the CMB, the slight boost of power for $k \lesssim k_\star$ will translate into a slight boost of the angular power spectrum of the CMB anisotropies (as compared to the inflationary prediction) for angular scales $\ell \lesssim \ell_\star$.

Another possibility is to set the initial conditions for perturbations deep in the contracting phase. Then, for both the deformed algebra and dressed metric approaches, one can choose a Minkowski vacuum state for all the wavenumbers, $v_{S(T),k}(\eta \to -\infty) = \exp(ik\eta)/\sqrt{2k}$. In the dressed metric, the standard power law spectrum is recovered for $k \gg k_\star$ for the very same reason as described above: the modes are not affected by the background 'curvature' during both the classical contraction and the quantum bounce. In the infrared limit, $k \to 0$, the modes are mainly affected by the background during contraction leading to a scale invariant power spectrum. In between, there is a range of modes which are not affected by contraction but by the bounce. In that range of wavenumbers, the primordial power spectrum exhibits oscillations with an envelope exhibiting a boost of the power. As shown in Figure 4, the prediction differs in the deformed algebra approach [21]. For modes such that $k > k_\star$, the shape of the primordial spectrum is mainly driven by Ωk^2. Since Ω is negative around the bounce, this leads to an exponential increase of the primordial power spectrum at short scales roughly given by $\mathcal{P}_T(k \gg k_\star) \propto \exp\left(k \int_{\eta_-}^{\eta_+} \sqrt{|\Omega|} d\eta_1\right)$ with $\eta_\pm$ defining the time laps around the bounce during which Ω is negative. For larger length scales, $k < k_\star$, the term Ωk^2 becomes subdominant in the differential equation satisfied by the Mukhanov-Sasaki variable. This regime is therefore very similar to the dressed metric approach previously discussed and the scale invariant behavior in the infrared limit as well as the oscillations for intermediate scales are recovered. A detailed comparison of both approaches was made in [20].

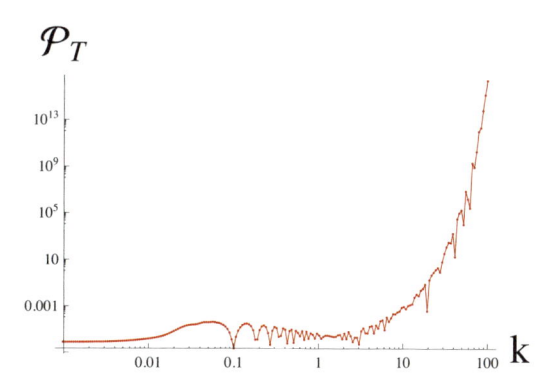

Fig. 4. Primordial power spectrum for tensor modes in the deformed-algebra approach (from [21]). The exponential increase is not necessarily a problem as (i) the observational window might fall out of this region, (ii) the spectrum has anyway a natural cutoff in the UV as the small-scale physics is not described by the primordial spectrum, (iii) backreaction should be taken into account when the amplitude becomes high. The spectrum for scalar modes was derived in [22].

Similar studies have been performed for the case of inverse volume (IV) corrections. This includes the derivation of an anomaly-free perturbation theory with IV corrections alone, and, with both holonomy and IV corrections [23]. However, the impact of the IV corrections on the bounce itself is not well understood and the primordial power spectra with such corrections has been computed during inflation only. Fortunately, an imprint appears on the largest scales for scalar and tensor modes in the form of a polynomial boost below the pivot scale k_0, $\mathcal{P}^{\text{IV}}(k) = \mathcal{P}^{\text{STD}}(k) \times (1 + \Gamma \delta_0 (k/k_0)^{-|\sigma|})$ [24]. Starting from such a predicted power spectrum, the IV parameters have been constrained using WMAP data on the CMB anisotropies showing that e.g. for $\sigma = 2$, the parameter δ_0 is constrained to be smaller than 6.5×10^{-5} at 95% of confidence level [25].

3.3. Measuring the Barbero-Immirzi Parameter

The above results are based on loop quantum cosmology with a *real-valued* Barbero-Immirzi parameter, inherited from the standard formulation of loop quantum gravity. Originally, the Ashtekar formulation of gravity as a gauge theory was however built with a complex-valued Barbero-Immirzi parameter, $\gamma = \pm i$, thus simplifying the constraints into being polynomials in the phase-space variables. Though γ plays no role at the classical level, it is of primary importance at the quantum level: $\gamma = \pm i$ makes the gauge group to be complex, rendering the quantization difficult. Quantization is usually performed with $\gamma \in \mathbb{R}$ for the gauge group is $SU(2)$, which is directly related to the discreteness of the spectra of geometric operators. The role of γ is then crucial in LQC since the discreteness of geometric operators plays an important role in the bounce scenario via the minimal area gap. Phenomenologically speaking, the value of γ fixes the value of ρ_c which could be measured by searching for the characteristic scale $k_\star = \sqrt{8\pi G \rho_c}$ in the CMB anisotropies. It was however argued that in the context of three-dimensional gravity, a natural choice would be $\gamma = \pm i$ which still leads to a consistent quantum theory [26]. This still has to be fully extended to four-dimensional gravity, but this shows that trying to experimentally probe the nature of the Barbero-Immirzi parameter is important.

The two (independent) helicity states of primordial gravitational waves are classically derived from a linearization of Einstein's equations around the inflationary background and subsequently quantized using a Fock scheme on curved spaces. The resulting primordial power spectra for the right-handed and left-handed gravitons are equal, $\mathcal{P}_{r/l} \propto (H/M_{\rm Pl})^2$ with H the Hubble parameter during inflation ('graviton' is used to denote a *Fock* quantization of tensor modes). The CMB angular power spectrum of the BB correlation is sourced by the sum of the two helicity states ($\mathcal{P}_{\rm T}$ in equation (4) is the sum $\mathcal{P}_r + \mathcal{P}_l$). The cross-correlations between temperature and B-modes (called TB), and between E- and B-modes (called EB) are however sourced by the *difference* of the two helicity states, $\mathcal{P}_r - \mathcal{P}_l$. Because $\mathcal{P}_r = \mathcal{P}_l$ by linearizing Einstein's equations and quantizing *à la* Fock, $C_\ell^{TB(EB)}$ are vanishing. However, it was argued that primordial gravitons may have a helicity-dependent behavior if linearization is performed in the Ashtekar formalism [27]. More precisely, it is argued that if the Barbero-Immirzi parameter is imaginary, the reality condition imposes that at the *quantum level*, left-handed and right-handed gravitons do not propagate similarly in an inflationary background, suggesting that linearized gravity may violate parity at the quantum level. (This helicity-dependent behavior only arises if γ has an imaginary part and at the quantum level. At the classical level or for a real-valued γ, there is no such parity breaking in linearized gravity.) If this is indeed the case, the TB and EB angular power spectra are non-zero if γ has a non-vanishing imaginary part while these spectra are zero if γ is real-valued.

Some C_ℓ^{TB} and C_ℓ^{EB} (with the C_ℓ^{BB} autocorrelation) are depicted in Figure 5, including lensing of CMB photons by large scale structures [28]. Dotted parts stand for negative values of TB and EB correlations which is an important piece of information since e.g. a negative C_ℓ^{TB} at $\ell \leq 15$ corresponds to more power in the right-handed gravitons. The amplitude of the BB autocorrelation is set by the value of the tensor-to-scalar ratio, r (equal to 0.05 in Figure 5). Introducing $\delta = \frac{\mathcal{P}_r - \mathcal{P}_l}{\mathcal{P}_r + \mathcal{P}_l}$ which amounts to the level of parity violation in the linearized gravitational sector, the amplitude of the TB and EB correlations is set by ($r \times \delta$). A reconstruction of r and δ is then possible from a measurement of C_ℓ^{BB}, C_ℓ^{TB} and C_ℓ^{EB}. The parameter δ is a direct measure of the level of parity breaking, and subsequently a direct test of a possible non-vanishing imaginary part of γ, as $|\gamma| = \left(1 \pm \sqrt{1 - \delta^2}\right)/|\delta|$ for the simplified case of a purely imaginary Barbero-Immirzi parameter.

For a future, highly-sensitive satellite mission dedicated to the CMB polarization, the measurements of polarized B-modes would be accurate enough for detecting at least 50% of parity violation at e.g. 95% of Confidence Level (C.L.) for $r = 0.2$

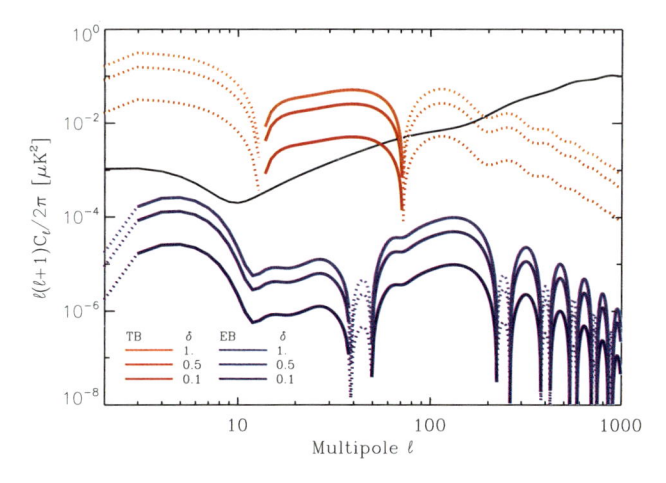

Fig. 5. CMB angular power spectra for the BB, TB and EB cross-correlations ($r = 0.05$) if γ is purely imaginary (from [28]).

(the uncertainties are dominated by the sampling variance). Similarly, measuring $C_\ell^{TB(EB)}$ consistent with zero would lead to an upper bound on δ, directly translated into an exclusion range for $|\gamma|$. For $r = 0.05$, the exclusion range at 95% C.L. is $0.66 \leq |\gamma| \leq 1.5$, and it is enlarged to $0.2 \leq |\gamma| \leq 4.9$ for $r = 0.2$ [28].

4. Lorentz Invariance Violation

Testing for quantum gravity usually assumes an access to gravitational phenomena for which the curvature becomes close to the Planck scale. In Ref. [29], it was first argued that one can also search for quantum gravity imprints by studying the propagation of particles whose energy is comparable to the quantum gravity energy scale (or even much below if the propagation distance is high enough). The basic idea is that discreteness is a genuine property of the quantum space-time. In the context of LQG, this can be understood from the *discreteness* of geometric operators as volume and area operators. This granularity fixes an invariant length scale in apparent contradiction with special relativity (as a boost can contract any length scale). Even though arguments showing that the discreteness of geometric operators in agreement with Lorentz invariance have been put forward (see [30], which argues that the discrete spectrum is observer invariant but the expectation values are not), this granularity idea has opened a wide area of quantum gravity phenomenology aiming at searching for Lorentz invariance violation or deformation as a tracer of quantum gravity. This rich phenomenology is encoded in the fact that the energy-momentum dispersion relation is modified $E \simeq p + m^2/2p \pm \xi(E^2/M_{QG})^n$, with M_{QG} the energy scale of quantum gravity, $\xi > 0$, and n usually chosen as an integer. Because of that, the group velocity for e.g. photons becomes momentum dependent. This means that two photons emitted at the same time but at different momenta would be received at two different times by a distant observer, as, (for $n = 1$), $\Delta v \simeq \xi \Delta k D/M_{QG}$ with Δk the momentum difference and D the distance from the emitter to the receiver. One should therefore look for energetic phenomena (thus Δk is close enough to the quantum gravity scale) and cosmological distances (for having a cumulative impact) for such an effect to be detectable.

If Lorentz invariance is indeed broken or deformed by quantum gravity, this could be described at an effective level. There are many different ways of implementing this idea, ranging from non-commutative space-time to effective field theories and nonlinear Poincaré symmetries. Here, we only mention a few which are closely related to LQG and refer the interested reader to [31] and references therein for details. In all the implementations discussed here, one arrives at a modified dispersion similar to the one mentioned above, with a potentially additional helicity dependence. One approach consists in analyzing the Hamiltonian of the electromagnetic field in a semi-classical state being an discrete approximation of the flat geometry, dubbed a weave [32]. Because the densitized triad operator enters the Hamiltonian for electromagnetism, its expectation value on the weave state is expected to receive loop quantum gravity corrections. The resulting modified dispersion relation for photons acquires a helicity-dependent correction $\omega_{\pm}^2 = k^2 \mp 4\chi k^3/M_{Pl}$ with $\chi \sim 1$. In such a case, photons would experience birefringence in vacuum modifying their polarization state. This effect has been investigated (albeit in the framework of effective field theory) in [33] and [34].

Another approach was put forward in [35]. The idea is that, classically, the action functional $S[A] = \int_\Sigma \mathcal{S}[A]$ can be used to define a slicing of the space-time. If one now considers a quantum setting, this slicing fluctuates around the classical neighborhood corresponding to space-time variations. The explicit calculation performed in [35] considers a Born-Oppenheimer state $\Psi_0[A]\chi[A,\phi]$ with Ψ_0 a semiclassical state peaking at the classical solution and ϕ a matter field. The expectation value of the densitized triad on such a semiclassical state, evaluated around the classical trajectory, is deformed to $E^{(0)}{}_i^a(x,t,\omega) = E^{(0)}{}_i^a(x,t)(1 - \alpha L_{Pl}\omega)$ with $E^{(0)}{}_i^a(x,t)$ the classical solution and ω to be interpreted as the energy of the matter field (in the sense that $\chi[t,\phi] \propto e^{-i\omega t}\chi_\omega[\phi]$). The time parameter t is defined from the action functional $\mathcal{S}[A]$. Since the triad is now ω-dependent, this defines an ω-dependent metric and thus a modified dispersion relation: $m^2 = \omega^2 - k^2/(1 - \alpha L_{Pl}\omega)$. A possible interpretation is that quantum gravity fluctuations lead to an effective frame in which momenta are measured [37]. Classically, the physical momenta p^a is measured in a local inertial frame fixed by the space-time manifold, $p_a = e_a^\mu \pi_\mu$ with π_μ interpreted as the generator of translations. Quantum fluctuations of the space-time itself would then lead to an effective frame $\tilde{e}_a^\mu$ which is nonlinearly related to e_a^μ with a π_μ dependence, $\tilde{e}_a^\mu = F(e_a^\mu, \pi_\mu)$. Since the physical momenta are now measured by $\tilde{p}_a = \tilde{e}_a^\mu \pi_\mu$, the transformation law for momenta would not be given anymore by the Lorentz matrices. In that case, one is therefore considering a deformation of the Poincaré symmetry since the relativity principle is preserved but the transformation rules are now nonlinear [36].

5. Black Holes

Black holes have been extensively studied in loop quantum gravity (see *Chapter 7*). As their macroscopic structure hopefully coincides (up to very small corrections) with the one predicted by general relativity, it is very hard to test LQG observationally using black holes. Recovering the correct value of the entropy is a very powerful consistency test but can hardly be considered as an experimental confirmation. The only way to observationally investigate LQG with black holes would probably be through their Hawking evaporation. As no evaporating black hole has been seen up to now, this is a possibility of future. However, a wide variety of phenomena, reviewed in [38], can in principle lead to primordial black holes.

The idea proposed in [39] is to search for possible LQG signatures in the spectrum of evaporating black holes. The state counting for black holes in LQG relies on the isolated horizon framework (that is a boundary of the underlying manifold considered before quantization). For a given area A of a black hole horizon, the states arise from a punctured sphere whose punctures carry quantum labels (see, e.g., [40]). Two labels (j, m) are assigned to each puncture, j being a spin half-integer with information on the area and m being its associated projection with information on the curvature. They satisfy the condition $A - \Delta \leq 8\pi\gamma\ell_P^2 \sum_{p=1}^N \sqrt{j_p(j_p + 1)} \leq A + \Delta$, where γ is the Barbero-Immirzi parameter of LQG, Δ is a 'smearing' parameter and p labels the different punctures. One may also add the closure constraint: $\sum_p m_p = 0$, which corresponds to a horizon with spherical topology.

In the past, it was postulated that due to quantum gravitational effects, the change in the area of a black hole should be proportional to a fundamental area, of the order of ℓ_{Pl}^2. It was then hoped that associated lines in the evaporation spectrum should appear and might reveal quantum gravity effects. However it was understood in [41, 42] that the situation is different in LQG because the spacing of the energy levels decreases exponentially with the energy. In [39], this issue was readdressed and it was shown that several different signatures can in fact be expected.

To investigate the evaporation in the deep quantum regime, a dedicated and optimized algorithm was developed. It is based on [43] and improved by a breadth-first search. To see if there is a measurable difference, the evaporation has been considered both according to the pure Hawking law and according to LQG. In each case, it was modeled by expressing the probability of transition between states as the exponential of the entropy difference modulated by the greybody factor. Those factors were computed beyond the optical limit by solving the quantum wave equations in the curved background of the black hole. Figure 6 shows that some specific lines associated with transitions occurring during the last stages of the evaporation can be identified in the LQG spectrum whereas the pure Hawking spectrum is naturally featureless.

Monte-Carlo simulations were performed to estimate the energy resolution and the number of black holes required for distinguishing between the different scenarios. At each step of the evaporation process, the energy of the emitted quantum was

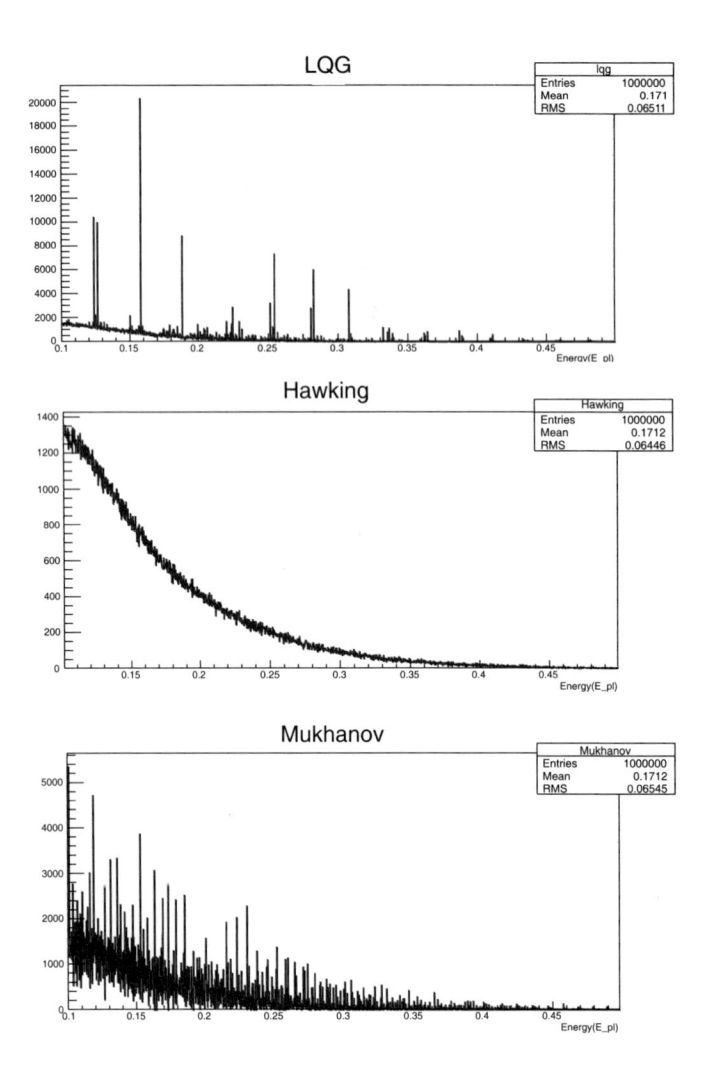

Fig. 6. Spectrum of emitted particles in LQG, in the pure Hawking case, and with an area proportional to the Planck area (Mukhanov), from top to bottom (from [39]).

randomly chosen according to the relevant statistics and to the (spin-dependent) greybody factor. A Kolmogorov-Smirnov (K-S) test was performed to quantify the distance between the cumulative distribution functions and used for a systematic study of possible discriminations between models. Figure 7 shows the number of black holes that would have to be observed for different confidence level in distinguishing between models, as a function of the relative error of the energy reconstruction. With either enough black holes or a relatively small error, a discrimination is possible, therefore showing to a clear LQG footprint in the evaporation spectrum. In this study, only emitted leptons were considered to avoid taking into account

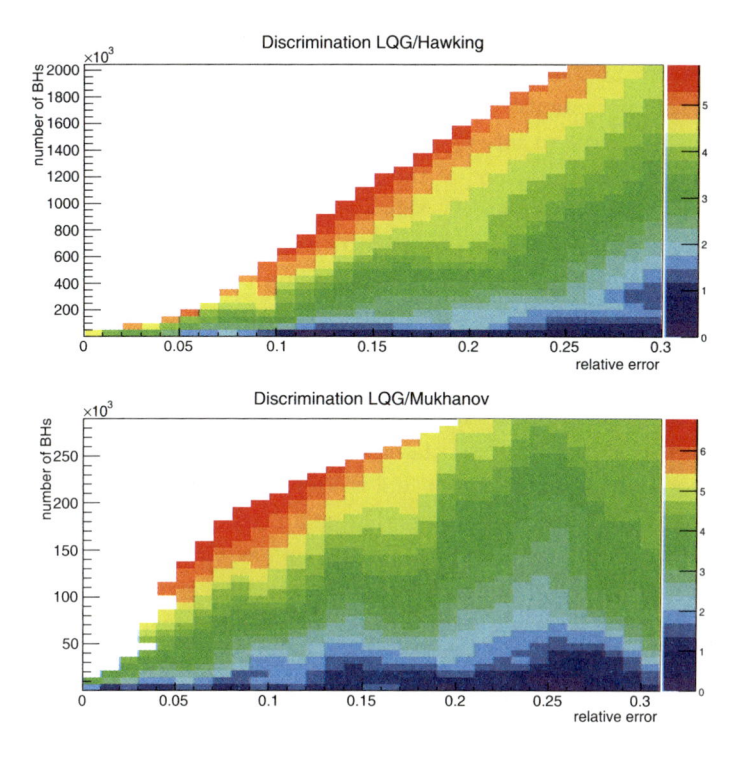

Fig. 7. Number of evaporating black holes that should be observed as a function of the error on the energy reconstruction of the emitted leptons for different confidence levels (the scale corresponds to the number of standard deviations). Up : discrimination between LQG and the Hawking hypothesis. Down : discrimination between LQG and the 'area proportional to the Planck area' hypothesis (from [39]).

complicated fragmentation effects. For a detector located close to the black hole, and due to the huge Lorentz factors, the electrons, muons and taus can be considered as stable.

There is another specific feature of the end-point of the evaporation process which can also be considered. In LQG, the last transitions take place at definite discrete energies associated with the final peaks in the mass spectrum whereas in the usual Hawking picture, the simplest way to implement a minimal mass is to perform a truncation of the standard spectrum to ensure energy conservation. This leads to the consequence that in the standard picture, the energy of the emitted quanta will progressively decrease and asymptotically approach to zero. This 'low-energy' emission associated with the end-point can be distinguished from the 'low-energy' particles emitted earlier in the evaporation process thanks to the dynamics. The time interval between consecutive emissions will increase with decreasing energies as E^{-3}. At 100 TeV, the mean interval is around 1 s. This specific feature of the "standard" spectrum is very different from the absence of low-energy particles expected in the LQG case.

A final possible test is associated with the pseudo-periodic 'large scale' structure of the area spectrum (see [43] and references therein). Most recent arguments suggest that this periodicity is damped for high masses. If, however, it was to remain, this would lead to interesting features. The area gap dA between peaks can be shown to be independent of the scale. As, for a Schwarzschild black hole, $dA = 32\pi M dM$ and $T = 1/(8\pi M)$, this straightforwardly leads to $dM/T = \text{const}$ where dM refers to the mass gap between peaks. This is the important point for detection: in units of temperature, the mass gap does *not* decrease for increasing masses. Any observable feature associated with this pseudoperiodicity can therefore be searched for through larger black holes. If primordial black holes are formed with a definite mass (as expected for example from phase transitions) and not with a continuous spectrum, their resulting emission can be shown [39] to exhibit potentially detectable features associated with this pseudo-periodicity.

A new proposal about statistics, holography, and black hole entropy in loop quantum gravity was suggested in [44]. The main change is that the degeneracy of area eigenvalues of LQG is now modified in a simple way by taking into account vacuum fluctuations in the near horizon region. The area spectrum will not be modified but instead of having basically a degeneracy of $2j + 1$ for each puncture state, we would now have $e^{a_j/4}$ (where a_j is the area eigenvalue, that is $8\pi\gamma\ell_p^2\sqrt{j(j+1)}$). Importantly, punctures should in this case be considered as indistinguishable bosons. The very same Monte Carlo simulation approach is being performed to account also for this new model.

6. Planck Stars

Recently another idea about black holes and possible observational consequences was pushed in [45]. The key insight comes first from lessons from quantum cosmology. In loop cosmology, the Friedmann equation is modified by quantum gravitational effects by a term determined by the ratio of ρ to a Planck scale density ρ_{Pl}. The quantum gravity regime seems to be reached when the energy density of matter reaches the Planck scale, $\rho \sim \rho_{Pl}$. The point is that this may happen well before relevant lengths l become Planckian. The bounce is due to a quantum-gravitational repulsion which originates from the Heisenberg uncertainty and does not happen when the universe is of Planckian size but instead happens when the energy density reaches the Planck density. Quantum gravity could become relevant when the volume of the universe is some 75 orders of magnitude larger than the Planck volume [46].

The analogy between quantum gravitational effects on cosmological and black-hole singularities has been successfully used to make a proposal as to how quantum gravity could also resolve the singularity at the center of a collapsed star. It is assumed that the energy of a collapsing star and any energy falling into the hole could condense into a highly compressed core with density of the order of the Planck density. If this is the case, the gravitational collapse of a star does not lead to a singularity but to an additional phase in the life of a star: a quantum gravitational phase where the gravitational attraction is balanced by a quantum pressure. A star in this phase is called a 'Planck star'. The key observation is that a Planck star can have a size $r \sim \left(\frac{m}{m_{Pl}}\right)^n l_{Pl}$ where m is now the mass of the star and n is positive. For instance, if $n = 1/3$ (as can be naively computed), a stellar-mass black hole would collapse to a Planck star with a size of the order of 10^{-10} cm, that is 30 orders of magnitude larger than the Planck length. The main hypothesis is that a star so compressed would *not* satisfy the classical Einstein equations anymore, even if huge compared to the Planck scale, because its energy density is already Planckian.

The event horizon is replaced by a 'trapping' — or rather, 'dynamical' — horizon [47] which looks like the standard horizon locally, but from which matter can eventually bounce out. The core, that is the 'Planck star', retains memory of the initial collapsed mass m_i. In particular, primordial black holes exploding today may produce a distinctive signal for this scenario. Let $m_f = am_i$ be the final mass reached by the black hole before the dissipation of the horizon. It was shown in [45], using arguments based on information conservation avoiding the firewall hypothesis, that the preferred value is $a \sim \frac{1}{\sqrt{2}}$. The whole observational scenario relies on the assumption that when the black hole reaches this mass it releases all its energy.

During the evaporation phase, the mass loss rate is given by $\frac{dm}{dt} = -\frac{f(m)}{m^2}$, where $f(m)$ is given above each threshold by $f(m) \approx (7.8\alpha_{s=1/2} + 3.1\alpha_{s=1}) \times 10^{24}$ g^{3}s^{-1}, where $\alpha_{s=1/2}$ and $\alpha_{s=1}$ are the number of degrees of freedom (including spin, charge and color) of the emitted particles. If $f(m)$ is assumed to be constant, this leads to:

$$m_i = \left(\frac{3t_H f(m_i)}{1-a^3}\right)^{\frac{1}{3}}. \qquad (15)$$

To account for the smooth evolution of $f(m)$ a numerical integration can be carried out and leads to $m_i \approx 6.1 \times 10^{14}$ g, and $m_f \approx 4.3 \times 10^{14}$ g. The value of m_i is very close to the usual value m_* corresponding to black holes needing a Hubble time to fully evaporate. This was expected as the process is explosive. The size of the black hole when it reaches m_f is the only scale in the problem and therefore fixes the energy of the emitted particles in this last stage. All quanta are assumed to be emitted with the same energy taken at $E_{\text{burst}} = hc/(2r_f) \approx 3.9$ GeV.

Most of the emitted gammas are not emitted with the energy E_{burst} but, instead, come from the decay of hadrons produced in the jets of quarks. If one assumes that the branching ratios are controlled by the internal degrees of freedom, the direct emission represents only a small fraction (1/34 of the emitted particles). To simulate this process, the 'Lund Monte Carlo' PYTHIA code was used to generate the mean spectrum expected for secondary gamma-rays emitted by a Planck star reaching the end of its life. The main point worth noticing is that the mean energy is of the order of $0.03 \times E_{\text{burst}}$, that is in the tens of MeV range rather than in the GeV range, with a high multiplicity of 10 photons per $q\bar{q}$ jet.

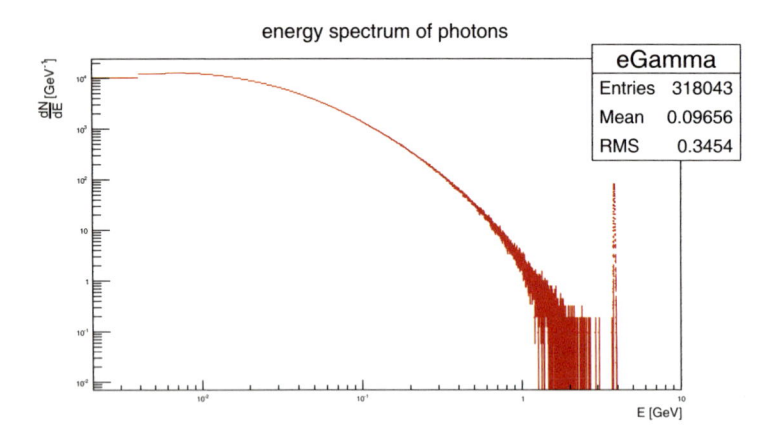

Fig. 8. Full spectrum of gamma-rays emitted by a decaying Planck star (log scales) (from [48]).

It is straightforward to estimate the number of photons $< N_{\text{burst}} >$ emitted during the burst. As for a black hole radiating by the Hawking mechanism, the particles emitted during the bursts (that is those with $m < E_{\text{burst}}$) are emitted proportionally to their number of internal degrees of freedom: gravity is democratic. The spectrum resulting from the emitted u, d, c, s quarks (t and b are too heavy), gluons and photons is shown in Figure 8. The little peak on the right corresponds to directly emitted photons that are clearly sub-dominant. By also taking into account the emission of neutrinos and leptons of all three families (leading to virtually no gamma-rays and therefore being here a pure missing energy), one obtains a total number of photons emitted of $< N_{\text{burst}} > \approx 4.7 \times 10^{38}$.

If one assumes a 1 m^2 detector, this leads to a maximum distance of detectability of $R \approx 205$ light-years. The 'single event' detection of exploding Planck stars is therefore *local* and only a tiny galactic patch around us can be probed. The signal is therefore expected to be *isotropic*.

If Planck stars reaching m_f were to saturate the dark matter bound their number within this detectable horizon would be

$$N_{\text{det}}^{\max} = \frac{4\pi \rho_*^{DM}}{3 m_f} \left(\frac{S < N_{\text{burst}} >}{4\pi N_{\text{mes}}} \right)^{\frac{3}{2}} \approx 3.8 \times 10^{22}. \tag{16}$$

However the usual constraint on primordial black holes $\Omega^{PBH} < 10^{-8}$ for initial masses around 10^{15} g basically holds and this leads to $N_{det} < 3.8 \times 10^{14}$, which is still a high number showing that the individual detection is not impossible.

It is possible to estimate the number of events observable in a time Δt corresponding to Planck stars that have masses between m_f and $m(\Delta t)$ at the beginning of the observation time, within the volume $R < R_{det}$. In this case, $m(\Delta t)$ is simply: $m(\Delta t) = \left(m_f^3 + 3f(m)\Delta t \right)^{\frac{1}{3}}$. The number of expected 'events' during Δt is given by

$$N(\Delta t) = \frac{\int_{m_f}^{m(\Delta t)} \frac{dn}{dm} dm}{\int_{m_f}^{m_{\max}} \frac{dn}{dm} dm} \Omega^{PBH} N_{\det}^{\max} \Omega_{sr}, \qquad (17)$$

where $m_{\max}$ is the maximum mass up to which we assume the mass spectrum dn/dm to be 'filled' by black holes and Ω_{sr} is the solid angle acceptance of the considered detector. An upper limit on the value of Ω^{PBH} can be taken conservatively at 10^{-8}. If one sets $m_{\max} = m_*$ and a density of a few percents of the maximum allowed density, that is $\Omega^{PBH} \sim 10^{-10}$, this leads to one event per day.

Could such events be associated with some gamma-ray bursts (GRBs) already detected? The long GRBs are well understood and have no link with Planck stars. Were Planck star explosions to be associated with some of the known GRBs, this would be with short gamma-ray bursts (SGRBs). Interestingly, SGRBs are the less well understood; the redshifts are not measured for a large fraction of them; they are known to have a harder spectrum and some of them do indeed reach the energies estimated here; and a sub-class of SGRB, the very short gamma ray bursts (VSGRBs), do exhibit an even harder spectrum and can be assumed to originate from a different mechanism as the SGRB time distribution seems to be bimodal. This does not mean that exploding Planck stars have been detected but this raises an interesting question.

Recently, the model has been developed in [49] and the resulting phenomenology was investigated in [50] and [51].

References

[1] BICEP2 Collaboration, Detection of B-mode Polarization at Degree Angular Scales, arXiv:1403.3985.

[2] A. Ashtekar and J. Lewandowsk, "Background independent quantum gravity: A status report," *Class. Quant. Grav.* **21** (2004) R53; C. Rovelli and F. Vidotto, *Covariant Loop Quantum Gravity* (Cambridge University Press (2014)); R. Gambini and J. Pullin, *A First Course in Loop Quantum Gravity* (Oxford, Oxford University Press (2011)); C. Rovelli, "Zakopane lectures on loop gravity," arXiv:1102.3660v5 [gr-qc]; P. Dona and S. Speziale, "Introductory lectures to loop quantum gravity," arXiv:1007.0402V1; T. Thiemann, *Modern Canonical Quantum General Relativity* (Cambridge, Cambridge University Press (2007)); C. Rovelli, *Quantum Gravity* (Cambridge, Cambridge University Press (2004)); L. Smolin, "An invitation to loop quantum gravity," arXiv:hep-th/0408048v3; A. Perez, "The spin-foam approach to quantum gravity," *Living Rev. Relat.* **16** (2013) 3.

[3] A. Barrau, T. Cailleteau, J. Grain and J. Mielczarek, "Observational issues in loop quantum cosmology," *Class. Quant. Grav.* **335** (2014) 053001; I. Agullo and A. Corichi, "Loop quantum cosmology," arXiv:1302.3833; G. Calcagni, "Observational effects from quantum cosmology," arXiv:1209.0473 [gr-qc]; M. Bojowald, "Quantum cosmology: Effective theory," arXiv:1209.3403 [gr-qc]; K. Banerjee, G. Calcagni and M. Martin-Benito, "Introduction to loop quantum cosmology," *SIGMA* **8** (2012) 016; M. Bojowald, *Quantum Cosmology* (Springer, New-York (2011)); A. Ashtekar and P. Singh, "Loop quantum cosmology: A status report," *Class. Quant. Grav.* **28** (2011)

213001; A. Ashtekar, "Loop quantum cosmology," *Gen. Rel. Grav.* **41** (2009) 707; M. Bojowald, "Loop quantum cosmology," *Living Rev. Rel.* **11** (2008) 4; A. Ashtekar, M. Bojowald and J. Lewandowski, "Mathematical structure of loop quantum cosmology," *Adv. Theor. Math. Phys.* **7** (2003) 233.

[4] E. Alesci and F. Cianfrani, "Loop quantum cosmology from quantum reduced loop gravity," *Europhys. Lett.* **111** (2015) 40002; M. Hanusch, "Projective structures in loop quantum cosmology," *J. Math. Anal. Appl.* **428** (2015) 1005; N. Bodendorfer, J. Lewandowski and J. Swiezewski, "A quantum reduction to spherical symmetry in loop quantum gravity," *Phys. Lett. B* **747** (2015) 18; C. Fleischhack, "Kinematical foundations of loop quantum cosmology," arXiv:1505.04400.

[5] I. Agullo, A. Ashtekar and W. Nelson, "The pre-inflationary dynamics of loop quantum cosmology: Confronting quantum gravity with observations," *Class. Quant. Grav.* **30** (2013) 085014; I. Agullo, A. Ashtekar and W. Nelson, "An extension of the quantum theory of cosmological perturbations to the Planck era," *Phys. Rev. D* **87** (2013) 043507; I. Agullo, A. Ashtekar and W. Nelson, "A quantum gravity extension of the inflationary scenario," *Phys. Rev. Lett.* **109** (2012) 251301.

[6] A. Barrau, M. Bojowald, G. Calcagni, J. Grain and M. Kagan, "Anomaly-free cosmological perturbations in effective canonical quantum gravity," arXiv:1404.1018.

[7] A. Ashtekar and D. Sloan, "Loop quantum cosmology and slow roll inflation," *Phys. Lett. B* **694** (2012) 108; A. Ashtekar and D. Sloan, "Probability of inflation in loop quantum cosmology," *Gen. Rel. Grav.* **43** (2011) 3519.

[8] C. Germani, W. Nelson and M. Sakellariadou, "On the onset of inflation in loop quantum cosmology," *Phys. Rev. D* **76** (2007) 043529.

[9] L. Linsefors and A. Barrau, Duration of inflation and conditions at the bounce as a prediction of effective isotropic loop quantum cosmology," *Phys. Rev. D* **87** (2013) 123509.

[10] A. Corichi and A. Karami, "On the measure problem in slow roll inflation and loop quantum cosmology," *Phys. Rev. D* **83** (2011) 104006.

[11] A. Ashtekar and E. Wilson-Ewing, "Loop quantum cosmology of Bianchi I models," *Phys. Rev. D* **79** (2009) 083535.

[12] L. Linsefors and A. Barrau, "Modified Friedmann equation and survey of solutions in effective Bianchi-I loop quantum cosmology," *Class. Quant. Grav.* **31** (2014) 015018.

[13] L. Linsefors and A. Barrau, "Exhaustive investigation of the duration of inflation in effective anisotropic loop quantum cosmology," arXiv:1405.1753.

[14] M. Zaldarriaga and D. Harari, "Analytic approach to the polarization of the cosmic microwave background in flat and open universes," *Phys. Rev. D* **52** (1995) 3276; U. Seljak and M. Zaldarriaga, "A line-of)sight integration approach to cosmic microwave background anisotropies," *Astrophys. J.* **469** (1996) 437.

[15] J. Grain and A. Barrau, "Cosmological footprints of loop quantum gravity," *Phys. Rev. Lett.* **102** (2009) 081301; J. Grain, T. Cailleteau, T. Cailleteau and A. Gorecki, "Fully loop quantum cosmology corrected propagation of gravitational waves during slow-roll inflation," *Phys. Rev. D* **81** (2010) 024040; J. Mielczarek, T. Cailleteau, J. Grain and A. Barrau, "Inflation in loop quantum cosmology," *Phys. Rev. D* **81** (2010) 104049.

[16] J. Grain, A. Barrau, T. Cailleteau and J. Mielczarek, "Observing the big bounce with tensor modes in the cosmic microwave background: Phenomenology and fundamental LQC parameters," *Phys. Rev. D* **82** (2010) 123520.

[17] A. Ashtekar, W. Kaminski and J. Lewandowski, "Quantum field theory on a cosmological, quantum space-time," *Phys. Rev. D* **79** (2009) 064030.

[18] T. Cailleteau, J. Mielczarek, A. Barrau and J. Grain, "Anomaly-free scalar perturba-

tions with holonomy corrections in loop quantum cosmology," *Class. Quant. Grav.* **29** (2012) 095010; J. Mielczarek, T. Cailleteau, A. Barrau and J. Grain, "Anomaly-free vector perturbations with holonomy corrections in loop quantum cosmology," *Class. Quant. Grav.* **29** (2012) 085009.

[19] T. Cailleteau, A. Barrau, J. Grain and F. Vidotto, "Consistency of holonomy-corrected scalar, vector and tensor perturbations in loop quantum cosmology," *Phys. Rev. D* **86** (2012) 087301; M. Bojowald and G. M. Paily, "Deformed general relativity," *Phys. Rev. D* **87** (2013) 044044.

[20] B. Bolliet, J. Grain, C. Stahl, L. Linsefors and A. Barrau, "Comparison of primordial tensor power spectra from the deformed algebra and dressed metric approaches in loop quantum cosmology," *Phys. Rev. D* **91** (2015) 084035.

[21] L. Linsefors, T. Cailleteau, A. Barrau and J. Grain, "Primordial power spectrum in holonomy corrected Ω loop quantum cosmology," *Phys. Rev. D* **87** (2013) 107503.

[22] S. Schander, A. Barrau, B. Bolliet, L. Linsefors and J. Grain, "Primordial scalar power spectrum from the Euclidean big bounce," arXiv:1508.06786.

[23] M. Bojowald, G. Hossain, M. Kagan and S. Shankaranarayanan, "Anomaly freedom in perturbative loop quantum gravity," *Phys. Rev. D* **78** 063547; T. Cailleteau, L. linsefors and A. Barrau, "Anomaly-free perturbations with inverse-volume and holonomy corrections in loop quantum cosmology," *Class. Quant. Grav.* **31** (2014) 125011.

[24] E. J. Copeland, D. J. Mulryne, N. J. Nunes and M. Shaeri, "Gravitational wave background from superinflation in loop quantum cosmology," *Phys. Rev. D* **79** (2009) 023508; J. Grain, A. Barrau and A. Gorecki, "Inverse-volume corrections from loop quantum gravity and the primordial tensor power spectrum in slow-roll inflation," *Phys. Rev. D* **79** (2009) 084015; M. Bojowald and G. Calcagni, "Inflationary observables in loop quantum cosmology," *JCAP* **03** (2011) 032.

[25] M. Bojowald, G. Calcagni and S. Tsujikawa, "Observational test of inflation in loop quantum cosmology," *JCAP* **11** (2011) 046; M. Bojowald, G. Calcagni and S. Tsujikawa, "Observational constraints on loop quantum cosmology," *Phys. Rev. Lett.* **107** (2011) 211302.

[26] M. Geiller and K. Noui, "A note on the Holst action, the time gauge, and the Barbero-Immirzi parameter," *Gen. Rel. Grav.* **45** (2013) 1733; M. Geiller and K. Noui, "Testing the imposition of the spin foam simplicity constraints," *Class. Quant. Grav.* **29** (2012) 135008; S. Alexandrov, M. Geiller and K. Noui, "Spin foams and canonical quantization," *SIGMA* **8** (2012) 055; J. Ben Achour, M. Geiller, K. Noui and C. Yu, "Testing the role of the Barbero–Immirzi parameter and the choice of connection in loop quantum gravity," arXiv:1306.3241 [gr-qc] (2013); J. Ben Achour, M. Geiller, K. Noui and C. Yu, "Spectra of geometric operators in three-dimensional LQG: From discrete to continuous," to be published in *Phys. Rev. D* (2013).

[27] C. R. Contaldi, J. Magueijo and L. Smolin, *Phys. Rev. Lett.* **101** (2008) 141101; J. Magueijo and D. M. T. Benincasa, *Phys. Rev. Lett.* **106** (2011) 121302; L. Bethke and J. Magueijo, *Phys. Rev. D* **84** (2011) 024014; L. Bethke and J. Magueijo, *Class. Quant. Grav.* **29** (2012) 052001.

[28] A. Ferté and J. Grain, "Detecting chiral gravity with the pure pseudospectrum reconstruction of the cosmic microwave background polarized anisotropies," *Phys. Rev. D* **89** (2014) 103516.

[29] G. Amelino-Camelia, J. Ellis, N. E. Mavromatos, D. V. Nanopoulos and S. Sarkar, "Potential sensitivity of gamma-ray burster observations to wave dispersion in vacuo," *Nature* **293** (1998) 763.

[30] C. Rovelli and S. Speziale, "Reconcile Planck-scale discreteness and the Lorentz-Fitzgerald contraction," *Phys. Rev. D* **67** (2003) 064019.

[31] F. Girelli, F. Hinterleitner and S. A. Major, "Loop quantum gravity phenomenology: Linking loops to observational physics," *SIGMA* **8** (2012) 098.
[32] R. Gambini and J. Pullin, "Nonstandard optics from quantum space-time," *Phys. Rev. D* **59** (1999) 124021.
[33] L. Maccione, S. Liberati, A. Celotti and J. G. Kirk, "New constraints on Planck-scale Lorentz violation in QED from the crab nebula," *JCAP* **10** (2007) 013.
[34] R. J. Gleiser, C. N. Kozameh and F. Parisi, "On low-energy quantum gravity induced effects on the propagation of light," *Class. Quant. Grav.* **20** (2003) 4375.
[35] L. Smolin, "Falsifiable predictions from semiclassical quantum gravity," *Nucl. Phys. B* **742** (2006) 142.
[36] J. Magueijo and L. Smolin, "Generalized Lorentz invariance with an invariant energy scale," *Phys. Rev. D* **67** (2003) 044017; J. Magueijo and L. Smolin, "Lorentz invariance with an invariant energy scale," *Phys. Rev. Lett.* **88** (2002) 190403.
[37] S. Liberati, S. Sonego and M. Visser, "Interpreting doubly special relativity as a modified theory of measurement," *Phys. Rev. D* **71** (2005) 045001; R. Aloisio, A. Galante, A. F. Grillo, S. Liberati, E. Luzio and F. Méndez, "Modified special relativity on a fluctuating spacetime," *Phys. Rev. D* **74** (2006) 085017; F. Girelli, S. Liberati, R. Percacci and C. Rahmede, "Modified dispersion relations from the renormlaization group of gravity," *Class. Quant. Grav.* **24** (2007) 3995.
[38] B. J. Carr, K. Kohri, Y. Sendouda and J. Yokoyama, "New cosmological constraints on primordial black holes," *Phys. Rev. D* **81** (2010) 104019.
[39] A. Barrau, T. Cailleteau, X. Cao, J. Diaz-Polo and J. Grain, "Probing loop quantum gravity with evaporating black holes," *Phys. Rev. Lett.* **107** (2011) 251301.
[40] A. Corichi, J. Diaz-Polo and E. Fernández-Borja, "Black hole entropy quantization," *Phys. Rev. Lett.* **98** (2007) 181301.
[41] C. Rovelli, "Black hole entropy from loop quantum gravity," *Phys. Rev. Lett.* **77** (1996) 3288.
[42] A. Ashtekar and K. Krasnov, "Quantum geometry and black holes," arXiv:gr-qc/9804039.
[43] I. Agullo *et al.*, "Detailed black hole state counting in loop quantum gravity," *Phys. Rev. D* **82** (2010) 084029.
[44] A. Ghosh, K. Noui and A. Perez, arXiv:1309.4563.
[45] C. Rovelli and F. Vidotto, "Planck stars," arXiv:1401.6562.
[46] A. Ashtekar, T. Pawlowski, P. Singh and K. Vandersloot, "Loop quantum cosmology of k=1 FRW models," *Phys. Rev. D* **75** (2007) 024035.
[47] A. Ashtekar and M. Bojowald, "Black hole evaporation: A paradigm," *Class. Quant. Grav.* **22** (2005) 3349.
[48] A. Barrau and C. Rovelli, *Phys. Lett. B* **739** (2014) 405.
[49] H. M. Haggard and C. Rovelli, "Black hole fireworks: Quantum-gravity effects outside the horizon spark black to white hole tunneling," arXiv:1407.0989.
[50] A. Barrau, C. Rovelli and F. Vidotto, "Fast radio bursts and white hole signals," *Phys. Rev. D* **90** (2014) 127503.
[51] A. Barrau, B. Bolliet, F. Vidotto and C. Weimer, "Phenomenology of bouncing black holes in quantum gravity: A closer look," arXiv:1507.05424.

Subject Index

$\bar{\mu}$-quantization, 190, 193

ADM, 31
algebraic dual, 72
anomaly free representation, 69, 87
area ensemble, 252
area gap, 10, 59
area operator, 55
Ashtekar variables, 33, 36, 56, 100, 105, 187, 249
Ashtekar-Lewandowski measure, 44, 251
Ashtekar-Lewandowski representation, 16, 43, 55, 143, 154
asymptotic safety, vii, 8, 165
asymptotics, 116
atom of space, 10, 108, 129, 130

B-modes, 294
Baratin-Oguri (BO) model, 135
Barbero-Immirzi parameter, 9, 24, 34, 36, 100, 105, 187, 247
Bekenstein–Hawking entropy, 242
BF representation, 16, 156, 174
BF theory, 99, 134
Bianchi models, 17, 201, 206
BICEP2, 281
BKL conjecture, 210
black hole evaporation, 297
black hole thermodynamics, 242
Bohr compactification, 188
Boltzmann equations, 288
bulk Hilbert space, 251

Bunch–Davies vacuum, 220

canonical quantization, 38
Casimir operators, 104
cellular decomposition, 101
charge network states, 84
Chern–Simons surface term, 248
Clebsch–Gordan coefficients, 97
CMB anomalies, 225
CMB observations, 24, 221
coarse graining, 158, 165
coherent spin networks, 114
combinatorics, 126
condensates, 145
connection variables, 33
constraint algebra, 36
continuum limit, 87, 130, 144, 154, 156
cosmological constant, vi, 99, 201, 202
cosmological perturbations, 185, 213, 214
cyclic representation, 46
cylindrical consistency, 44, 154, 164, 171
cylindrical functions, 42

deformed constraint algebra, 290
densitized triad, 34
diffeomorphism constraint, 36, 157
diffeomorphism invariance, 69, 91, 158
diffeomorphism invariant Hilbert space, 73

Diophantine equations, 254, 257
Dirac observables, 191
Dirac quantization, 12, 32, 52
dressed metric, 218, 282
dynamical triangulation, 101, 119, 125, 145

E-modes, 294
effective spacetime description, 193
embedding method, 195
Engle-Pereira-Rovelli-Livine (EPRL) model, 13, 102, 135, 142
entanglement entropy, 269
Euclidean gravity, 76, 83

Feynman amplitudes, 127
finite triangulation constraint operators, 85
FLRW spacetimes, 17, 183, 185, 197
frame fields, 33
Freidel-Krasnov (FK) model, 13, 102

gamma-ray bursts, 303
gauge invariant Hilbert space, 47, 54
gauge theories, 36
Gauss law, 36
Gelfand–Naimark–Segal theorem, 46
generalized connection, 42
generating functions, 254
geometric operators, 55
geometric phase, 145
Gowdy models, 17, 201, 210
graph Hilbert space, 52
Gross-Neveau model, 7
group averaging, 70, 71, 75, 190
group field theory, 14, 125, 173

Hamiltonian constraint, 36, 157, 191, 214
Hawking area theorem, 241
Hawking radiation, 20–22, 273
heat kernel, 116, 273
helicity, 294

high performance computing, 193
higher space-time dimensions, 37
holography, 265
holonomy, 40, 103
holonomy corrections, 292
holonomy-flux algebra, 38, 39, 43, 48, 249
Husain–Kuchavr (HK) model, 82
hybrid quantization, 210, 219

inductive limit Hilbert space, 153
inflation, 203, 205, 215, 220
intertwiner, 54, 106, 111
inverse volume corrections, 292
isolated horizons, 244

kinematical Hilbert space, 49
Koslowski–Sahlmann representation, 46

large-N limit, 141
lattice gauge theory, 40, 93
laws of black hole mechanics, 242
length operator, 62
Lewandowski Marolf habitats, 75
Lewandowski–Okolow–Sahlmann–Thiemann (LOST) Theorem, 45, 64
Lewandowski-Marolf habitat, 70, 80
logarithmic corrections, 271
loop quantum cosmology, 183
Lorentz invariance violation, 295

matrix models, 125
melonic sector, 141
Minkowski theorem, 108
modified Friedmann equation, 194
Mukhanov-Sasaki variables, 214, 287

n-point functions, 128
non-Gaussianity, 226
number theory, 254

parameterized field theory, 82

parametrized field theory, 12
partition function, 126
Penrose metric operator, 112
Peter-Weyl theorem, 49
phase space dependent diffeomorphisms, 87
phase structure, 143, 168
phase transition, 143, 146, 168
Planck stars, 23, 300
Plebanski action, 101, 134
Ponzano–Regge model, 94
Ponzano-Regge model, 98, 125
pre-inflationary dynamics, 213, 221
primordial power spectrum, 291
pseudo-manifolds, 138

QFT on quantum spacetimes, 217, 273
quantum bounce, 191
quantum configuration space, 42, 55
quantum Einstein's equations, 52
quantum geometry, v, 112, 249, 262
quantum group, 138
quantum polyhedra, 110
quantum shift, 85
quantum spacetimes, 213

random lattice, 127
reduced quantization, 32
refined algebraic quantization, 157
Regge action, 97
Regge gravity, 116, 127
renormalization group flow, 143, 153, 162

scalar and tensor modes, 215
Schwinger-Dyson equations, 146
semiclassical black holes, 242, 259
semiclassical spacetime, 117
simplicial complex, 133
simplicity constraint, 101, 106, 168
solvable loop quantum cosmology, 197

spherically symmetric models, 213, 273
spin foams, 12, 98, 133, 159, 166, 268
spin network observables, 128
spin networks/spin-nets, 10, 49, 73, 75, 106, 126, 145, 166, 168, 251
Stone-von Neumann theorem, 10, 39
string theory, vi
strong curvature singularities, 204
strong observables, 72
superselection, 73
surface Hilbert space, 251

tensor models, 125
tensor networks, 166, 169
tetrahedron, 97
Thiemann procedure, 71
Thiemann trick, 95
Thiemann's Hamiltonian constraint, 76
topological field theory, 99
truncation method, 195
Turaev-Viro model, 125
two-complex, 102, 134

uniform regulator covariance, 78
uniform Rovelli Smolin (URS) topology, 79
unitary equivalence, 39
unitary representations, 103
Unruh temperature, 21, 263

vertex amplitude, 107, 116
volume operator, 59

weak coupling limit, 83
weakly isolated horizons, 245
Wheeler-DeWitt theory, 191
widely spread, squeezed, non-Gaussian states, 193
Wigner $6j$ symbol, 97

ZAMO's, 260